IPP Studies in the Frontiers of China's Public Policy

Series Editor
Feng Zhang, South China University of Technology, Guangzhou, China

IPP Studies in the Frontiers of China's Public Policy combines original research and theoretical innovation to provide fresh insights into the fast-changing landscape of China's public policy. Books in the series, written by scholars based inside China or commissioned by the IPP, draw on local Chinese experiences to generate empirical findings and theoretical insights. The field of China's public policy is broadly defined to include all aspects of the country's social, economic, technological and foreign policies. The series encourage interdisciplinary approaches to public policy, with a special but not exclusive focus on the study of south China, especially the Guangdong, Hong Kong, Macau Greater Bay Area, one of the most vibrant regions of growth and innovation in China.

Yang Lijun

Contest for Citizenship and Collective Violence During China's Cultural Revolution

palgrave
macmillan

Yang Lijun
South China University of Technology
Guangzhou, China

IPP Studies in the Frontiers of China's Public Policy
ISBN 978-981-97-0905-2 ISBN 978-981-97-0906-9 (eBook)
https://doi.org/10.1007/978-981-97-0906-9

The original submitted manuscript has been translated into English. The translation was done using artificial intelligence. A subsequent revision was performed by the author(s) to further refine the work and to ensure that the translation is appropriate concerning content and scientific correctness. It may, however, read stylistically different from a conventional translation.

Cover credit: Photo12/Ann Ronan Picture Library/Alamy Images

This Palgrave Macmillan imprint is published by the registered company Springer Nature Singapore Pte Ltd.
The registered company address is: 152 Beach Road, #21-01/04 Gateway East, Singapore 189721, Singapore

Paper in this product is recyclable.

To the Memory of My Father Yang Yurong (1936–2001)

Preface to the English Edition

Twenty years after the original Japanese publication and five years after the Chinese translation, the English edition of this book has finally arrived. As the author, I deeply appreciate Palgrave for publishing the English version and enabling this work to find its way to more readers. I sincerely thank all who lent their time and expertise to make this English edition possible. In particular, I want to recognize Peter Cuthbert, whose dedication through the long translation process has been invaluable in bringing this book to an English-speaking audience.

This book delves into China's Cultural Revolution of the 1960s, attempting to answer a multitude of related questions:

- How did criticism of "capitalist-road authorities" within the Communist Party during the Cultural Revolution lead to a sweeping mass factional struggle and widespread armed conflict?
- What underlying reasons caused followers loyal to the same leader to divide into two incompatible factions during the Cultural Revolution?
- In the midst of the Cultural Revolution's turmoil, what were the primary motivations driving the divided factions—loyalty, faith, interests, or political identity?
- To what extent were participants in the Cultural Revolution active agents in their actions, and did they have any choice in their involvement?

- What dominant factors influenced and shaped participants' behavior during the Cultural Revolution?
- How did participants strategically use these influential factors to enhance their own power and invincibility throughout the Cultural Revolution?
- What were the perceptions and implications of the cult of fanatical leadership during the Cultural Revolution, and how did it impact the course of events?
- Did the widespread god-making movement primarily stem from genuine faith, or did calculated political strategy drive it?
- How did individuals involved in the Cultural Revolution develop and implement competing strategies amidst the fervent ideological slogans and campaigns?
- How can we gain insight into the nature and causes of collective violence during the Cultural Revolution? Was it solely a consequence of the movement's disorder, or were other underlying factors at play?

These interconnected questions are essential for understanding the behavior of political leaders, faction leaders, and the masses during the Cultural Revolution. The book examines them through the lens of the distribution of and competition for citizenship, which arose from China's unique institutional setting after the founding of the People's Republic. It argues that, as a movement, the Cultural Revolution was a competition for citizenship within the state–society framework. In this competition, obtaining political rights (expressed as acquiring political identity during the Cultural Revolution) became paramount. The Cultural Revolution can be viewed as a contest to embody the state's idea of a "good citizen." However, as the state held a monopoly on defining good citizenship criteria and competition methods, which were subject to adjustments based on shifting state goals and power structures, it introduced uncertainty and internal dynamics to the competition.

In tackling these questions, I avoid viewing the Cultural Revolution as merely a historical event confined to a specific time. Instead, I take a broader approach using comparative institutionalism to explore the underlying factors behind its phenomena. Recognizing the magnitude of the Cultural Revolution's impact, I emphasize the importance of learning from it to prevent similar catastrophes in the future. By examining it institutionally, we can draw relevant lessons for contemporary China. Moreover, I also acknowledge the autonomy of participants' behavior in the Cultural Revolution, considering it not only a political movement but also a social one. Thus, I approach it as a politically mobilized social movement and utilize the social movement approach to explain collective behavior. Through this book, I aim to engage in a dialogue with scholars in social movement studies.

In the 1960s, the world experienced a period of intense upheaval, with various social movements taking shape in different countries. Alongside the Cultural Revolution in China, significant events like the anti-Vietnam War protests and civil rights movement in the United States, the May Storm in France, student/worker protests across Europe, and new left-wing movements in Japan. These movements echoed and intersected with each other, captivated by the ideology of Mao Zedong and the struggle for global hegemony between the United States and the Soviet Union during the Cold War. Despite varying focal points, they shared common traits. Dissatisfaction with and challenges to existing orders, widespread public participation, and the pursuit of citizenship were at the core of these movements. In this context, the Cultural Revolution in China was not an isolated occurrence but rather an offshoot of the larger global social movements of the time.

Entering the 1970s, significant changes occurred in the global power dynamics with Nixon's visit to China and the détente between China and the Soviet Union. The world order gradually shifted, and both China and the Soviet Union initiated reforms in their systems by the end of the 1970s and mid-1980s, respectively. The Cold War eventually came to an end with the fall of the Berlin Wall and the dissolution of the Soviet Union. Consequently, the fervent social movements that had swept through countries began to subside. While some East Asian countries witnessed street protests advocating for democracy during the third wave of democratization in the 1980s, these movements were more localized and regional rather than global in nature.

The world has recently seen a resurgence of social movements akin to those in the 1960s. These movements have taken various forms in many countries, encompassing protests such as Occupy Wall Street, Black Lives Matter, Occupy Central in Hong Kong, strikes in France and the UK, anti-immigrant protests in Sweden, and more. According to the Carnegie Endowment, as of July 6, 2023, over 400 significant antigovernment protests erupted worldwide, over 132 countries have experienced significant protests, 23% of significant protests have lasted longer than more than three months, 135 significant economic antigovernment protests have occurred since 2017.

Across the globe, there is a growing surge of passion and enthusiasm for participating in the public sphere. These movements reflect the challenges posed by the troubled international order in recent years and the unequal distribution of wealth resulting from capital-led globalization since the end of the Cold War. While each country faces unique developmental stages and issues, and protest manifestations vary, they all share a common goal—the struggle for citizenship. Some movements prioritize political rights, while others focus on civil or social rights. This study of China's 1960s competition for citizenship offers valuable insights for understanding contemporary social movements and highlights the enduring significance of the universal quest for citizenship, spanning generations and inspiring meaningful transformations in societies worldwide.

Guangzhou, China
July 2023

Yang Lijun

Preface to the Chinese Edition

Some fifteen years have passed since the Japanese edition of this book was first published, and now, with great joy and gratitude, I am honored to witness the publication of the Chinese edition by the esteemed Chinese University of Hong Kong Press. In these intervening years, the field of foreign studies on the Chinese Cultural Revolution has flourished, yielding many works that have enriched our understanding of this complex period. I am deeply grateful to the scholars and friends who inspired me to think deeply about the subject of the Cultural Revolution many years ago. It is truly humbling to see my book considered worthy of republication in Chinese, thereby making it accessible to both Chinese readers and researchers worldwide.

My journey into researching the Cultural Revolution began during my studies in Japan, and the Japanese edition of this book, originally published in 2003 by Ochanomizu Shobo, represents an abridged version of my doctoral dissertation. The theory of "state–society relations," the theory of the public sphere, and the discussion of citizenship, which were widely discussed in academic circles at the end of the twentieth century, greatly influenced my study of the Cultural Revolution. Since then, there has been a significant change in the scholarly focus of the relevant research areas. This may make my book seem a bit "outdated." However, I firmly believe that its analytical framework remains contemporary in its perspective on factionalism and collective violence during the Cultural Revolution. From a practical standpoint, the analysis of the intervention of public power in the private sphere, political control, and competition for political loyalty is still valid, not only for the Cultural Revolution but also for the analysis of Chinese society today. Thus, in this Chinese edition, I have maintained the original analytical framework, preserving the essence of the Japanese edition. The only modifications made were the addition of research reviews on significant works in Cultural Revolution studies from the past decade and supplementary discussions on the central analytical concept of "citizenship," as suggested by reviewers.

In addition, I have introduced two new chapters in this Chinese edition. Chapter 7 delves into the competitive strategy of citizenship, while Chapter 9 explores the intricate relationship between collective violence and social order during the Cultural Revolution. These two chapters are based on papers published in the Japanese academic *journals Asian Studies* and *Historical Review* in 2005 and 2007, respectively. However, I have revised and condensed these papers, incorporating subsequent reflections and insights that emerged after the Japanese edition's publication. The final chapter of the Japanese edition was completed in 2001. Now, after some time, many of the situations described have changed, but the basic issues analyzed and the root causes of the problems have not, so I have adhered in the Chinese version to the original format, albeit taking into account that it reflected the situation in China in 2001.

During the writing process of this book, I experienced the profound loss of my beloved father. It remains a lifelong regret that, as a student in a foreign land, I could not be by his side during his final days. My mentor, Professor Takashi Mitani, also passed away a few years ago. Most of the professors who led me on the academic path, such as Yuzo Kato, Kazuko Mori, Amako Satoshi, Ryosei Kokubun, Hiroko Sakamoto, Shinichi Sato, Tang Liang, and Katsuji Nakagane, are now beyond or approaching retirement age. Former fellow students of mine are now the backbone of the academy. On this occasion of the publication of the Chinese edition, I extend my heartfelt gratitude to all the scholars and friends mentioned in the postscript of the Japanese edition and the numerous friends who supported and assisted me during my studies abroad. I would also like to thank the original publisher, Ochanomizu Shobo, and the Japan Society for the Promotion of Science, which provided financial support for the publication; it was because of these two organizations that my work as a new young scholar was published, and that my book was awarded the

Masayoshi Ohira Memorial Academic Prize in 2005. Their assistance and recognition helped motivate my continued academic work.

Many people have contributed to the publication of the Chinese version of this book. Professor Zhao Xiaoliang translated it with great patience and enthusiasm, notwithstanding her busy teaching and research schedule; without her efforts, the Chinese version would not have been possible. This book was originally planned to be published by Zhejiang Publishing Group, which paid for the translation, but unfortunately, it could not do so for various reasons. I am grateful to Ms. Wang Libo, who diligently edited the manuscript despite the challenges. I also express my gratitude to Dr. Liu Ji, a former colleague at the Institute of Public Policy of South China University of Technology, for his assistance in facilitating the translation process.

Three anonymous academic reviewers meticulously reviewed the text and provided valuable insights and suggestions, which played a pivotal role in shaping the Chinese version. If the Chinese version has been improved over the original, it is entirely due to these individuals' comments. I thank the Chinese University of Hong Kong Press for publishing this book and Dr. Ye Minlei and Dr. Yang Yanni for their hard work during the publication process.

Lastly, I wish to express my deepest gratitude to my husband, Professor Zheng Yongnian. Throughout the writing of this book, our continuous discussions and his invaluable insights have greatly enriched its content. His unwavering support and intellectual collaboration have been indispensable and have significantly shaped the Japanese and Chinese versions.

Finally, I sincerely hope that a future will soon arrive when limitations or restrictions no longer confine the study of the Chinese Cultural Revolution.

Guangzhou, China
2018

Yang Lijun

Contents

List of Figures

List of Tables

CHAPTER 1

Introduction: Central Themes and a Review of the Literature

The Great Proletarian Cultural Revolution (hereinafter referred to as the "Cultural Revolution"), also known as the "Ten-Year Catastrophe," took place in the People's Republic of China during the decade from 1966, with the publication of the May 16 Notification to the death of Mao Zedong and the crackdown of the Gang of Four in October 1976. This book, however, takes the Ninth Congress of the Chinese Communist Party (hereinafter referred to as the "CCP") up to April 1969 as the subject of analysis from the perspective of the struggle for citizenship. This is because even though the ten years of the Cultural Revolution were characterized by a consistent philosophy of class struggle and an emphasis on political movements, there were significant differences between the first three years and the last seven years in terms of the form of the movement. While the first three years emphasized the autonomy of movement proponents, a bottom-up movement, the last seven years emphasized political mobilization, returning to a top-down form under the party's leadership as before the Cultural Revolution. Undoubtedly, the movements that took place during the first three years of the Cultural Revolution had two prominent features: the first was the formation, fragmentation, and reorganization of various factions, and the second was collective violence.

Y. Lijun, *Contest for Citizenship and Collective Violence during China's Cultural Revolution*, IPP Studies in the Frontiers of China's Public Policy, https://doi.org/10.1007/978-981-97-0906-9_1

Researchers have conducted many studies from different perspectives to analyze the causes of these two significant phenomena. This book will attempt to revisit these phenomena from an institutionalist perspective, drawing on previous research findings.

In this introduction, the author will explain the problematic issues encountered in researching this book and will proceed to organize and analyze studies on the Cultural Revolution and relevant social movements.

1 Questions Raised

There is little doubt in academic circles that the formation of numerous factions characterized the first three years of the unprecedented Cultural Revolution, their frequent changes and reorganizations (referred to here as the "fluidity of factions"), and the escalation of collective violence in the process. The concept of "faction," as used in this book, does not refer to the term "political faction" commonly used in the field of political science but instead to the spontaneous formation of social organizations by the masses during the Cultural Revolution. The fluidity of factions can be broadly divided into three categories. The first category is the fluidity of factions in the political arena. In other words, during a certain period of the Cultural Revolution, a faction may become a central player in the political arena because of the support of Mao or other state agents. Still, with the reorientation of the movement or the downfall of the political figures behind the faction, other factions will soon replace its position. The second category is the fluidity of the factions, i.e., they are constantly in flux as they unite, divide, disintegrate, and reorganize in response to changing political situations. Although many factions existed during the Cultural Revolution, it was not uncommon for them to disintegrate today, be absorbed by other factions, or even die out tomorrow. The third category is the fluidity of members within factions. In other words, factions during the Cultural Revolution had a loose internal structure. Membership within the same faction was not fixed, with new members joining and old members leaving at any time. Even if a faction could exist for a long time and its name remained the same, the members of the faction would be very different from when it was first established. These three significant fluxes constituted the daily state of social and political life during the Cultural Revolution.

The changing factions were accompanied by massive collective violence. Many scholars in Europe and the United States have compared the collective violence during the Cultural Revolution to the "Great Purge" in the Soviet Union in the 1930s.[1] Japanese scholar Mitsuyuki Kagami believes that the Cultural Revolution and the civil war between the nationalists and the communists in the 1940s were large-scale "civil unrest in which compatriots fought each other."[2] The Great Purge in the Soviet Union resulted in the killing of some 13 million people and the exile of many intellectuals to Siberia.[3] In contrast, the number of people killed during the Cultural Revolution is still difficult to calculate accurately due to the lack of information.[4] But from the facts revealed by research,[5] it can be said that the collective violence during the Cultural Revolution was no less harmful than the Soviet purges.

Generally speaking, violence can be classified as individual, collective, and state, depending on the subject. The violence studied in this book is not inter-individual violence, nor is it perpetrated by the state apparatus, such as the army or the police, but is collective violence since the subjects of the acts are members of factions.

Collective violence during the Cultural Revolution can be divided into collective violence within homogeneous factions (fighting between different factions within the rebellion), collective violence between heterogeneous factions (fighting between conservatives and rebels), collective violence committed against class enemy (black five, "anti-party and anti-socialist group," etc.), and collective violence committed by members of factions against agents of the state. Maintaining a stable social order and preventing conflicts between various interest groups were initially the most basic functions of the state; as long as the government functions properly, it is difficult for one social group to exercise violence against other social groups in any setting. Therefore, when studying the collective violence during the Cultural Revolution, it is necessary to ask what role the state played. In other words, were there factors in the state system that induced violence before it occurred? Did the state take measures to stop the violence when it happened? Or did it encourage violence and contribute to its expansion and social deterioration?

One of the four types of collective violence mentioned above is violence exercised by mass organizations (primarily by rebel groups) against party and government cadres, i.e., by social groups/factions against state agents. Why, then, did the state allow such violence to take place? This question is the key to understanding the Cultural Revolution. To understand factional conflict and collective violence during the Cultural Revolution, looking at two aspects besides focusing on and exploring inter-factional conflict is essential. The first is the friction within state institutions, especially the power struggle between state leaders and policymakers. The second is the impact of that friction on the conflict among social factions. To sum up, it is the opinion of the author that the factional conflicts and collective violence during the Cultural Revolution were closely related to the policy differences and power struggles within the state. It can even be said that the conflicts and violence that started at the impetus of state leaders were rewarded, tolerated, and tacitly approved by the leadership. Moreover, changes in the subject matter of factional conflict and violence are also inextricably linked to conflict among state leaders. Therefore, to analyze the causes of the two significant phenomena of factional change and collective violence, it is necessary to develop a theory that can effectively explain collective violence in the Cultural Revolution. This theory demands careful consideration of two critical elements: the direct actors within the social factions and the indirect actors within the state. It requires a thorough understanding of the changes occurring within each of these factors and an examination of their interaction. By doing so, we can better understand how this interaction drove the progression of the Cultural Revolution. Of course, this does not simply mean reading the interaction between the state and society but putting the two in the same institutional space, the setting or "playing field" where the Cultural Revolution took place and evolved. The author believes that the study of the Cultural Revolution must be preceded by an analysis of the institutional spatial setting, which constrains and regulates proponents' behavioral characteristics and how state and society interact. Furthermore, the author also believes that the two components of "temporal and spatial" are significant in studying the Cultural Revolution. The "temporal" component is easy to understand and refers to the timeline of the Cultural Revolution. It is essential because the Cultural Revolution was a dynamic process in which participants and goals of the movement were constantly changing. Likewise, changing was the participants' perception of the Cultural Revolution and the strategies and methods developed in the participation process. "Old athletes" was a term used in the late period of the Cultural Revolution, often used by proponents of the Cultural Revolution to refer to themselves or to tease each other. Here, the so-called old athletes have experienced political movements for some time and have developed the skills to protect themselves in the political environment. The "spatial" component refers to the institutional spatial setting mentioned earlier and the geographical space where the events occurred, having unfolded differently in urban and rural areas. This book attempts to incorporate the "temporal" and "spatial" components of the unfolding of the Cultural Revolution into the theoretical constructs and the behavioral characteristics of the proponents of the Cultural Revolution. The author will focus on the changes in the state and society, their interaction, and their impact on the Cultural Revolution's process. Before discussing this book's research methodology and theoretical framework, it is necessary to examine the studies of the Cultural Revolution to date. As the studies are overwhelming, this book limits the literature review only to that relevant to exploring collective action in the Cultural Revolution.

2 The History of Cultural Revolution Studies and Their Different Interpretations

Due to the differences in academic and political environments, the studies of the Cultural Revolution in China and abroad are very different. This section will discuss the history of Cultural Revolution studies in China, Japan, Europe, and the United States, respectively.

2.1 *Research on the Cultural Revolution in China*

"The Cultural Revolution occurred in China, but its research is conducted abroad" is a common saying among Chinese researchers. This is precisely the case. The main reason is that various forbidden areas exist in studying the Cultural Revolution in China. The official explanation of the Cultural Revolution is set out in the *Resolution on Certain Historical Issues Since the Founding of the People's Republic of China* (hereinafter referred to as "the Resolution") issued in 1981 by the Central Committee of the Communist Party of China (hereinafter referred to as "the CCP Central Committee"). It states, "The Cultural Revolution was wrongly launched by leaders and exploited by counter-revolutionary groups. Counter-revolutionary groups used it to bring about civil strife that brought deep disaster to the party, the state, and people of all ethnicities." This official interpretation rejects the Cultural Revolution in its entirety, both to maintain the legitimacy of the party and to "eliminate the continuing impact of the factionalism of the Cultural Revolution on real life."[6] Such an interpretation significantly constrains the scope and possibilities of research on the Cultural Revolution. The state further reinforces this constraint by restricting access to archival materials and interfering directly or indirectly with the media. For example, in 1988, the state issued a notice requiring special approval for publishing books related to the Cultural Revolution.[7] This publication ban continues today, more than fifty years after the start of the Cultural Revolution, and still confines scholarly research on the Cultural Revolution. Against the backdrop of constraints on multifaceted studies of the Cultural Revolution, many scholars have shifted their research perspective from theoretical analysis to historical narratives. As a result, most domestic studies of the Cultural Revolution in China focus on exposing inside stories of significant events taking place at the time. Moreover, such studies are always based on the party's official narrative and thus can hardly be described as value-neutral and objective.

The earliest work on the history of the Cultural Revolution in China is *A Ten-Year History of the Cultural Revolution* by Yan Jiaqi and Gao Gao.[8] In this work, to sort out the history of the Cultural Revolution, the authors used openly published government documents and printed content issued by various factions as the two main types of historical material. Gao Gao later admits that due to political constraints made it impossible to analyze the history of the Cultural Revolution in depth.[9] It is the author's opinion that this self-evaluation is very objective and appropriate. In addition, Wang Nianyi's *The Age of Great Turmoil* also provides a detailed overview of the history of the Cultural Revolution. Though arousing the attention of the academic community,[10] this book remains in the rut of emphasizing historical facts without advancing theories.

While researchers have been sorting out the Cultural Revolution from the perspective of history, literary scholars have been relatively quick to think about it from an artistic perspective. "Scar literature" mainly exposes the inhuman cruelty of the Cultural Revolution, which emerged as early as 1976, just after the end of the Cultural Revolution, and received much attention at the time. Scar literature resonated with and was supported by many who experienced the Cultural Revolution, but it did not last long. One of the reasons for its cooling down was that as time went on, enthusiasm waned, and people began to look at scar literature dispassionately. Many began to think that fictional literature was not a good way to reflect on the Cultural Revolution. Another reason was that to maintain the prestige of the ruling party and social stability, the Central Committee of the CCP began to advocate a "forward-looking" attitude toward the Cultural Revolution. It took indirect measures to obstruct the development of scar literature. Despite this, there is still an intense desire among those who lived through the Cultural Revolution to record their painful experiences.[11] For example, the famous writer Feng Jicai documented the Cultural Revolution's history in interviews for almost a decade, from the mid-1980s to 1996. In the 1990s, even the younger generation who had not experienced the Cultural Revolution showed increasing interest, as most were searching for the truth behind what happened during the Cultural Revolution.[12] For example, Wu Wenguang and others interviewed five Red Guards from 1992 to 1993 and compiled the interviews into a document. In 1997 after the return of Hong Kong, with the increasing openness of China, scholars had more opportunities for overseas exchanges. At the same time, scholars conducting research overseas also used Hong Kong as a platform to publish Chinese-language works on the Cultural Revolution.

Overall, among the Chinese Cultural Revolution books published from the end of the last century to the present, just like previous Chinese Cultural Revolution books, they mainly focus on exploring historical truth through the excavation of historical materials. There are works that serve as panoramic studies of the Cultural Revolution, such as Bu Weihua's *"Smashing the Old World": Havoc of the Chinese Cultural Revolution (1966–1968)*[13] and Shi Yun and Li Danhui's *When the "continuous Revolution" Goes Awry: From the Anti-Lin Biao Campaign to the Anti-Deng Xiaoping Campaign (1972–1976)*.[14] There are also studies of the Cultural Revolution from local or industry perspectives, such as Ma Jisen's *The Cultural Revolution in the Foreign Ministry*,[15] and Gao Shuhua and Cheng Tiejun's *The Cultural Revolution in Inner Mongolia: Oral History of a Rebel Leader*.[16] Works that focus on the violence of the Cultural Revolution include *The Massacres during The Cultural Revolution*, edited by Song Yongyi,[17] and *The Myth of Blood—A Chronicle of the 1967 Cultural Revolution Massacres in Daoxian, Henan Province*, by Tan Hecheng.[18] Historiographical collections of significant value include *Self-Criticism Books during the Cultural Revolution*, published by the China Cultural Revolution History Press.[19] A collection of oral histories published by Bu Dahua, Qin Xiao, Li Dongmin, Kuai Dafu, and other figures of the Cultural Revolution are presented in *Memories and Reflections: One of the Oral Histories of the Red Guard Era*.[20] There is also a series of Cultural Revolution data collections published by Sisyphus Cultural Publishing in Taiwan.[21] In comparison with previous periods, the most significant changes in Cultural Revolution historiography during this period are seen in the more detailed use of historical materials and the expansion of research objects and themes.

In addition to historical investigation, there are also several books that have attracted attention for their attempts to provide theoretical analysis of the Cultural Revolution. The first is Xu Youyu's *Rebels of All Stripes: A Study of Red Guard Mentalities*.[22] The second is Tang Shaojie's *An Episode of the Cultural Revolution: The 1968 Handred-day War at Tsinghua University*.[23] The third is Yin Hongbiao's *Footprints of the Missing: Thoughts of the Youth during the Cultural Revolution*.[24] Xu Youyu's book, *Rebels of All Stripes: A Study of Red Guard Mentalities*, published at the end of the last century, explains the causes of behavioral characteristics and regional differences of the Red Guard movement. The author absorbed the interpretations of social conflict that dominated English-language Cultural Revolution research at the time, emphasizing the role of the official revolutionary and class struggle ideological indoctrination among young people before the Cultural Revolution. Xu argues that the Red Guards, amid the Cultural Revolution, were a significant force in the rebellion. He pointed out that the seemingly incomprehensible rebellious behavior of the Red Guards during the Cultural Revolution, with its regular sectarian divisions and fierce factional struggles, was the result of three factors, mainly the combination of official ideological indoctrination before the Cultural Revolution, Mao's skillful use of tactics during the Cultural Revolution, and the outbreak of pre-existing contradictions in the Chinese social and political system at the time.[25] Although not explicitly stated, the author's efforts to add the state element to the social conflict theory paradigm that has dominated mainstream studies of the Cultural Revolution since the 1980s can be seen. It is also an academic dialogue between a local Chinese scholar and the Western-dominated Cultural Revolution studies. Before this, there was little scholarly back and forth between Chinese academics and the overseas world of Cultural Revolution studies due to various factors such as language barriers and lack of information.

Tang Shaojie's study of factional struggles at Tsinghua University focuses on describing the course of the Cultural Revolution at the university from June 1966 to August 1968. The author's study of the factional struggle at Tsinghua University, which was once at the center of the political stage during the Cultural Revolution, helps to clarify how the power struggle at the top of the central government interacted and mutually conflicted with factional behavior. In particular, the author's study of the factional organizational structure of Tsinghua University and how this structure influenced rebel movements across the country is very illuminating. However, in his explanation of the causes of the factional struggles and violence of the Cultural Revolution, Tang attributes them to "fear"[26] without further analysis of the underlying institutional factors that produced such a "fear" phenomenon. This lack of analysis can be considered a shortcoming of the work. In addition, in 1972, William Hinton, an American scholar, also published a book based on interviews with people involved in the sectarian struggle at Tsinghua University entitled *The Hundred-day War: The Cultural Revolution at Tsinghua University*.[27] The interviews, conducted shortly after the end of the Hundred-days War, provide a clear picture of how the participants in the movement perceived the Cultural Revolution, their motivations for participating, and the various interests and stakes wrapped up in the language of "revolution." These two excellent works complement each other and provide a good understanding of the multiple aspects of the Cultural Revolution at Tsinghua University. In addition to first-hand historical material in this book, the author uses William Hinton's interviews as a source of second-hand research. However, this book's focus is different from that of Hinton's historical—presentation in that, based on Hinton's historical research, this book analyzes how interactions between the state and society form factional behavior under the institutional spatial configuration.

If Xu Youyu's and Tang Shaojie's studies focus mainly on the behavior of the participants in the Cultural Revolution (though heretical thought trends in the Cultural Revolution are not the focus of these two scholars' studies, both deal with it to varying degrees), then Yin Hongbiao's *Footprints of the Missing: Thoughts of the Youth during the Cultural Revolution*, published in 2009, is a masterpiece that explores the ideological changes of the movement's participants. Understanding heretical thinking during the Cultural Revolution is critical to understanding the Cultural Revolution as a social movement. It is no less essential to study heretical thinking than to explore the rebellious behavior of the participants in the Cultural Revolution. Unfortunately, in Cultural Revolution studies in China and abroad, there is a lack of systematic and in-depth studies on heretical thinking during the movement. So far, the only books on this subject are mainly historical collections.[28] This may be related to the fact that mainstream Cultural Revolution studies were dominated by overseas scholars' "outside-in perspectives" which often had to face language and cultural barriers. Yin Hongbiao's study focuses on the various popular youth ideological trends from the left and right from 1966 to 1967 and analyzes the political and social environment in which they emerged. The author argues that the proliferation of heretical youth trends from 1966 to 1968 was mainly due to the "Great Democracy" advocated by the Cultural Revolution, which gave young people "the opportunity to express themselves, form groups, and publish newspapers and journals without the approval of the party and government leaders, on the premise that they supported the Cultural Revolution. The proliferation of heretical thinking after 1968 was made possible because young people sent to the countryside could break free from official ideological control, forming "folk villages of thought."[29] Yin Hongbiao's analysis partly reflects the facts of the time but is too superficial. The author believes this is mainly because Yin focuses too much on the political context of the ideological level of a particular group but does not pay enough attention to the interaction and competition between the state and society and between different social groups in terms of ideas or ideologies. As mentioned by the author, although the right to express oneself without the approval of party and government leaders appeared to have gained a certain degree of freedom and autonomy during the Cultural Revolution, it was not in actuality. Regarding literary inquisition, the Cultural Revolution was the worst period in the history of the People's Republic of China. In addition, the practice of being sent down to the countryside was already in place before the Cultural Revolution. Still, it did not lead to developing an ideology different from the official one. In the author's view, the various heretical ideologies of the Cultural Revolution resulted from interaction and competition between the state and society and between different social groups, and not merely the expression of various grievances due to the availability of a more lenient environment. With this in mind, it is also necessary to introduce the institutionalist view of the interaction between the state and society, which the author will refer to later in studying the thinking of young people during the Cultural Revolution.

In summary, although it has been more than 50 years since the end of the Cultural Revolution, the study of the Cultural Revolution is still in its infancy in China. Apart from a few works, a comprehensive study has yet to go deeper. Although many researchers have tried to explore the Cultural Revolution from different angles, progress has been slow, and it will take quite some time to achieve results. There are many reasons for this. First, it is a time-consuming project to uncover and organize historical material, and there is still a long way to go to collect and organize historical material on the Cultural Revolution. At the same time, due to the restrictions on the research of the Cultural Revolution in China, the recorded material related to examining the Cultural Revolution in various archives is basically closed. In 2015, when the author requested access to Cultural Revolution materials at an archive in an inland city, the author learned that since 2013, archival materials related to all movements after the founding of the People's Republic of China, including the Cultural Revolution, had been banned from viewing. This is more restrictive than when the author accessed the Cultural Revolution archives at the exact location in 2003. Second, any significant new research area requires the collective efforts of many researchers, who explore it from different perspectives and gradually form a pluralistic research paradigm through mutually complementing other research findings, thus producing new academic results. This necessary academic process, however, has yet to be practiced in the field of Cultural Revolution studies. Finally, as mentioned above, the study of the Cultural Revolution is still an academic "no-go area" in China, with many political restrictions, not only in terms of collecting, organizing, and analyzing historical materials on the Cultural Revolution but also in terms of limiting researchers' access to documentation. It also restricts the freedom of scholars to conduct research from different perspectives, thus hindering the emergence of new research paradigms.

However, apart from the reasons mentioned above, a more critical factor that makes it difficult to deepen the study of the Cultural Revolution lies in the problem of methodology. A methodology must be scientific, meaning analyzing a given socio-political phenomenon must be value-neutral and objective.[30] It can be said that the most significant controversy in Cultural Revolution studies is the methodological issues associated with values.[31] In other words, it becomes the question of who is doing the research. Chinese researchers of the Cultural Revolution can be broadly divided into two categories. The first category is the so-called Cultural Revolution generation that experienced it firsthand and recorded their experiences. The second group is the generation that grew up during the Reform and Opening-Up period and did not experience the Cultural Revolution firsthand. These two generations have similar perceptions of the Cultural Revolution. However, the perceptions of the generation that experienced the Cultural Revolution firsthand can be very different. Although understandably, the victims denied the

Cultural Revolution, and the beneficiaries affirmed it, the problem is that during the long process of the Cultural Revolution, there was no clear boundary between victims and the beneficiaries. As the direction of the movement changed, today's victims could become tomorrow's beneficiaries, or conversely, today's beneficiaries could become tomorrow's victims. So, even if we study the same period, the evaluation of the Cultural Revolution changes due to the different experiences of individual researchers or the same researcher's evaluation of different periods of the Cultural Revolution as it unfolded. The evaluation of the Cultural Revolution varies from one period to another by the same researcher. The same situation exists among researchers who did not experience the Cultural Revolution. For example, researchers have reached different conclusions when analyzing the Cultural Revolution from the perspective of democracy and human rights. "Some scholars see the Cultural Revolution as an extreme example of disregard for democratic human rights, while others emphasize the existence of legitimate elements in the Cultural Revolution-style democracy."[32]

For an event as multifaceted as the Cultural Revolution, it would be inevitably one-sided and superficial to judge it based on positive or negative values. Even if researchers cannot avoid being influenced by their values, they will not be able to make progress in studying the Cultural Revolution if they only regard their values as legitimate or conduct research solely to justify them. Therefore, it is challenging to develop a convincing analysis without changing the status quo of researchers relying solely on their values to conduct research. One cannot expect to produce a theory that can effectively explain the Cultural Revolution in such a manner.

2.2 *Cultural Revolution Studies in Japan*

As early as the Cultural Revolution was underway, Japanese academics began to conduct much research on it, which was rare among scholars from other countries. However, after the end of the Cultural Revolution, Japanese researchers turned their attention to the ongoing Reform and Opening-Up period. This was especially true in the 1990s when the study of the Cultural Revolution almost disappeared in Japan. The reason for this is not only that the dramatic changes brought about by the reform and opening policies have drawn the attention of scholars, making contemporary China a more relevant and urgent topic of study, but also because of the profound effect the Cultural Revolution had on the late 1960s and early 1970s Japanese intellectual community. The Cultural Revolution eventually became a subject that no one wanted to touch in the Japanese field of Chinese studies. At the time, the differences in perceptions of the Cultural Revolution were hotly debated among Japanese intellectuals. This debate eventually led to an organizational split for previously pro-China intellectuals within certain social groups, including the Sino-Japanese Friendship Association. Those who praised the Cultural Revolution saw it as a Mao-led experiment in creating an egalitarian society by eliminating the three "significant differences"—between workers and peasants, urban and rural areas, and intellectual and manual labor. By contrast, many deniers of the Cultural Revolution considered it to be a power struggle at the top. In the late 1960s, praise for the Cultural Revolution far overshadowed its denial. This phenomenon was related to the two countries' lack of communication and information. Likewise, in 1960 the Anpo Protests Movement in Japan (where the Socialist Party of Japan, the Communist Party of Japan, and Japanese revolutionary forces such as labor unions, prevented the revision of the Japan-US Security Treaty in the Diet). At its peak, the protest saw hundreds of thousands of people marching.[33] Also very relevant to this background is the ongoing Zenkyoto Movement (Japan's All-Campus Joint Struggle movement, commonly known as the Zenkyoto, demanded the democratization and autonomy of school management), which started at Waseda University in the late 1960s and later moved to the University of Tokyo and expanded to major national and private universities. The Zenkyoto movement found a line of willing resonance in China's Cultural Revolution, and the ideas of the Cultural Revolution profoundly influenced the Zenkyoto movement in Japan, where Mao Zedong's ideas were revered, and a portrait of Mao Zedong once hung at the main entrance of the University of Tokyo. This shows the breadth and depth of the influence of the Cultural Revolution on Japan at that time. Later, this misjudgment of the Cultural Revolution became a sore point for many Japanese academics, and some even stopped their academic work as a result.[34] For these reasons, after the end of the Cultural Revolution, the study of the Cultural Revolution in Japan was left in a relatively withered state.

Borrowing from Satoshi Amako's inductive approach, the methodologies of Japanese scholars who studied the Cultural Revolution from its completion to the end of the last century include the Power Struggle

Theory represented by Mineo Nakajima, the Wholistic Approach of Cultural Revolution illustrated by Minoru Takeuchi, and the Theory of the Deterioration of Ideas expressed by Mitsuyuki Kagami.[35]

First, a discussion on the Power Struggle Theory. This type of research believes that the Cultural Revolution had nothing to do with culture but was a power struggle over policy decisions between the Maoist and the Liu Shaoqi factions in the Chinese Communist Party's top leadership. The policy differences between the two sides in politics, economics, national defense, and foreign affairs eventually evolved into a power struggle. Its representative, Mineo Nakajima, explained that "the fundamental meaning of the Cultural Revolution ...... is nothing more than a power struggle"[36] and is still the consensus of many researchers.

Second, the methodology of the Wholistic Theory of the Cultural Revolution. Minoru Takeuchi emphasizes that "the Cultural Revolution was originally a whole, ... but also had multiple layers, with five or seven levels—(1) ideological, (2) institutional, (3) power, (4) social authority, (5) social ethos, (6) literary and artistic, and (7) various academic levels—superimposed."[37] This interpretation emphasizes the Cultural Revolution's diverse, multifaceted, and complex nature but is not conducive to lending any understanding. Rather, it reveals the confusion that Minoru Takeuchi and his Japanese contemporaries felt when they tried to understand the Cultural Revolution but failed to do so.

Third, the Theory of the Deterioration of Ideas. Mitsuyuki Kagami focuses on the ideological significance of the Cultural Revolution, which he sees as "a process in which idealistic notions succumb to the realities of power struggle and thus deteriorate."[38]

Based on a generalization of Japanese studies of the Cultural Revolution, Satoshi Amako proposes an interpretation of the Cultural Revolution that resonates with ideas and grievances. He argues that the study of the Cultural Revolution should focus both on Mao's theory of a unified, independent, democratic, and strong China and the political practice of realizing his grand idea, as well as on the social reality of unmet social aspirations arising from the hierarchy of status in terms of class, employment, and education. In his study, the Cultural Revolution is the resonance of both.[39] Of course, Amako's *Resonance Theory of Ideas and Grievances*, published in 1999, coincided with the searing trend of social conflict theory seen at the time in Cultural Revolution studies in the English-speaking world.

After the turn of the century, the rise of a new generation in Japanese Cultural Revolution studies brought a paradigm shift and a tiny peak in Cultural Revolution studies.[40] The chapter on the methodology of Cultural Revolution studies in this book (Chapter 2) was published in 2001 as the research paper, *Contest for Citizenship and Collective Violence in China's Cultural Revolution-An Institutional Approach.*[41] In this paper, the author attempts to introduce the concept of citizenship into Cultural Revolution studies and proposes a theoretical framework that observes the interaction between the state and society in the same institutional spatial configuration. Previous studies of the Cultural Revolution in Japan have usually adopted a historiographical approach, focusing on collecting and organizing historical materials. The author argues that the explanatory approach of social movements as movements of the Cultural Revolution is equally helpful in its understanding. Introducing the research methods of sociology, especially social movements, into the study of the Cultural Revolution was one of the author's attempts in this thesis back then. In 2003 the Japanese version of this book, *The Cultural Revolution and Social Structure of China—Distribution of Citizenship and Collective Violence*, was published.[42] This was followed in 2008 by Jun Konno's book *Chinese Society and Mass Mobilization—Political Power and the People in the Mao Zedong Era*[43] and Shinichi Tanikawa's book *The Dynamics of the Cultural Revolution in China* [44] in 2011. Although the focus of these two books differs slightly from that of the author, they both draw on the research methods of social movements and emphasize the interaction between the state and society in studying the Cultural Revolution.

In addition, among the works on the Cultural Revolution in Japan related to the author's topic, two collections of research essays are also worthy of attention. One is *The Setback of Modern China: An Examination of the Cultural Revolution*, published in 1985 by the research group The Cultural Revolution and the Change of Chinese Socialism, chaired by the previously mentioned Mitsuyuki Kagami.[45] The other is *The Chinese Cultural Revolution Revisited*, published in 2003 by The Cultural Revolution Research Group chaired by Ryosei Kokubun.[46] As mentioned earlier, China's Cultural Revolution not only influenced the left-wing movement in Japan conceptually but also sent shockwaves through Chinese studies. The decade-long research group chaired by Mitsuyuki Kagami aims to reconceptualize and reinterpret the Cultural Revolution from the literature. It can be seen as an attempt to confront the field of Chinese studies in Japan after the shocks of the Cultural Revolution. One of the results of this attempt is the book *Setbacks in Modern China: An Examination of the Cultural Revolution*. While the book's findings focus on the state's philosophy of governance and cultural background that led to the Cultural Revolution, some chapters focus on the movement's participants. For example, Tsuneto Yamamoto explores how the social discontent caused by the policies before the Cultural Revolution influenced the participants' behavior by examining the "half-work/half-education" policy before the Cultural Revolution.[47] Ichie Watanabe explores the causes of the Hunan rebellion movement and the social basis behind the "Shengwulian" (the Chinese-character acronym for the Hunan Provincial Proletarian Revolutionary Great Alliance Committee) ideology by examining the rural decentralization movement of intellectual youth before the Cultural Revolution.[48] This is very much integrated with the social conflict theory perspective of the English-language literature, which will be discussed later. While most of the studies on movement participation among young people in the Cultural Revolution have so far focused on urban full-time schools, only a few articles have examined the motivations and behavioral characteristics of their involvement in the Cultural Revolution from the perspective of the 1960s half-work/half-study system, and the intellectual youth decentralization movement. Therefore, these two papers are still relevant today. The research group chaired by Ryosei Kokubun had its beginnings in the 1990s when, as a result of China's institutional reforms and the restructuring of state-owned enterprises, a large amount of material that had once been stored in the archives of government departments or state-owned enterprises reached the public domain through the used-book market. Tsuneto Yamamoto appropriately assesses the research group's collection of essays, *Chinese Cultural Revolution Revisited*. He argued that although the editor, Ryosei Kokubun, emphasizes that the purpose of the book is to shed light on the Cultural Revolution from the perspective of the state and society interaction by providing case studies of the Cultural Revolution which focused on central government departments (e.g., the party, the PLA, the Ministry of Foreign Affairs), namely, those directly related to the centers of power, it also observes cases of those organizations on the periphery of power, such as democratic parties, universities, factories, and ethnic minority areas. On the whole, this effort does not achieve its desired goal. Analytical studies on government departments tend to favor state-level interpretations, while essays on the periphery of power neglect relevant interactions with the state.[49] Nonetheless, the collection of essays built on solid historical analysis and a selection of cases covering representative departments of the unfolding Cultural Revolution remains valuable in Japan's history of Cultural Revolution studies.

2.3 Cultural Revolution Studies in Europe and the United States

Compared with the stagnant situation in China and Japan, studying the Cultural Revolution in Europe and the United States has made significant progress in theoretical interpretation and empirical analysis. Overall, although Western scholarship on the Cultural Revolution is characterized by the times and influenced by the individual values of its researchers, both the methodology and the analysis perspective are rich and varied.

By the end of the last century, research on the Cultural Revolution in Europe and the United States could be broadly divided into two categories. The first category focuses on state policies, national leaders, and political elites and adopts a top-down perspective, which this book calls the "state-centric approach." Another type of research focuses on the group and factional level and adopts a "bottom-up" approach, referred to in this book as a "socio-centric approach." After the turn of the century, the approach to the Cultural Revolution began to focus on the political process of forming collective action, referred to in this book as the "political constructionist approach."

The State-Centric Approach to Research

There are three approaches to the study of state-centrism: the Mao-centric Theory, which emphasizes Mao's philosophy, ideals, extraordinary charisma, and personality traits and psychological elements; the line/ Power Struggle Theory, which highlights the opposition between Mao and other leaders; and the Policy Theory, which emphasizes aspects of state policy.

First, the Mao-centric perspective. Scholars have focused differently on the interrelationship between Mao and collective action during the Cultural Revolution. In the early days of the Cultural Revolution, overseas scholars could only learn about China through official public documents of the Chinese government and the press due to the lack of other information channels. Under such circumstances, scholars were quickly attracted by the idealistic values of the Cultural Revolution. Some scholars of this period tended to romanticize the Cultural Revolution in terms of

Mao's stated intentions for launching it. Richard Pfeiffer, for example, argues that Mao launched the Cultural Revolution in pursuit of revolutionary purity, that it was a genuine revolution, and that in doing so, Mao intended to create a new structure of rule.[50] In addition, many researchers focused on Mao's ideology, arguing that the collective action during the Cultural Revolution resulted from Mao's mass mobilization strategy. Harry Harding argues that the task of the Cultural Revolution was primarily Mao's attempt to prevent or stop the growing bureaucratic tendencies within the party by expanding mass participation in politics.[51] For these scholars, the collective action during the Cultural Revolution resulted from Mao's mobilization strategy.

In addition, some scholars have borrowed Max Weber's concept of the leader's transcendental charisma to analyze the irrationality of collective action during the Cultural Revolution. Anne F. Thurston, for example, points out that Mao's hypnotic charisma transformed the rational masses into irrational ones and that his immense influence over his admirers far exceeded the masses' ability to reason and to judge in the pursuit of interests. Hence, the mass behavior rooted in fanatical worship and loyalty inevitably took on an irrational character. Thurston believes this was the root of the Chinese people's fanatical worship of and loyalty to Mao Zedong during the Cultural Revolution.[52] Wang Shaoguang further points out that transcendental charisma is challenging to sustain permanently. Once Mao's transcendental charisma disappeared, the entire movement fell into anarchy and was highly susceptible to factional conflict.[53]

Other studies have adopted a social psychological approach to explain Mao's motivations for the Cultural Revolution and its collective violence. Robert Lifton, for example, suggests that the psychological element of Mao's activist response to death in his later years played a significant role in his motivation for the Cultural Revolution. In Lifton's view, Mao's last years were marked by a dual psychological crisis of fear of his death and anxiety over the situation in China after his death. Although Mao knew that death was inevitable, he hoped that his revolutionary spirit would not die with his death and that this goal could only be achieved by revolutionizing Chinese society as a whole.[54] Moreover, Ross Terrill argues that the psychological element of Mao, who was outside of reality and buried in his own ideal world, played a crucial role in launching the Cultural Revolution.[55]

Lucian Pye's research represents the field of studies that explains the Cultural Revolution from a psychological perspective. Pye is one of the quintessential figures in applying social psychology to analyze Chinese political culture. He points out that China's more than one hundred years of humiliating modern history have internalized the psychology of "resentment" as part of Chinese political culture and constitutes the inner cause of collective violence, including the Cultural Revolution. Pye argues, "In terms of political emotions, both resentment and the hypothesis of the enemy have played a vital role in modern Chinese politics. There is no other country where resentment plays a more important role in political culture than in China."[56] According to Pye, if one considers the importance of the emotional element of resentment in political culture, it is easy to understand why conflicts between leaders and social groups are more likely than in any other country to turn to collective violence when they occur.

In addition, sociologist Albert J. Bergesen, who views the Cultural Revolution as a massive state-initiated political persecution campaign, advocates for "political witch hunts" to explain the reasons for the Cultural Revolution. In other words, the unification of the state requires not only loyal admirers but also enemies. If there were no enemies, they would be created by purges, trials, denunciations, arrests, imprisonment, etc. The state sanctifies its existence through the ritual of triumph over the "enemy," which is the process of political witch-hunting. This process is a punitive ritual, as opposed to a rewarding ritual that honors heroes. In the course of repeated punitive rituals, the state's interests are internalized as part of the daily life of the masses. Bergson argues that this political ritualization of enemy creation was responsible for the widespread collective violence of the Cultural Revolution.[57]

Second, the line/Power Struggle Theory perspective. Before the Cultural Revolution, many overseas researchers used to believe that there was a relatively high level of consensus in the leadership of the CCP. However, the outbreak of the Cultural Revolution shattered this myth. After the Cultural Revolution, researchers began to trace the history of the struggle for power in the pre-Cultural Revolution leadership and to explore how this struggle influenced the occurrence of the Cultural Revolution. The line/Power Struggle Theory is a new approach, and it is in this context that the theory emerged.[58] Roderick MacFarquhar traces the causes of the power struggle of the party leadership during the Cultural Revolution back to before the Great Leap Forward. According to his research, power struggles among party leadership had occurred since at least the late 1950s and had been growing, leading to the outbreak of the Cultural Revolution. The power struggle among the top leaders, in turn, influenced their subordinate factions, triggering a chain of struggles between different factions and between the central and local levels, which further contributed to the process of the Cultural Revolution.[59] Similarly, Lowell Dittmer argues that the Cultural Revolution was the inevitable result of the power struggle between Mao and Liu Shaoqi and the factions they led.[60]

Third, the policy theory perspective emphasizes elements of national policy. The most representative of this category is Lynn White's research. White argues that collective violence during the Cultural Revolution and its aftermath was no accident but the interaction of three state policies in place before the Cultural Revolution. The first policy element was the division into social status components and classes, the implementation of which created new social groups and contributed to the formation of group consciousness (labeling groups). The second element was the excessive power given by the state to the leaders of work units (i.e., the danwei and state agents), which created a relationship of dependence of subordinates on the leadership so that people consciously fought those whom the leaders told them to fight. This dependency makes it easy for party cadres to gain the masses' support and mobilize them to attack the targets of struggle. The third element was the legitimization of violence in political campaigns. In the seventeen years since the state's founding before the Cultural Revolution, the Communist Party launched mass political campaigns to give a political basis for legitimizing collective violence. White believed these three policy elements were highly institutionalized before the Cultural Revolution and became part of the Communist system, which was the root cause of its collective violence.[61]

Although the elements emphasized in the state-centered theory above, such as Mao's extraordinary charisma, the different philosophies of governance among state leaders and their different interpretations of the communist ideology, and the various state policies before the Cultural Revolution, were undoubtedly important sources of collective action of the Cultural Revolution. These research approaches emphasizing the state element can be considered "top-down." Such explanations treat the vast number of participants in the Cultural Revolution as passive and manipulated objects. They tend to underestimate the sense of participation of ordinary individuals and their behavioral characteristics in the movement.

Moreover, "top-down" explanations fail to address the Cultural Revolution's essential feature, the involvement in collective violence. Although it is indisputable that the Cultural Revolution was initiated directly by Mao Zedong and that Mao bears inescapable responsibility for its occurrence, it would not have happened without the active response of the masses. Similarly, while White points out that state policies triggered the Cultural Revolution, he cannot explain why the policy factors that brought about social instability contributed to the massive collective violence in the Cultural Revolution in 1966 and not at other times. A state-centric approach to research cannot effectively explain these types of questions.

The Socio-Centric Approach to Research

The socio-centric approach to research emerged as a critique of state-centrism. Its main feature is the emphasis on social conflict. It no longer sees many ordinary participants as objects manipulated by leaders such as Mao Zedong and other senior party leaders but as active participants in the movement and as essential players in taking the Cultural Revolution forward. Such studies focus on analyzing participants' personalities and psychology, the conflict of interests that existed before the Cultural Revolution, and their different ideals to explore their motivation for involvement. Some of the representative studies are explained below.

Hong Yung Lee criticizes the view of the Cultural Revolution as a mere product of power and policy struggle at the central decision-making level, arguing that the leadership's decisions somehow reflected the interests of various social groups. In his study of the 1966–1968 Cultural Revolution, Lee examines not only the political relations between multiple actors at the level of the political elite (e.g., the military, the government, the party, Mao, and the Central Cultural Revolution Group), but also the factionalism in mass organizations, thus inferring the relationship between the two levels of the political elite and the masses. According to Lee's research, mass participation in the movement resulted from mobilization and manipulation by Mao and other senior party leaders. As the movement developed, splits within the party organization gradually weakened their control over the masses. As a result, the social contradictions implicit in the pre-Cultural Revolution era surfaced, and the mass movement began to develop in its own direction. In this context, the masses saw the instructions of Mao and the CCP Central Committee as a tool to pursue their interests, either by choosing instructions that suited them or by arbitrarily distorting the instructions of Mao and other leaders to justify their actions.[62] Lee's study points out how top political figures such as Mao and others manipulated political groups so that conflicts between mass factions were manifestations of elite-level conflicts. It also showed how the Red Guards of various factions influenced top politics and the process of the Cultural Revolution.

For scholars of socio-centrism, the Red Guards were a group of particular interest. Some researchers have traced the behavioral characteristics of the Red Guards and the causes of the factional struggle back to the pre-Cultural Revolution social environment. David Raddock, for example, analyzes the relationship between individual psychological growth and political development from a social psychological perspective. He points out that the socialization of the family before the Cultural Revolution was closely related to young people's attitude toward political participation at the time.[63] Anita Chan, Stanley Rosen, Jonathan Unger, and other scholars cite the situation in Guangzhou province as an example to analyze the motivations and behavioral characteristics of youth political participation in the pre-Cultural Revolution education system.[64] Stanley Rosen tries to identify the structural relationship between the Red Guard faction and the education system before the Cultural Revolution and finds a pattern of behavior. That is, students from favored family backgrounds were more motivated to join the Communist Youth Organization, the more likely they were to enter prestigious or better middle and high schools, the more likely they were to attend college before the Cultural Revolution, and the more active they ultimately were in the Cultural Revolution. He also pointed out that since the essential criteria for university admission were "grades, class origin (*chusheng*), and political performance," the question of which of the three took precedence created divisions among young people before the Cultural Revolution, thus laying the institutional foundation for the factional struggle in the Red Guard movement later. In addition, regarding the relationship between class origin and faction, Stanley Rosen and Hong Yung Lee see the conflict between the conservatives and the rebels in the Red Guards as a collective conflict between the privileged and non-privileged classes. However, Hong Yung Lee argues that most conservatives are students with good class origins, and most rebels are students with bad class origins,[65] a view Stanley Rosen does not fully share. Based on a questionnaire survey of 2000 former Red Guards in Guangzhou, Rosen points out that the children of the red five class origins (revolutionary soldier, revolutionary cadre, worker, revolutionary martyrs, peasant, and poor and lower-middle peasant) should not be considered as a whole. That there were differences between the children of cadres and those from worker/peasant families. According to Rosen's research, children from worker/peasant families had two options. One was to join the children of cadres as a member of the red five to support the origin theory and defeat the "Black gangs" and the black five (landlord, rich peasant, counter-revolutionary, bad element, rightist).[66] The other option was to join the rebel group as a member of the non-privileged class to criticize the privileged class. In addition, regarding the family origin of rebel group members, unlike Hong Yung Lee's conclusion that the rebels were made up of children of bad family backgrounds, Rosen found that there was a large number of rebel group members with the family backgrounds of intellectual, capitalist, and old company employee who were regarded as being politically in between the red five and the black five.[67]

Unger also analyzes the factional conflicts of the Red Guards from the educational system's perspective before the Cultural Revolution. Using the Guangzhou example, he explores the changes in China's education system from the 1950s to the 1960s. He identifies the paradox of the disproportionate increase in high school students in the early 1960s and the gradual decrease in university students due to severe restrictions. He thus argues that reduced access to education and a deteriorating job market led to fierce competition among students. This influenced the behavior of young people during the Cultural Revolution, and the conflicting interests around schooling and employment led to factional conflicts among students.[68]

Anita Chan focuses on the different paths of socialization of the Red Guard generation and the relationship to factional differentiation and conflict seen in the Cultural Revolution. She finds that different types of pre-Cultural Revolution political activists constituted the members of various factions in the Cultural Revolution. According to her research, although pre-Cultural Revolution adolescents received the same ideological education, they acted differently because their types of heroes, ideas, and literary imaginations were different. "Pre-Cultural Revolution adolescents had tendencies to think and behave differently, and the Cultural Revolution merely provided a stage for them to express themselves."[69]

As mentioned earlier, the central theme of the socio-centric approach analyzes the masses' motivation to participate in the Cultural Revolution. It considers the masses as prominent participants in the movement and focuses on their political and economic interests and educational backgrounds before the Cultural Revolution. The studies following this approach have explained some of the questions that the state-centered approach has failed to answer and provided a valuable perspective for understanding the sectarian divisions and conflicts during the Cultural Revolution. However, there are still questions that need to be examined. For example, why did potential conflicts before the Cultural Revolution erupt in 1966? Why did participants acting in their interests persecute and even massacre ordinary people who had nothing to do with the socio-political system that caused their discontent? How did the sources of factionalism and conflict shift from potential to reality? Socio-centric studies explain the causes of factions and conflicts, but the explanations for the reasons for an act cannot account for the act itself. We cannot see from these explanations why and how the potential (for violence) became a reality in 1966.

To explain the above, it is necessary to look beyond social factors to the role played by the leaders/elites. Without the opportunity provided by the leaders/elites at the level of the Cultural Revolution, these features of factional division and conflict would have remained in a state of potential. It was the Cultural Revolution launched by Mao that provided an opportunity for factional and conflict factors to be transformed from possibility to reality, and without the stage-setting of the Cultural Revolution, potential social disputes would not have erupted. Moreover, the variables of factional differentiation, reorganization, transformation, and escalation of violence cannot be explained without analyzing factors at the leader/elite level. Likewise, it is difficult to explain the severity of the escalation of factional conflict in the Cultural Revolution and the phenomenon of the often-violent clashes that took place by considering only some specific conflicts of interest between factions.

Political Constructionist Research Methods

Influenced by the development of social movement theories and the discussion of the relationship between the state and society in related fields, overseas scholars of the Cultural Revolution since the turn of the century have gradually tried to put the "state" element, which was secondary to "social conflict theory," back into the analytic framework of the Cultural Revolution. In general, there are two significant trends in the past decade or so of research on collective action during the Cultural Revolution. The first trend continues the methods of the previous century, taking collective behavior itself as the object of analysis, exploring its causes, behavioral characteristics, and the political logic behind it. The second trend is to take the political and cultural factors behind collective behavior as the object of analysis, exploring how collective behavior is shaped by official political and cultural influences, and how it produces a political culture that is out of sync with the official one—heterodox ideologies—in the process of following and interacting with official political culture. It also attempts to uncover the political and cultural diversity hidden beneath seemingly rigid official political ideologies. In contrast to the first category, which explores the characteristics and causes of factionalization and violence, the second focuses more on the political rituals that shape collective action, the symbols, and political discourses or "scripts" of political culture. This author believes that both types of studies belong to the realm of political constructionism. The difference, however, lies in that the first type of study is inclined to analyze the political process in political constructionism. The latter type is more inclined to explore cultural constructs in political constructionism.

In the first category, Andre Walder's research on the Red Guard movement[70] and Su Yang's book *Collective Killing in Rural China during the Cultural Revolution* are representative. In Walder's study of the Red Guard movement at Peking University, he disagrees with scholars who argue that the class origin or the different interests of the beneficiary and non-beneficiary groups before the Cultural Revolution led to the division between conservatives and rebels. He contends that the reason why some of the Red Guards joined the conservatives and others joined the rebels was related to the political choices of the participants in the movement at the beginning of the Cultural Revolution. He finds that the ambiguity of the political message of the central leadership at the beginning of the movement in 1966 led some to join the conservatives, who were aligned with the party, and others to join the rebels, who challenged the party's authority. In his subsequent research with Dong Guoqiang on the Nanjing rebel movement, Walder continues to follow this perspective in analyzing how ambiguous directives from the top in political movements create local factional divisions and how factional behavior escalates collective action in collusion with local political struggles.[71] This interpretation of political choice in an uncertain political environment has significantly contributed to understanding the causes of the differentiation of Cultural Revolution factions, but it also leaves some issues open. First, if the political choice of participants at the beginning of the Cultural Revolution determined which faction to join, what factors usually influenced the political choice of participants? Was this political choice the result of some contingent political judgment? Or were other factors behind it that led to its inevitable outcome? Furthermore, do participants have to make certain political choices? Or why do participants not wait and see rather than actively engage in an action when the direction of political development is unclear? Although political choice in an uncertain political environment can explain the factionalization and the factional struggles of the Cultural Revolution, it cannot explain the underlying reasons for the broad and deep spread of the movement.

Su Yang's book *Collective Killing in Rural China during the Cultural Revolution* is one of the only scholarly explorations of collective violence during the Cultural Revolution to date. Using county records (published Xianzhi volumes) and unpublished internal documents, the book provides a detailed and in-depth examination of the collective killings that took place in rural areas between 1967 and 1968. Su Yang rejects the social conflict theory, which views the collective violence of the Cultural Revolution as a result of social discontentedness leading to collective action and radicalization. Also rejected is the state-centered explanation, which attributes the causes of collective violence exclusively to state policy advocacy or encouragement by high-level state agents. He argues that collective killings are not a direct product of state policy but rather a combination of state policy advocacy and state control failures. He divides the capacity of state control into informational and organizational management, arguing that some local communities attach great importance to the centrally promoted framework of class struggle and conflict, which is a successful form of informational control but fails to anticipate the punishment for the extreme violence that only organizational control can achieve.[72] Su Yang argues that the local conditions and the localized process play a mediating role between state policies and the results of collective killings. The difference between these two mediating conditions leads to the geographical disparity in mass killings. Among them, the different policies implemented by the state in the process of regime re-establishment in other regions, the intensity of grassroots control by local party organizations, the framework of conflict created by local governments or political elites, the distance relationship between communities and power centers, and disputes among village clans all create different political processes, which result in different outcomes in local acts of collective killing. Here, Su Yang takes rural communities as the object of analysis, though he is more inclined to study the impact of state policies, social control systems, and actions of state agents than social collective action. He does not directly explore the deeper level of the participants' motivations in collective action or the conflict of interests among village clans. Of course, the lack of inquiry into the motivations and behavioral characteristics of actors at the social level has much to do with the difficulty of access to historical materials. It is precisely because of this lack of inquiry at the social level that Su Yang's study, despite the persuasiveness of its theoretical formations in explaining why and how mass killings occur and why there are geographical differences, is not able to fully explain exactly what motivations led local political elites to actively direct mass killings. In one case provided by Su Yang, a local political elite organized mass killings when fully aware of the central government's discouraging attitude. From this phenomenon, it is clear that the motives for organizing mass killings are not very convincing when explained purely in terms of the lack of political control, the consolidation of power bases by local elites, and the claiming of credit from superiors. His analysis does not show a necessary connection between mass killings and disputes among village clans. Furthermore, Su Yang's analysis of data from county records (Xianzhi volumes) on mass killings reveals that the places where mass killings took place were often poor and remote. Su Yang thus argues that distance and remoteness led to the inaccessibility of state control, which allowed local party organizations with relatively high intensity of control to organize mass killings based on their policy interpretation. This explanation has merit, but if one looks at it from the perspective of "the mountains are high, and the emperor is far away," distance can also keep localities out of the movement or reduce participation. For a local elite in a poor rural area, to what extent the political performance of radicalization has affected their official career is also a question that needs further investigation.

As already mentioned, the main feature of the second type of Cultural Revolution studies is to explore the logic and the features of collective action from the perspective of political culture. In his book, *The Cultural Revolution at the Margins*,[73] Wu Yiqing criticizes that the studies of the Cultural Revolution have been too focused on the violent and destructive nature of collective action to effectively explain the impact of the Cultural Revolution on the political and economic changes in China after the Reform and Opening-Up period. He argues that the roots of China's changing political and economic system from the late 1970s to the 1980s should be traced back to the grassroots rebel movement in the Cultural Revolution, which rationally pursued political and economic interests. By examining the controversies over the "bloodline theory" and the "family origin" debate, the socioeconomic demands from disaffected segments of Shanghai's laboring population, and the Shengwulian (a loose rebel coalition that mounted a serious challenge to the national trend toward political recentralization in Hunan Province), Wu argues that the political chaos within the top leadership at the height of the Cultural Revolution provided an opportunity for the disillusioned groups of the pre-Cultural Revolution to express their grievances and fight for their political and economic benefits, and they turned a critique of individuals, namely, "capitalist roaders within the Party," to a critique of the class system, the employment system, the distribution system, and other political systems. Through this work, Wu tries to reveal an alternative history of the Cultural Revolution and makes an essential contribution to the field of Cultural Revolution studies by emphasizing the multifaceted nature of the Cultural Revolution and also advocates the use of critical thinking in its study. However, as Wu mentions at the beginning of his work, his research is heavily influenced by the recent debate on the Cultural Revolution on the Chinese Internet. The discussion of the "two Cultural Revolution theories"[74] (one being the Cultural Revolution of official power struggle and the other being the Cultural Revolution of the people against the bureaucrats) that originated in the late Cultural Revolution period and have been fermenting on the Chinese Internet for nearly two decades have influenced the author's view.

Like Wu Yiqing, Yang Guobin also attempts to construct a relationship between collective action during the Cultural Revolution and contemporary political developments, though from a different perspective and approach. In his book, *The Red Guard Generation and Political Activism in China*,[75] Yang tries to use performance theory in social movement theory, as well as Victor Turner's theory of ritual process and Weber's concept of the routinization of charisma to explain the political behavior of the Red Guards during and after the Cultural Revolution. His work explores how the Red Guard movement and the subsequent experience of being sent to the countryside have seriously affected and shaped the life trajectory of the Red Guard generation. On this basis, his work adds some perspective on the relationship between the political protest movement in contemporary China and the generation of the Red Guards. Yang sees the factional struggle in the Red Guard movement as a competition around revolution and loyalty, and the violence is the inevitable result of the radicalization of that competition. He argues that this occurred because the people began to display and perform political loyalties under the interplay of a political culture that sanctified the revolution in a vague and uncertain political process. That is, Yang sees the factional behavior of the Red Guard movement as a performance in the political arena, arguing that the performance of the revolution and the rituals and sacred scripts of the political culture was very influential in the occurrence and development of the Red Guard movement. However, as the sacred scripts reached violence, their backlash also created heresy. He argues that when the sacralization of the political script reached its zenith was also its weakest period and that its repercussion led directly to the subsequent de-sacralization of the political script and the creation of new values.

Daniel Leese also sees the collective action of the Cultural Revolution as a "revolutionary performance."[76] Leese uses communication and hermeneutic approaches to analyze the phenomenon of rhetorical and ritual worship in reverence of Mao during the Cultural Revolution. In his book, *Mao Cult - Rhetoric and Ritual in China's Cultural Revolution*, the author not only analyzes the roles played by Mao Zedong himself, Lin Biao, and the large number of participants in the movement in shaping, promoting, and suppressing the Mao cult, but also explored the functions of instigation, compliance, and acceptance. Furthermore, the author analyzes specific phenomena such as the "Little Red Book" and the "loyalty dance" to elucidate the utilitarian and instrumental nature behind the Mao cult. Unfortunately, the author did not provide a convincing answer to the question of why the Mao cult had such a powerful function. The author sees the Mao cult as "a deliberate political device to ensure party unity and an unintended consequence of the relationship between 'patron and client' in a political system where the rules of promotion are not clearly defined."[77] This explanation may partially explain the phenomenon of the Mao cult but not its widespread and fanatical nature.

The trend to explore the multifaceted nature of the Cultural Revolution has emerged in recent years. Pang Lai Kwan's book, *The Art of Cloning: Creative Production during China's Cultural Revolution*, focuses on the analysis of the multifaceted and contradictory nature of the Cultural Revolution and the tension between the various roles in the subject formation, social mimesis, and propaganda of political culture at the time. The author calls it "the dialectical and contradictory nature of the Cultural Revolution."[78] The Cultural Revolution involved a dialectical tension between individuality and collectivity, as there was both extreme homogeneity and some space for expressing particular personalities. The exemplars created to guide the masses were intended to promote conformity and exclusivity, but the mass copying and imitation of these exemplars often led to their distortion and diversification. The propaganda of the Cultural Revolution contained inherent contradictions and dialectics, such as inciting the masses to rebel while also trying to control them. While the repetitive cultural propaganda reinforced ideology, the popular folk culture of writing big-character posters and performing loyalty dances also spread widely. Additionally, while the violent content was used to incite revolution in the propaganda culture of the Cultural Revolution, the aesthetic values promoted in the propaganda culture distorted the political information emphasized by the state. Pang argues that the aesthetic power generated by literature during the Cultural Revolution brought tremendous energy to the politics of the time and pushed the ideals and violence of the Cultural Revolution to extremes. Revealing the diversity and dialectic of political culture during the Cultural Revolution from a highly discursive perspective and guiding the reader to think about the Cultural Revolution from different angles are the most successful aspects of Pang's work. But, in constructing the two aspects of the dialectic, the author often neglects to analyze the relevance of those aspects. For example, when the author refers to the pervasive role that art plays in the Cultural Revolution's process of mass reproduction, its aesthetic dissemination, and the element of distortion in the process of cultural reproduction, she often forgets the extremely politicized institutional environment that governed everyone's behavior at the time, and also fails to distinguish the element of time in the Cultural Revolution. For instance, big-character posters improved calligraphic skills, artistic appreciation, and cultivation of calligraphy as a hobby are products rather than another dialectical aspect of writing big-character posters.

The interpretation of the Cultural Revolution in terms of the relevance of the state and society has been a common feature of the field of Cultural Revolution studies over the last decade. Under this commonality, some scholars have focused on the political process by which acts took place. In contrast, others have resorted to phenomenological and semiotic approaches to analyze the many phenomena of collective action in the Cultural Revolution. Diverse perspectives and analytical approaches have opened up the study of the Cultural Revolution. However, this author argues that these works still need to explain the depth and breadth of what happened during the Cultural Revolution. In other words, whether it is the factionalization and struggles of the Cultural Revolution, the phenomenon of the Mao cult, or the ideological debates such as heretical thinking, the breadth of these phenomena and the depth of their involvement in the Cultural Revolution are the main characteristics of the movement. If the Cultural Revolution was a "revolutionary performance," why did most Chinese people participate in it and pay the price with their lives? Does the mere radicalization of the revolutionary performance explain the massive collective violence?

Therefore, as mentioned above, analyzing state and social relevance is essential to provide a thorough account of the two major issues of factionalization and collective violence. For the author, however, focusing on both causes does not mean combining them mechanically to analyze how each affects factions and conflict. If this is done, it will be easier for a new explanation to go beyond the explanations offered by each of the three approaches: state-centric, socio-centric, and political constructionism. A unique and effective explanatory approach must capture the state and society under the same institutional system. The author refers to this as the institutional spatial configuration in which the state and society coexist. This so-called "institutional spatial configuration" refers to the institutional factors that go beyond the single elements of the state and society and have a regulatory effect on the single behavior of the state or society and the interaction between them. The institutional environment influences and regulates the conduct of actors in a given polity, the formation of inter-organizational relations, and the way they interact. It can also be understood simply as the institutional environment that regulates the relationship between the state and society. Understanding the institutional spatial configurations that governed state–society relations during the Cultural Revolution is crucial to studying the Cultural Revolution, as the author will elaborate on the analytical framework in the next chapter.

For the author, an approach to explaining the state and society interaction needs to focus on the institutional spatial configurations that exist between the two, examining how they, bound by the same institutional spatial configuration, interact and what influences this institutional spatial configuration exerts on the interface between them. Common to all the studies mentioned above, whether state-centric, socio-centric, or political constructionist, is neglecting the fact that leaders/elites and ordinary participants coexist in the same institutional spatial configuration. Both leaders/elites and the masses participating in the movements are influenced by a particular institutional spatial configuration, which determines the respective behavioral characteristics of leaders/political elites and participants and the patterns of interaction between the two. Similarly, changes such as factionalization and reorganization and the collective violence during the Cultural Revolution can be closely related to institutional spatial configurations.

This book aims to develop an institutionalist approach to explaining the phenomenon of faction formation, fragmentation, reorganization, and collective violence in the Cultural Revolution. Simply put, the institutionalist approach examines the state and society within the same institutional spatial configuration and analyzes how the characteristics of their individual behavior and the resulting interactions under this institutional spatial configuration affected the collective action of the Cultural Revolution. In explaining the interaction between the state and society, the author attempts to construct the relationship between the two using the concept of citizenship. As will be elaborated in the next section, the author argues that the way the state distributes citizenship determines the way members of society compete for it and that the concept of citizenship is both an important tool for understanding the interaction between the state and society and an adequate explanation of the collective action of participants in the Cultural Revolution. In contrast to the previous studies, which considered the Cultural Revolution a political movement, the author believes that the Cultural Revolution was a social movement of political mobilization in which various interest groups took advantage of political opportunities to compete for citizenship. In the course of the struggle for citizenship, the behavior and changes of each group were influenced by the institutional factors of how citizenship was distributed under the leadership of the CCP. In this way, it is possible to analyze the distribution of and competition for citizenship under the same institutional spatial configuration and to grasp both state and social elements in an integrated manner.

As conceived by the author, the analytical framework of institutionalism consists of four main elements. First, the institutional spatial configuration of the state and society; second, the system of distribution of citizenship and the behavioral characteristics of the population competing for citizenship; third, the relationship between institutionalization and collective violence at the three levels of the state, society, and state–society interactions; and fourth, how the above three institutional elements created interest quantified by loyalty, and the role such a national psychological construct played during the Cultural Revolution. Of these four elements, we can consider the institutional spatial configuration of the state and society as the first dimension of the institutionalist analysis framework. The second dimension is the citizenship distribution system, the competition rules, and the degree of state/society institutionalization. The psychological construct that interest is measured by loyalty can be interpreted as the behavioral characteristics caused by the institutional factors at the above two dimensions.

In relation to the above four elements, the author attempts to explain the factionalization, restructuring, and collective violence of the Cultural Revolution by analyzing the following questions: In what institutional spatial configuration of the state and society did the Cultural Revolution unfold? How did the state's exclusive and hierarchical approach to the distribution of citizenship within the broader context of the institutional spatial configuration of the state and society determine the behavioral characteristics of members of society competing for citizenship and influence collective action during the Cultural Revolution? How did the de-institutionalization of the state, society, and state–society interactions during the Cultural Revolution intensify inter-factional competition for interests and allow for repeated violence, culminating in a general outbreak of national armed struggle? When this book understands the Cultural Revolution as a politically mobilized social movement for citizenship, it has set up its participants as rational actors. In this respect, there is a significant difference from the state-centric view of the movement's participants as fanatical admirers motivated by Mao's charisma. However, it is undeniable that fanatical admiration for Mao was a widespread social phenomenon during the Cultural Revolution. So, when participants in the Cultural Revolution are assumed to be rational actors, what exactly should be understood about fanatical leader worship? At the same time, why would rational participants lead such irrational behavior as intergroup brutalization? These would be the questions that the author would like to answer through an institutionalist approach to analysis.

Since the concepts and methods used in this book are inspired by research on social movements conducted in Europe and the United States, a brief overview of Western analysis, represented by the United

States, is in order before specifying the definition of citizenship and institutionalist research methods.

3 Methodology of Social Movement Research

There are so many studies of social movements in European and American sociology that a comprehensive survey is burdensome and unnecessary and will be limited here to the literature relevant to the research methods of this book.[79]

3.1 Definition of Social Movements

The study of social movements focuses on a series of issues related to analyzing the occurrence, development, and results of social movements. Social movements will be discussed here in terms of internal and external causes.

Research on the internal factors of social movements focuses on analyzing the following three major questions. First, what kind of protest behavior do the movement participants adopt? Is it violent, emotional, and irrational, or a rational and more moderate protest? Second, are participants voluntary, or are they mobilized to participate? Third, does the movement aim to achieve individual interests or pursue collective values, beliefs, and goals beyond individual interests?[80]

Research on the external factors of social movements focuses on analyzing the relationship between social movements and the existing political system. Generally, social movements are defined as acts of protest by the weak against the strong. Specifically, the weak, through collective action, demand that the existing political system meet their needs. Therefore, the researcher is most interested in analyzing how the existing political system affects the occurrence of the movement. Moreover, since social movements are anti-establishment, explaining how social movements affect existing political institutions is also a research focus.

Like many concepts in social sciences, there is no universally accepted definition of social movements. However, the understanding that social movements are a form of collective behavior has become a consensus. For example, Charles Tilly believes that social movements are collective actions carried out in pursuit of collective goals, taken by people who come together with a shared sense of grievance, hope, and interest.[81]

Sidney Tarrow, based on a review of the history of social movement studies, proposes that social movements are sustained interactions between people with common goals and cohesion, and elites, opponents, and those in power, in which the former express protest against the latter. In other words, social movements have four characteristics: collective challenge, common purpose, solidarity, and sustained interaction.[82]

3.2 Research on the Intrinsic Elements of Social Movements

Studying the factors inherent in social movements has undergone a complex process before reaching today's results. In the classical studies of social movements, sociologists and political scientists saw social movements as an irrational and emotional collective action in response to the crisis of the social system and the breakdown of social order. The French sociologist Emile Durkheim pointed out that mass movements, especially the violence that occurs in collective rituals of fanatical leader worship, result in the collective's total subjugation of the individual. The collective overtakes the individual due to the sway of the collective consciousness. Durkheim's ideas have profoundly influenced modern social science and were repeatedly invoked until the 1970s in analyzing the irrational dimension of social movements. This is because the fascist movements that took place in the first half of the twentieth century, especially in their fanaticism, largely influenced scholars' perceptions of social movements.[83]

Although fascist movements died out after the war, for a long time, social movements were seen as "dysfunctional" phenomena triggered by economic crises or social disruptions. It was in the 1960s when researchers approached social movements from a historical perspective, that social movements began to be seen as constant social phenomena, and the rational factors behind them began to enter the picture. E.P. Thompson defined social movements as acts of collective bargaining through riots.[84] By analyzing the French Revolution, Tilly points out that social movements are not exclusively interruptions of everyday political life and that even violent acts of protest contain stages in which goals are reached rationally and peacefully.[85]

In the 1970s, the study of social movements was influenced by the economist Mancur Olson, and the understanding of social movements as rational choice became mainstream. Olson's masterpiece, *The Logic of Collective Action,* was not a study of social movements, but the

Rational Choice Theory discussed therein has influenced many sociological researchers. Olson analyzed collective action through the lens of "cost–benefit." According to his research, collective action can produce "public goods," so to obtain greater benefits at lower costs, participants will use "selective incentives" to decide the behavior to be taken. That is, participants may choose not to participate in an action if they feel that they will not benefit from participating or if they will benefit from not participating. In short, participants act rationally by calculating the costs and benefits. This is known as the "free-riding principle of public goods."[86]

As mentioned earlier, according to the Rational Choice Theory, social movements result from people acting rationally based on calculated costs and benefits and collective actions taken to protect and promote their interests.[87] This theory differs from the previous studies in that it emphasises on emotional, destructive, and irrational elements of collective action, rather than on the participants' motivations and norms of behavior, why they participate, and on what basis of behavioral norms they act. Even so, several issues remain for further analysis. For example, it fails to account for why people choose to take to the streets at a particular time and not at another.

The 1970s were the heyday of research on social movements. Influenced by Rational Choice Theory, scholars actively formed a new paradigm for social movement research and explored methodological approaches in conducting research from different perspectives. The study in this period can be broadly divided into three phases. The first phase focused on the relatively institutionalized forms of protest in the process of social movements. It examined how forms of protest are institutionalized along with the successive occurrences of movements.[88] The second phase studied violence in the process of social movements and suggested that violence can occur intentionally by participants or as a result of interactions between participants and political institutions.[89] The third phase took the perspective from the object of study. Unlike the previous two phases, research in this phase distinguished between two types of social movements: unorganized collective actions (e.g., sudden events) and organized social movements.[90]

In summary, a common feature of the research approaches associated with Rational Choice Theory is the emphasis on the fact that participants act in pursuit of their interests in a given situation. However, a new contradiction arises between Rational Choice Theory and the definition of a social movement as a collective action. That is, the individual decides whether or not to participate in a movement. As a collective member (such as a factory, a school, a company, a religious group, an ethnic group, etc.), the individual does not live in isolation. It is not difficult to imagine that their decisions cannot be based on individual interests alone but must include elements of loyalty and obligation to the collective. Many researchers have recognized that social movements cannot be discussed in isolation from the collective context when studying social movements as collective action. In this context, to address new issues, researchers began to explore new methodologies, and the focus of research has also shifted accordingly.

In the early 1980s, research in this new methodology focused on analyzing the causes of collective action. In contrast to the rational choice theories of the past, research in the 1980s focused on exploring the role of emotional factors in collective action. This was not, however, a revival of the classical collective action research represented by Le Bon. This research phase focused on new meaning, identity elements, and the cognitive framework that induces and legitimizes collective action.[91] The change results from the influence of advanced research of the time, such as the linguistic theory of French sociology, the symbolic meaning of anthropology, and the development of social psychology since the 1960s.

Social movement researchers in the 1980s identified a variety of "movement subcultures" and "movement networks" within social movements. That is, movement participants are not independent individuals as described by Rational Choice Theory but rather socialized groups. Such movement subcultures and movement networks were not inevitable products of social development but were created by the efforts of movement organizers. In addition, the development of the French sociolinguistics theory at the time gave new life to studying social movements. Under its influence, attention began to be paid to the role of ideology in movements, transmitted through language to influence people's thinking, judgment, and, consequently, their behavioral traits.

The research paradigm of the 1980s emerged as a critique of Rational Choice Theory. It was characterized by the assertion that it is not enough to interpret social movements in terms of individual rational choice but that a combination of "culture," "language," "networks," etc., is needed. Nevertheless, researchers do not see this new research paradigm as a rejection of Rational Choice Theory but rather as a complement. The process by which participants organize themselves through the perception of shared interests and values is considered a social movement. In other words, the study of social movements must focus on the rational and emotional elements and the collectivization process that translates them into action.

Since social movements are collective acts, sorting out the following two relationships is crucial. One is the relationship between social movements and organization, and the other is the relationship between social movements and mobilization. Social movements are often perceived as the weak against the strong and cannot confront those in power if the participants lack mechanisms for mutual coordination and a sense of cooperation. This begs the question: do social movements emerge spontaneously, or are they mobilized?

Generally, most collective acts before the nineteenth century to seize food and resist taxes were not mobilized nor planned well in advance. As such, movement leaders only existed after the movements arose. Since movements in this period were primarily sudden reactions to policies implemented by those in power, movement leaders emerged in the course of the movement in response to current events. In other words, the social movements of the period were not yet institutionalized and organized, and thus the masses disbanded as soon as the movements ended. Social movements after the mid-nineteenth century began to be organized, but they were more loosely organized, and there was a considerable distance between them and modern organizations. Into the twentieth century, as demonstrated by environmental activism, social movements began to take on a high degree of organization. However, social movements that lacked organization, including riots, remained viable. From this phenomenon, organized social movements and unorganized social movements can coexist.

Under what circumstances, then, are social movements manipulated by organizations? Some social movement scholars, influenced by Robert Michels' "iron law of oligarchy,"[92] place a high value on the role played by organized participants in social movements. In contrast, other scholars have argued that spontaneous movements have more impact. While the former advocate that the energy of the masses can only be unleashed if they are organized, the latter counters that the role of organizing is only to reduce the disorderly nature of movements. For example, the British historian Eric Hobsbawm is pessimistic about spontaneous collective action, noting that collective action lacking organization can hardly succeed.[93] In contrast, Piven and Cloward argue that spontaneous behavior is instead prone to success. The reason is that governments cannot predict when, where, and in what form spontaneous behavior will erupt and, therefore, cannot take effective emergency measures. Instead, organized social movements evolve protest behavior into a part of everyday life, which results in less energy for the movement to break out.[94]

In response to these two schools of thought, Tarrow proposes a compromise[95] that distinguishes and explores social movements from their organization, specifically, that they are ongoing, conflictual interactions between the challengers and oppressors of society. He points out that social organization brings together people with common goals and their coordinated struggle in protest against political elites, authority, and other rival organizations. Spontaneous collective action is susceptible to chaos due to private motivations such as the interests of the participants. Conversely, it can be argued that the two main features of personal motivation and disorganization make it possible for movements to reach rational outcomes based on the participants' rational choice of action. In contrast, organized collective action, while more orderly, is highly likely to radicalize in pursuing rational goals and lead to irrational outcomes. Of course, it is difficult to think of social movements and their organization as separate in practice; spontaneous movements often create organizations as they proceed, and existing organizations may also become the movement's mainstay. This is because movements do not only arise from outside the existing institutions but may also evolve from within them. Tarrow's views mentioned above have widely influenced the study of contemporary social movements.

3.3 *Research on External Factors of Social Movements—The Relationship with Politics*

Having sorted out the research on social movements above, this section focuses on collating and discussing research on the relationship between social movements and politics.

The success or failure of social movements is closely related to politics, but the inclusion of the political element in the study of social movements occurred after the 1970s. This is because, after the 1970s, researchers began to generally recognize that the study of social movements must put the state element at the center.[96]

In the social sciences, political scientists were the first to pay attention to the study of social movements. But most researchers initially approached social movements from a non-political perspective, i.e., as the result of alienation and normative disorder, not directly related to existing institutions. During the modernization and economic development in the 1960s, the relative deprivation theory focused on analyzing the relationship with social movements.[97] The relative deprivation theory, which analyzes the causes of social movements from the perspective of social malcontent caused by relative deprivation, has dramatically impacted the study of social movements. However, the theory focuses on the psychological rather than political factors arising from economic outcomes. Political factors were discussed but were outside the framework of the study. In the context of pursuing economic modernization at the time, the relationship between the economy and social movements was the focus of analysis.

The types of social movement research in the 1960s differed more or less between the United States and Europe. American researchers tended to conduct comparative multinational studies of social movements and were particularly inclined to gather analysis of collective action using data from questionnaires in their sociological research methods. This had the unintended consequence of underestimating the relationship between social movements and politics. In contrast, European scholars studying social movements emphasized the relationship between social movements and politics. Tarrow illuminates why focusing on political factors in social movement research is essential.[98] By examining the relationship between the communist movement and the peasant movement in post-war Italy, Tarrow concludes that political parties played a significant role in the process leading to the decline of the peasant movement.[99] In addition, Parkin analyzes the role of the Labor Party in the New Left Movement in Britain, noting that the New Left Movement was critical in promoting the Labor Party.[100]

Social movement research in the 1970s, while continuing the methodological approach of the 1960s, had begun to focus on the political element in those movements. Researchers such as Barnes and Tilly pointed out the vital link between collective action and political participation, i.e., the tendency of participants in social movements to be active in institutional politics.[101] However, with the emergence of the resource mobilization theory after the 1970s, the focus of research on social movements in this period changed accordingly, shifting from the individual as a participant in the movement to the organization of the movement. Most researchers of resource mobilization theory explained the relationship between movements and mobilization and movements and organizations from an economic perspective. Still, they needed to pay more attention to the political factors associated with the movements. McCarthy and Zald, however, see movements as political phenomena and introduce essential concepts such as entrepreneurship and competition from the field of economics. They argued that social movements were an "industry" that needed to be run with entrepreneurial leadership. Moreover, as an industry, there is no way to avoid competition among organizations for participants. However, economic concepts such as entrepreneurship, competition, and management take on a political meaning here and are used as political terms because, in the industry of social movements, both entrepreneurs and organizations are politicized.[102]

The resource mobilization theory research paradigm influenced social movement research in the 1970s, and researchers at the time achieved great success in explaining the relationship between social movements and politics within the framework of this paradigm. Notable among these are Gamson's *The Strategy of Social Protest,* and Piven and Cloward's, *Poor People's Movements*. The former analyzes how political alliances and state policies influence the outcomes of social movements,[103] while the latter sees social movements as acts of protest in which organizers use their resources to confront institutional authority.[104] Both emphasize how the political resources available to the participants of a movement can transform the potential to bring a movement into reality. While their studies also focus on the mobilization of resources, in contrast to other studies of resource mobilization theory, resources refer not to the organization of the movement itself and the entrepreneurial spirit of the organizers but to politics and existing institutions.

In general, research in the 1960s and 1970s addressed the political factor in social movements but did not consider it an independent variable. In the 1980s, researchers summarized the elements that influenced social movements and proposed a new research paradigm which focused on the political element. This research paradigm, although not the theory of a particular school of thought headed by a particular figure, was able to grasp the "political opportunity structure" of collective action, and thus to extract the political system, especially the state, as an independent variable to analyze its relationship with a given social movement.

This paradigm looks into the connections between the state and collective action by equally stressing both the strategies of social movements and their promoting elements. The theory of political opportunity structure has greatly influenced the study of this book, and its main ideas are briefly described below.

The question of how the state influences the rise of collective action, why the state cannot suppress social movements, and why the masses can accumulate resources to form organizations against authority are among the questions studied in political opportunity structuralism.

The concept of political opportunity structure emerged in the 1970s but has developed since the 1980s.[105] Some scholars have studied collective action in cities and found that the political opportunity structure strongly influences social movements' occurrence and outcome.[106] Although there are differences among scholars who study political opportunity structure, there are common concerns about the extent to which political institutions are open or closed, the stability of political alliances, the existence of political coalitions and alliance groups, the attitudes of the ruling class toward collective action, and the extent to which the development of collective action is tolerated.

In summary, among the political factors influencing social movements, the political opportunity structure attaches the most significant importance to the following three points. The first is the open and closed nature of the political system. The degree to which a political system is open and closed determines how collective protest, as an unusual form of political participation, is accepted, tolerated, rejected, and repressed. Of course, the same political system has different degrees of openness and closeness at different times. Thus, social movements under the same political system can take different forms and patterns depending on the time. The second point is the impact of changes in domestic political alliances, i.e., the waxing and waning of party power and the alternation of power between different parties, on social movements' occurrence, development, and outcome. The third point is the impact of conflicts within the political elite on social movements' circumstances, actions, and outcomes. Policy differences and power struggles within the leadership often provide opportunities for social movements to emerge. Social movements have easy access to support from some forces within the leadership, and some forces within the leadership need social support during a power struggle.

Moreover, the political opportunity structure influences social movements, and social movements can also react to the political system. As mentioned earlier, conflicts within the leadership provide opportunities for social movements to occur, and in turn, the occurrence of movements creates opportunities for the leadership and political elites. Included here are two levels of meaning: on one level, as mentioned above, some forces within the leadership use social movements to expand their power; on another level, the occurrence of social movements provides the leadership with the opportunity to suppress anti-establishment social forces. In addition, once a social movement occurs, it is autonomic and is influenced by its formation of opportunity. That is, the movement's development is influenced by the state of progress of the social movement itself and other movements occurring simultaneously. In addition to other relevant elements, they may not necessarily move forward as the organizers originally intended. In other words, social movements interact with existing institutions and other social movements occurring simultaneously.

3.4 *The Relevance of Western Social Movement Literature and Cultural Revolution Studies*

The theories mentioned above of social movements were developed by studying social movements in Europe and the United States and are effective in providing theoretical guidance in understanding social movements in Western countries. However, when the Cultural Revolution is understood as a movement, although consistent with European and American social movements in form, it is not possible to apply the above theories intact because the Cultural Revolution is different from Western social movements in many ways.

It is necessary to briefly explain the main features of Chinese social movements in the Maoist era and their relationship to the struggle for political power.

As mentioned earlier, social movements are generally defined as acts of protest in which the weak express their demands to the existing political system through collective action. However, in Mao-era China, where explicit opposition groups to the regime were not allowed to exist, almost all social movements can be seen as acts of protest that take advantage of the political opportunities provided by the state to fight for more citizenship within the state-arranged configuration of institutional space. Moreover, because China does not allow for public acts of protest, social movements in China are accompanied mainly by political struggles and political mobilization within the state.

Simply put, once a power struggle occurs between state agents, especially those at the policymaking level, they may mobilize social forces to gain an advantage. The social movement, at this point, is no longer a spontaneous act of protest but can be positioned as a politically mobilized social movement. On the one hand, mobilized political participation is an excellent opportunity for social groups to reap political and economic benefits. On the other hand, as the example of the rise and fall of the Red Guards shows, it is self-regulating once a social movement occurs. It may advance differently than intended by the mobilizers and organizers. Thus, once a movement is derailed, it allows the state to suppress anti-social forces. In other words, in politically mobilized social movements, it is only with the support of the state that social groups can enter the political arena and that their collective action in pursuit of interests can gain legitimacy. However, if this precondition is recovered, their actions gain legitimacy, and the organizations themselves may be judged illegal, making it difficult for a social movement to continue.

Precisely because the social movements in China during the Mao era had the above characteristics, it is natural that the theories of European and American social movements cannot be fully applied to explain China's problems. However, the views of European and American social movements can provide us with effective analytical perspectives and methodologies for analyzing Chinese issues. As such, this book will apply the research results of European and American social movements, such as the Cultural Revolution studies, Rational Choice Theory, resource mobilization theory, and the theory of political opportunity structure, and introduce the concept of citizenship and the theory of the public sphere in the next chapter. On this basis, this book will develop an institutional approach to studying the Cultural Revolution.

4 The Structure of this Book

Before discussing the research methodology, a brief description of the book's structure is given.

This book consists of eleven chapters. In addition to this introduction, the first chapter will first sort out the existing concept of "citizenship," clarify its definition, and then put forward an institutionalist approach to studying the Cultural Revolution.

Chapters 3 and 4 will go back to the pre-Cultural Revolution period to describe the institutional context in which the Cultural Revolution occurred. Chapter 3 will focus on reorganizing China's institutional spatial configuration during the Mao era and its impact on the Cultural Revolution. It will also examine the positioning of the Cultural Revolution in the historical context of nation-building. Chapter 4 will focus on the pre-Cultural Revolution system of citizenship distribution and its formation process. It will also explain how this system shaped and how the masses competed for citizenship and thus influenced collective action during the Cultural Revolution.

Chapters 5–7 discuss the collective action of the various factions during the Cultural Revolution. This book maintains that there were three "sub-movements" during the first three years of the Cultural Revolution and that the process of the Cultural Revolution unfolded as these three sub-movements shifted. In the process of transferring the sub-movements, the subjects and goals of the movements were different. The Old Red Guards (*lao hongweibing*) of the high schools in Beijing, the *Jinggangshan* Corps(*jinggangshan bingtuan)* of Tsinghua University, and the Shanghai Workers Revolutionary Rebel General Headquarters (*gongzongsi*) in Shanghai were the main rebel social groups of the three sub-movements. They were dedicated to advancing state goals while taking advantage of the opportunity to pursue their group's interests during the different stages of the sub-movement. Chapters 5–7 are devoted to the analysis of these factions and seek to clarify three main issues: (1) the process and causes of their formation, differentiation, and decline; (2) the driving forces of each sub-movement; and (3) the reasons for the shift of the sub-movements. In addition to these three common issues, each chapter has its focus. Chapter 5 focuses on the socialization of political struggle in the CCP leadership combined with the politicization of competing interests in society and how this led to the rise of factions. Chapter 6 focuses on how changes in state policy, the political struggle between top leaders of the central leadership, and the goals of the various factions in society created a split and led the Cultural Revolution into massive chaos. Chapter 7 analyzes the Shanghai Workers Revolutionary Rebel General Headquarters (*gongzongsi*), focusing on Shanghai's position in the country, the status of the January Revolution in Shanghai, and its comparison with other regions. By analyzing from a new perspective, we will see the process of establishing the Three-In-One Regime in Shanghai and the role played by Zhang Chunqiao, Wang Hongwen, and the Shanghai Workers Revolutionary Rebel General Headquarters in the reconstruction process drew different conclusions from those of previous studies.

Chapter 8 takes the rise and development of the rebellion movement at the Dongfanghong Textile Factory in Yuci, Shanxi Province, as an example to explore the three sub-movement periods described in the earlier three chapters. As opposed to the factional disputes at the center of the political arena, this study examines how politics and the struggle for citizenship were carried out within the institutional spatial configuration in local-level factional organizations. In contrast to the previous chapters, this chapter will look at the factional leaders who play a mobilizing and leading role in the factional competition, focusing on what kinds of people and what kinds of reasons or circumstances create factional leaders, how these factional leaders organize resources to lead factional struggles in a multilateral political environment, and how their strategies and methods interact with power struggles between state agents. The author attempts to analyze the textile mill rebellion to see the macro from the micro, to contrast the periphery of the political movement with that of the center. Furthermore, the author tries to understand more clearly the inter-factional struggle for citizenship by focusing on the behavior of the central faction leaders and how they organized competition to stop the loss of citizenship in the context of a changing distribution system or acquire more citizenship under the new rules.

Based on distinguishing different types of violent behavior that occurred during the three years from 1966 to 1969, Chapter 9 analyzes the reasons behind collective violence triggered by factional struggles.

Chapter 10 focuses on the relationship between collective violence and social order in the Cultural Revolution. While the collective violence of the Cultural Revolution was a remarkable phenomenon in terms of its longevity, scope, and destructiveness, it is worth considering that this extraordinary act of collective violence did not bring about a total collapse of social order. What, then, was it that maintained order in the Cultural Revolution? Was there a Cultural Revolution-style method of maintaining order? What is the relationship between the collective violence of the Cultural Revolution and social order?

The final chapter summarizes the book's main arguments and examines what it means to adopt the concept of competitive citizenship to study social movements in the Reform and Opening-Up period. Specifically, what impact the Cultural Revolution had on reform-era nation-building, and how did the changes in the state and society institutional spatial configuration that accompanied reform-era state-building transform collective action and how citizenship was pursued?

Notes

1. Lynn T. White III, translated by Yue Jing Lun, "Explaining Revolution in the Peak", (jieshi gaochao zhong de geming), *Journal of Social Sciences* (Cultural Revolution special issue) published by The City University of Hong Kong, 1996, no. 7, pp. 1–20.
2. Mitsuyuki Kagami, "General Introduction: The Tragedy in the Cultural Revolution", in Mitsuyuki Kagami, ed., *Frustration in Contemporary China: An Examination of the Cultural Revolution Setbacks in Modern China* (Institute of Asian Economic Research, 1986), pp. 3–20.
3. Lynn T. White III, op cit., p. 4.
4. Some researchers claim that the number of people killed during the Cultural Revolution reached 20 million. Kagami Mitsuyuki, op. cit., "General Introduction: The Tragedy in the Cultural Revolution", p. 3.
5. For facts about collective violence during the Cultural Revolution, see the following documents: Guangxi Cultural Revolution Chronology, edited by Guangxi Cultural Revolution Chronology Group, Guangxi People's Publishing House, 1992, pp. 53, 111 (guangxi wenge dashi nianbiao); Wen Yu, *China's Left Scourge (Beijing zuohuo)* (Beijing Chaohua Publishing House, 1992), p. 444.
6. Wang Shaoguang, "Expanding the Horizons of Cultural Revolution Studies" ("tuozhan wenge yanjiu de shiye", *wenhuadageming: shishi yu yanjiu*), in Liu Qingfeng *The Cultural Revolution: Facts and Analysis* (The Chinese University of Hong Kong Press, 1996), p. 516.
7. Id. p. 516.
8. Yan Jiaqi and Gao Gao, *A Ten-Year History of the Cultural Revolution (wenge shinian shi)* (Tianjin People's Publishing House, 1986).
9. Zeng Huiyan, "Interview with Gao Gao" ("caifang Gao Gao"), *The Nineties Monthly*, April 1987, No. 67.

10. Wang Ningyi, *The Age of Great Turmoil* (*dasongluan de niandai*) (Henan People's Publishing House, 1988).
11. Feng Jicai, *Ten Years of One Hundred People* (*yibaigeren de shinian)* (Jiangsu Literature and Art Publishing House, 1997).
12. Wu Wenguang, *Revolutionary Scene 1966 (geming xianchang 1966)* (Times Publishing, 1994).
13. Bu Weihua, *"Smashing the Old World": Havoc of the Chinese Cultural Revolution (1966–1968),* (*"zalan jiu shijie: wenhuadageming de dongluan yu haojie")* (The History of the People's Republic of China, Volume 6) (The Chinese University of Hong Kong Press, 2008).
14. Shi Yun and Li Danhui, *When the "Continuous Revolution" Goes Awry: From the Anti-Lin Biao Campaign to the Anti-Deng Xiaoping Campaign (1972–1976)* (*nanyi jixu de "jixugeming": cong pilin dao pikong)* (The History of the People's Republic of China, Volume 8) (The Chinese University of Hong Kong Press, 2008).
15. Ma Jisen, *The Cultural Revolution in the Foreign Ministry (waijiaobu wenge jishi)* (The Chinese University of Hong Kong Press, 2003).
16. Gao Shuhua and Cheng Tiejun, *The Cultural Revolution in Inner Mongolia: Oral History of a Rebel Leader(neimenggu wenge fenglei:yiwei zaofanpai lingxiu de koushushi)* (Hong Kong: The Mirror Books Press, 2007).
17. Song Yongyi, ed., *The Massacres during The Cultural Revolution (wenge datusha)* (Hong Kong: Open Magazine Press, 2002).
18. Tan Hecheng, *The Myth of Blood—A Chronicle of the 1967 Cultural Revolution Massacres in Daoxian, Henan Province (xue de shenhua: gongyuan 1967nian hunan daoxian wenge datusha jishi)* (Hong Kong: Tian Xing Jian Press, 2010); English edition: Translated by Stacy Mosher and Guo Jian, *The Killing Wind—A Chinese County's Descent into Madness during the Cultural Revolution* (New York: Oxford University Press, 2017).
19. Shi Shi, ed., *Self-Criticism Books during the Cultural Revolution (wengezhong de jiantaoshu)* (Hong Kong: The China History Press, 2011).
20. Mi Hedu, ed., co-authored by Bu Dahua et al., *Memories and Reflections: One of the Oral Histories of the Red Guard Era* (Hong Kong: The China Books Ltd., 1999).

21. A series of Cultural Revolution sourcebooks compiled by John Sisyphus and published in Taiwan by The Sisyphus Culture Publishing in 2016. The books in the series categorized by speeches given at mass rallies by major Cultural Revolution leaders such as Jiang Qing, Zhou Enlai, Wang Hongwen, Yao Wenyuan, Zhang Chunqiao, and Kang Sheng is a very important historical source when studying the behavioral characteristics of the major leaders of the Cultural Revolution.
22. Xu Youyu, *"Rebels of All Stripes: A Study of Red Guard Mentalities"(xingxingsese de zaofan:hongweibing jingshensuzhi de xingcheng yu yanbian)* (The Chinese University of Hong Kong Press, 1999).
23. Tang Shaojie, "*An Episode of the Cultural Revolution: The 1968 Handred-Day War at Tsinghua University" (yiyezhiqiu:qinghuadaxue 1968 nian "bairidawudou")* (The Chinese University of Hong Kong Press, 2003).
24. Yin Hongbiao, *Footprints of the Missing: Thoughts of the Youth during the Cultural Revolution(shizongzhe de zuji:wenhuadageming qijian de qingniansichao)* (The Chinese University of Hong Kong Press, 2009).
25. Xu Youyu, op. cit., pp. 16–17.
26. Tang Shaojie, op. cit., pp. 260–271.
27. William Hinton, Translated by Toru Haruna, *Hundred Days of War: The Cultural Revolution at Tsinghua University* (Heibonsha, 1976).
28. Among them is the commentary on heretical thinking in the collection of materials co-edited by Chinese American scholar Song Yongyi and Sun Dajin, *The Cultural Revolution and Its Heretical Thouths* (Hong Kong: Field House, 1997) which can be regarded as an excellent work in this field.
29. Yin Hongbiao, *Footprints of the Missing: Thoughts of the Youth during the Cultural Revolution*, 550–551 pp.
30. H. Gerth and C. W. Mills, eds., *From Max Weber: Essays in Sociology* (New York: Oxford University Press, 1958); A. M. Henderson and T. Parsons, eds. *Weber; The Theory of Sociology and Economic Organization* (New York: The Free Press, 1964); Reinhard Bendix, *Max Weber, An Intellectual Portrait* (New York: Garden City, 1960).

31. The different value judgments of scholars on the Cultural Revolution are fully reflected in Liu Qingfeng's comment on Cui Zhiyuan's and others' postmodernist affirmation of the Cultural Revolution: "(omitted) The most striking of these are undoubtedly those arguments that try to give the Cultural Revolution some kind of rationality or even advancement. First, based on their claim that they oppose one of the starting points of the supposed official Chinese Communist Party's ideological hegemony to completely repudiate the Cultural Revolution, they talked about the formal justification of Mao's 'great democracy' and 'justified rebellion,' claiming their validity and necessity for China's future development; or denying that the Cultural Revolution was a catastrophe for workers, peasants and soldiers, and even intellectuals, arguing that it was only a disaster for capitalists and privileged elements. The second is to invoke Western postmodern theory to reconstruct the Cultural Revolution, ignoring the huge gap between the West and China's social development, and thus to camp the 'post-modern' character of the Cultural Revolution phenomenon. In a positive sense, these arguments show a certain plurality and openness. Unfortunately, however, most of them are based on selective memory and are therefore quite different from the facts and experiences of the Cultural Revolution as experienced by most Chinese people. Some claim that these arguments are ignoring the immense suffering of the entire nation, which cannot be said to be without merit. ...... The Cultural Revolution has its many faces, including values that on the surface seem to warrant affirmation. But it was the pushing to extremes of those seemingly unquestionable ideals and values that produced the evil and savagery of the Cultural Revolution." (Liu Qingfeng, "Re-Questioning the Historical Question", (*dui lishi de zaifawen*) edited by Liu Qingfeng, op. cit., p. x).

    For Cui Zhiyuan's explanation of the Cultural Revolution, see the following paper, "The Gains and Losses of Mao's Theory of the Cultural Revolution and the Reconstruction of Modernity"("Mao Zedong 'wengelilun' de deshi yu 'xiandaixing'de chongjian", op. cit. *Journal of Social Sciences* (Cultural Revolution special issue), pp. 49–74. Revised version: Cui Zhiyuan, *The Second Thought Liberation and Institutional Innovation (dierci sixiangjiefang yu zhiduchuangxin)* (Hong Kong: Oxford University Press, 1997), pp. 333–364.

32. Most of the research papers written by Chinese researchers since the 1980s are included in the following two volumes: Liu Qingfeng, ed., *The Cultural Revolution: Facts and Analysis* (The Chinese University of Hong Kong Press, 1996); The City University of Hong Kong, The Centre for Contemporary Chinese Studies, ed., *Journal of Social Sciences* (Cultural Revolution special issue) (The City University of Hong Kong Press, 1996), Spring, No. 7.
33. Hashizume Osaburo, "'The Red Guards and the Zenkyoto'—Also on the New Left in Japan in the 1960s", in Liu Qingfeng, ed., *The Cultural Revolution: Facts and Analysis*, pp. 289–304.
34. On this point, refer to the following studies: Kazuko Mori, *Modern Chinese Politics* (Nagoya University Press, 1993), pp. 1–16; Furumaya Tadao, "The Cultural Revolution and Japan", in Makoto Ikeda et al. *Japan–China Relations in the World* (vol. 1) (Legal Culture Society, 1996), pp. 180–198; Mitsuyuki Kagami, "The Cultural Revolution and Modern Japan", in Liu Qingfeng, ed., *The Cultural Revolution: Facts and Analysis*, pp. 305–318.
35. Satoshi Amako, *China, Changing of a Socialist Power* (Tokyo University Press, 1992), pp. 94–95.
36. Mineo Naka Jima, *Chinese History, Society, and International Relations* (Chuo Koronsha, 1982), pp. 80–103.
37. Minoru Takeuchi. *Document Modern History 16: Cultural Revolution* (Heibonsha, 1973).
38. Mitsuyuki Kagami, "General Introduction: The Tragedy in the Cultural Revolution", op. cit., pp. 16–17.
39. Satoshi Amako, *History of the People's Republic of China* (Iwami Shinsho, 1999), pp. 77–81.
40. For an overview of Japanese studies of the Cultural Revolution in recent years, see Tsuneto Yamamoto, "Rethinking of The Chinese Cultural Revolution: From the Perspective of Research History", in *Being Nearby* (June 2007), pp. 21–32.
41. Yang Lijun "Contest for Citizenship and Collective Violence in China's Cultural Revolution: An Institutional Approach", *Asian Studies*, vol. 49, no. 1 (January 2001), pp. 86–107.

42. Yang Lijun, *The Cultural Revolution and Social Structure of China: The Distribution of Citizenship and Collective Violence* (Ochanomizshobo, 2003).
43. Jun Konno, *Chinese Society and Mass Mobilization: Political Power and the People in the Mao Zedong Era* (Ochanomizushobo, 2008).
44. Shinichi Tanikawa, *The Dynamics of the Cultural Revolution in* China (Ochanomizushobo, 2011).
45. Kagami Mitsuyuki ed., *Frustration in Contemporary China: An Examination of the Cultural Revolution* (Asian Economic Institute).
46. Ryosei Kokubun, ed. *The Chinese Cultural Revolution Revisited* (Keio University Press, 2003).
47. Tsuneto Yamamoto, "The Policy Trinity of Labor, Education, and Deportation in 1960 and Its Destruction: The Focus on the Half-Work/Half-Education System", in Mitsuyuki Kagami ed., *Frustration in Contemporary China: An Examination of the Cultural Revolution*, pp. 147–226.
48. Ichie Watanabe, "The Cultural Revolution in Hunan and the "Shengwulian", in Mitsuyuki Kagami, ed., *Frustration in Contemporary China: An Examination of the Cultural Revolution*, pp. 105–145.
49. Tsuneto Yamamoto, "Chinese Cultural Revolution Revisited: From the Prespective of Research History", *Being Nearby*, no. 6 (2007), pp. 25–27.
50. Richard Pfeiffer, "The Pursuit of Purity: Mao's Cultural Revolution", *Problems of Communism*, vol. 18, no. 6 (1969), pp. 12–15.
51. Harry Harding, *Organizing China: The Problem of Bureaucracy, 1949–1976* (Stanford: Stanford University Press, 1981).
52. Anne F. Thurston, *Enemies of the People: The Ordeal of Intellectuals in China's Great Cultural Revolution* (New York: Knopf, 1987).
53. Wang Shaoguang, *Failure of Charisma: The Cultural Revolution in Wuhan* (Hong Kong: Oxford University Press, 1995). In Chinese, Wang Shaoguang, *Reason and Madness: The Masses in the Cultural Revolution* (Hong Kong: Oxford University Press, 1993).
54. Robert Lifton, *Revolution Immortality: Mao Tsetung and the Cultural Revolution* (New York: Vintage Books, 1968).

55. Ross Terrill, *Mao: A Biography* (New York: Harper and Row, 1980).
56. Lucian W. Pye, *The Spirit of Chinese Politics: A Psychocultural Study of the Authority Crisis in Political Development* (Cambridge, MA: MIT Press, 1992), p. 67.
57. Albert J. Bergesen, “A Durkheimian Theory of ‘Witch-Hunts’ with the Chinese Cultural Revolution of 1966–1969 as an Example”, *Journal of the Scientific Study of Religion*, vol. 17, no. 1 (1978), pp. 19–29.
58. Lucian Pye, *The Dynamics of Chinese Politics* (Cambridge, MA: Oelgeschlager, Gunn & Hain, 1981); Frederick Teiwes, *Leadership, Legitimacy and Conflict in China: From a Conflict in China: From a Charismatic Mao to the Politics of Succession* (Armond, New York: M. E. Sharpe, 1984). Both Chinese and Japanese scholars often use their models of line and power struggle.
59. Roderick MacFarquhar, *The Origins of the Cultural Revolution*, three volumes (New York: Columbia University Press, 1974, 1983, 1966).
60. Lowell Dittmer, *Liu Shao-ch’i and the Chinese Cultural Revolution: The Politics of Mass Criticism* (Berkeley: University of California Press, 1987).
61. Lynn T. White III, *Policies of Chaos: The Organizational Cause of Violence in China’s Cultural Revolution* (Princeton, NJ: Princeton University Press, 1989).
62. Hong Yung Lee, *The Politics of the Chinese Cultural Revolution: A Case Study* (Berkeley, CA: University of California Press, 1978).
63. David Raddock, *Political Behavior of Adolescents in China: The Cultural Revolution in Kwangchow* (Tuscon, Ariz: The University of Arizona Press, 1977).
64. Jonathan Unger, *Education Under Mao: Class and Competition in Canton Schools, 1960–1980* (New York, NY: Columbia University Press, 1982); Anita Chan, Stanley Rosen and Jonathan Unger, “Students and Class Warfare: The Social Roots of the Red Guard Conflict in Canton”, *The China Quarterly*, no. 83 (September 1980), pp. 397–446.
65. Hong Yung Lee, “The Radical Students in Kwangchow during the Cultural Revolution”, *The China Quarterly*, no. 64 (December 1975), pp. 645–683.

66. "Black Gangs" refers to the counter-revolutionary reactionaries of the Cultural Revolution, and usually refers to people in power who have been overthrown.
67. Stanley Rosen, *Red Guard Factionalism and the Cultural Revolution in Guangzhou* (Boulder, Colorado: Westview Press, 1982).
68. Jonathan Unger, op. cit., 1982.
69. Anita Chan, *Children of Mao: Personality Development and Political Activism in the Red Guard Generation* (Seattle: University of Washington Press, 1985).
70. Andrew G. Walder, *Fractured Rebellion: The Beijing Red Guard Movement* (Cambridge: Harvard University Press, 2009).
71. Dong Guoqiang and Andrew G. Walder, "Factions in a Bureaucratic Setting: The Origins of Cultural Revolution Conflict in Nanjing", *China Journal,* no. 65 (January 2011); 1–25; Dong Guoqiang and Andrew G. Walter, "Foreshocks: Local Origins of Nanjing's Qingming Demonstrations of 1976", *China Quarterly*, vol. 220 (2014), pp. 1092–1110.
72. Su Yang, *Collective Killings in Rural China during the Cultural Revolution* (New York: Cambridge University Press, 2011). (Translated by Song Xi, Collective *Killings in Rural China during the Cultural Revolution)* (Hong Kong: The Chinese University of Hong Kong Press, 2017).
73. Wu Yiqing, *The Cultural Revolution at the Margins* (Harvard University Press, 2014).
74. About the discussion of the "two Cultural Revolution theories", see Xiao Xidong, "One or Two Cultural Revolution?", in Luo Jinyi and Zheng Wenlong, eds., *Beyond the Shoah—A Rediscussion of the Cultural Revolution* (Taiwan: Feng Yun Forum Press Ltd. 1997), pp. 136–172 (Xiao Dongxi, "lianggewenge, huo yigewenge?", Luowenyi, Zhengwenlong bian, *Haojiezhiwai: zailun wenhuadageming*).
75. Yang Guobin, *The Red Guard Generation and Political Activism in China* (Columbia University Press, 2017).
76. Daniel Leese, Translated by Qin Hosheng et al., *Mao Cult: Rhetoric and Ritual in China's Cultural Revolution (chongbai Mao:wenhuadageming Zhong de yancichongbai he yishichongbai)* (Hong Kong: The Chinese University of Hong Kong Press, 2017). English edition, Daniel Leese, *Mao Cult: Rhetoric and*

Ritual in China's Cultural Revolution (Cambridge: Cambridge University Press, 2011).

77. Id., p. 14.
78. Pang LaiKwan, translated by Li Cho-Kiu Joseph, *The Art of Cloning—Cultural Production and Practice during the Cultural Revolution (fuzhi de yishu:wengeqijian de wenhuashngchan he shijian)* (Hong Kong: The Chinese University of Hong Kong Press, 2017), p. 23. English version, Pang LaiKwan, *The Art of Cloning: Creative Production During China's Cultural Revolution* (Verso, 2017).
79. For the discussion of the history of research on social movements in Europe and the United States, draw on the following literature: Aldon D. Morris and Carol McClurg Mueller, eds., *Frontiers in Social Movement Theory* (New Haven: Yale University Press, 1992).
80. Sidney Tarrow, *Struggle, Polities, and Reform: Collective Action, Social Movements, and Cycles of Protest*, Western Societies Program Occasional Paper No. 21 (2nd edition) (Center for International Studies, Cornell University, 1991).
81. Charles Tilly, *From Mobilization to Revolution* (Englewood Cliffs, NJ: Prentice-Hall, 1978), p. 84. (Translated by Horie fukashi, *Theory of Political Change*, Ashishobo, 1984); Charles Tilly, *The Contentious French,* (Cambridge, MA: Harvard University Press, 1986), p. 3.
82. Sidney Tarrow, *Power in Movement: Social Movements, Collective Action and Politics* (New York: Cambridge University Press, 1994), pp. 1–6.
83. For example, sociologist Bergson interpreted the Cultural Revolution in terms of a "hunt for political dissidents" as a result of its influence.
84. E. P. Thompson, "The Moral Economy of the English Crowd in the Eighteenth Century", *Past and Present*, vol. 50 (1971), pp. 76–136.
85. Charles Tilly, "Collective Violence in European Perspective", in Hugh D. Graham and Ted R. Gurr, eds., *Historical and Comparative Perspectives* (Washington, DC: US Government Printing Office, 1969), p. 10.

86. Mancur Olson, *The Logic of Collective Action* (Cambridge, CA: Harvard University Press, 1965). (Translated by Ida Hiroshi et al., *Theory of Collective Action*, Minervashobo, 1983).
87. The analysis of social movements using rational choice theory is represented by Samuel L. Popkin, *The Rational Peasant: The Political Economy of Rural Society in Vietnam* (Berkeley and Los Angeles: University of California Press, 1979); Anthony Oberschall, *Social Conflict and Social Movements* (Englewood Cliffs, NJ: Prentice Hall, 1973); John D. McCarthy and Mayer N. Zald, "Resource Mobilization and Social Movements: A Partial Theory", *American Journal of Sociology*, vol. 82 (1977), pp. 1212–1241; W. A. Gamson, *The Strategy of Social Protest* (Homewood, III: Dorsey, 1975).
88. For example, Douglas Hibbs, "Industrial Conflict in Advanced Industrial Societies", *American Political Science Review*, vol. 70 (1976), pp. 1033–1058; Walter Korpi and Michael Shalev, "Strikes, Power, and Politics in the Western Nations", in I. M. Zeitlin, ed., *Political Power and Social Theory* (Greenwich, Conn.: JAI, 1980).
89. For example, Douglas Hibbs, *Mass Political Violence: A Causal Analysis* (New York: Wiley, 1973); Ted Robert Gurr, *Why Men Rebel* (Princeton, NJ: Princeton University Press, 1970).
90. For example, Peter Eisinger, "The Conditions of Protest Behavior in American Cities", American *Political Science Review*, vol. 67 (1973), pp. 11–28; Michael Lipsky, "Protest as a Political Resource", *American Political Science Review*, vol. 62 (1968), pp. 1144–1158; Mayer N. Zald and John D. McCarthy, *Social Movements in an Organizational Society* (New Brunswick, NJ: Transaction, 1987).
91. For example, Bert Klandermans, Hanspeter Kriesi, and Sidney Tarrow, eds., *From Structure to Action: Comparing Social Movement Research Across Cultures* (Greenwich, Conn.: JAI, 1988).
92. Michels developed the famous "iron law of oligarchy" theory. He proposed that in any organization there are only a few individuals who hold authority, and therefore the monopoly of power by these individuals inevitably makes the organization hierarchically inclined. Robert Michels, *Political Parties: A Sociological Study of the Oligarchical Tendencies of Modern Democracy* (New York: The Free Press, 1962).

93. Eric J. Hobsbawm, *Primitive Rebels* (Manchester, England: Manchester University Press, 1959). (Translated by Aoki tamotsu, *Primitive Rebels: Millennial Kingdomism and Social Movements*, Chugong Shinshu 239, 1971; Translated by Mizuta Hiroshi, et al., *The Simple Rebels: A Social History of Ideas*, Shakaishisousha 1989).
94. France Fox Piven and Richard A. Cloward, *Poor People's Movements: Why They Succeed, How They Fail* (New York: Vintage, 1979).
95. Tarrow, op. cit., 1991, p. 18.
96. For example, Charles Tilly, ed., *The Formation of National States in Western Europe* (Princeton, NJ: Princeton University Press, 1975); Peter B. Evans, Dietrich Rueschemeyer, and Theda Skocpol, eds., *Bringing the State Back In* (New York: Cambridge University Press, 1985).
97. Ted Robert Gurr, "A Causal Model of Civil Strife: A Comparative Analysis Using New Indices", *American Political Science Review*, vol. 62 (1968), pp. 1104–1124. Gurr makes the point that people are motivated to engage in antisocial behavior when they perceive a gap between their actual standard of living and their expected standard of living and make judgments that they are being denied their rights. Once given the opportunity by the external environment, political, social, and cultural, they will engage in it.
98. Charles Tilly, *The Vendee* (Cambridge: Harvard University Press, 1964); "Collective Violence in European Perspective", in Hugh D. Graham and Ted Robert Gurr, eds. op. cit., 1969, pp. 5–45.
99. Sidney Tarrow, *Peasant Communism in South Italy* (New Haven: Yale University Press, 1967).
100. Frank Parkin, *Middle Class Radicalism* (Cambridge: Cambridge University Press, 1968).
101. Samuel Barnes, Max Kaase, et al., *Political Action: Mass Participation in Five Western Democracies* (Beverly Hills, CA: Sage Publications, 1979); David Snyder and Charles Tilly, "Hardship and Collective Violence in France: 1830–1960", *American Sociological Review*, vol. 37 (1972), pp. 520–532.
102. Zald and McCarthy, op. cit., 1987.
103. William A. Gamson, *The Strategy of Social Protest* (Homewood, Ill: Dorsey, 1975).

104. Piven and Cloward, op. cit., 1979.
105. Doug McAdam, *The Political Process and the Development of Black Insurgency* (Chicago: University of Chicago Press, 1982); Herbert Kitschelt, "Political Opportunity Structures and Political Protest: Anti-Nuclear Movements in Four Democracies", *British Journal of Political Science*, vol. 16 (1986), pp. 57–85; Mary F. Katzenstein and Carole Mueller, *The Women's Movements of the United States and Western Europe: Consciousness, Political Opportunity, and Public Policy* (Philadelphia: Temple University Press, 1987); Tarrow, op. cit., 1994.
106. For example, James Button and others have shown the following. Social movements in American cities in the 1960s brought a variety of political opportunities and, as a result, local elites sought more financial support from the federal government. In addition, scholars such as Anthony Leeds, who have studied urban social movements in Brazil, have pointed out that the analysis of psychological factors alone is not comprehensive and that it is also important to analyze political factors such as the political consciousness and political preferences of the average urban citizen and the power structure of the state. James Button, *Black Violence: Political Impact of the 1960s Riots* (Princeton, NJ: Princeton University Press, 1978); Susan Welch, "The Impact of Urban Riots on Urban Expenditures", *American Journal of Political Science*, vol. 19 (1975), pp. 741–760; Anthony Leeds and Elizabeth Leeds, "Accounting for Behavioral Differences: The Political Systems and the Responses of Squatters in Brazil, Peru and Chile", in John Walton and Louis Masotti, eds., *The City in Comparative Perspective* (New York: Wiley, 1976), pp. 193–248; John Walton, "Urban Political Movements and Revolutionary Change in the Third World", *Urban Affairs Quarterly*, vol. 15 (1979), pp. 3–22.

CHAPTER 2

Institutional Approaches to the Cultural Revolution

In the Introduction, the author reviewed the history and current state of research on the Cultural Revolution, mainly focusing on research in China, as well as in Europe, America, and Japan. The author discussed the achievements and shortcomings of scholars from different regions in their research. The author also discussed the progress made in Western social movement studies and argued that we could study and understand the Cultural Revolution by placing it within the literature on social movements. However, this does not mean simply copying and mechanically interpreting the literature on Western social movements to explain the Cultural Revolution, as this social movement in China was constrained by China's unique cultural, political, economic, and social conditions.

In this chapter, the author further introduces the concept of "citizenship competition" and discusses why this concept can be used to explain the Cultural Revolution, especially the violent phenomena that occurred. Similarly, the concept of "citizenship" comes from the West. We need to redefine it based on China's political and social practices to make it more applicable to the Chinese reality. After defining "citizenship" in China, this chapter focuses on building an institutional framework to explain why the citizenship competition within this framework led to the Cultural Revolution and its violent behavior. The author's intention is clear: to construct a new conceptual framework of institutional spatialism

Y. Lijun, *Contest for Citizenship and Collective Violence during China's Cultural Revolution*, IPP Studies in the Frontiers of China's Public Policy, https://doi.org/10.1007/978-981-97-0906-9_2

to explain the Cultural Revolution and its violence based on the research on the Cultural Revolution and social movements in various countries. In other words, this book's main body of explanatory framework is the institutional spatial configuration, which integrates some concepts from previous social movement studies and Cultural Revolution studies. The author believes this conceptual framework has more explanatory power than some existing conceptual frameworks in describing and analyzing the violence of the Cultural Revolution while complementing other concepts in other areas, such as the origins of the Cultural Revolution.

1 On the Concept of Citizenship

1.1 Definition of Citizenship in Sociological Studies

The first researcher to use the concept of citizenship to explain social movements was the British sociologist T. H. Marshall. In 1949, in a lecture entitled *Citizenship and Social Class*, Marshall used the idea of citizenship to describe the class struggle under the capitalist system.[1] Marshall divided the process of acquiring citizenship into three stages: civil citizenship, political citizenship, and social citizenship.[2] According to him, civil citizenship arose in the eighteenth century. It included some rights necessary for individual freedom, such as personal freedom, freedom of speech, freedom of thought, freedom of belief, the right to own property, and the right to access justice. Political citizenship emerged in the nineteenth century and included the right to participate in the political process as a member of the body politic. Social citizenship has developed since the twentieth century. It consists of a range of rights, such as economic welfare and security, or what we today call the rights of the modern welfare state.

Of particular note, Marshall points out the tension between citizenship and capitalist society, the profound contradiction between the two, which is why class struggles and social movements occur. However, he also pointed out that the conflict between capitalism and citizenship was institutionalized, and various rules regulated the tension between the two.

Although Marshall's discussion of citizenship and its relation to social class was confined to Britain, the impact of the concept of citizenship was significant, and it has since been used by many researchers, including Reinhard Bendix, Ralf Dahrendorf, A. H. Halse, S. M. Lipset, David Lockwood, B. S. Turner, and many others.[3] Michael Mann was one of the more significant contributors to developing the concept of citizenship. Mann believes that the manifestations of citizenship are related to the domination strategies of the ruling class. He argues that the main political systems in modern history include the liberal, reformist, authoritarian monarchist, fascist, and authoritarian socialist and that the manifestations of citizenship vary from one political system to another.[4]

Discussing Mann's arguments about the manifestations of citizenship in the Soviet-style authoritarian socialist system is necessary here because of its usefulness for our study.[5] Mann believes that the authoritarian-style socialist system is mainly manifested in implementing a "high-pressure rule" system, which controls every corner of society through two means: the infrastructure capacities produced by the twentieth century and ideology.[6] In practice, the "high-pressure rule" is often based on negotiation. Still, the right to negotiate is not granted to groups outside of state power (out-groups) but is limited to groups within the regime. Since open opposition was never allowed, other groups, provided they were not rival classes or antigovernment groups, had the right to participate in the regime and form alliances and "patron–client" relationships with the regime, making it possible to exert influence on the government. Thus, although antigovernment social movements were rarely seen because they were not officially permitted, political struggles such as purges and factional struggles within the regime constantly occurred. Under the authoritarian socialist system, the state did not provide citizens with civil rights, nor did it even deliver political rights in the true sense of the word. However, the state provided more social rights than any democratic state, from daily necessities to housing, medical care, education, and recreation, almost every right in life.

1.2 Citizenship as an Intermediary between the State and Society

In Mann's account of the manifestations of citizenship in authoritarian socialist institutions, as described above, the element of the state is placed at the center of the discussion. In other words, Mann is concerned with how the state allocates citizenship. Mann's focus on the element of the state has influenced many researchers in the field of Chinese studies. For example, when studying how the state's monopoly of political, economic, and social resources in Mao-era China affected the formation of state–society relations, Andrew Walder empirically analyzes the authoritarian relations within Chinese state-owned enterprises from the perspective of industrial organization theory. He points out that the state monopoly of resources caused individuals to rely heavily on the state and even state agents (factory directors, etc.) in various ways.[7] Although his study provides a highly effective viewpoint for understanding state–society relations in the Mao era since the focus is on the state, the perspective when considering the issues is inevitably top-down.

As the Introduction mentions, understanding the Cultural Revolution requires a perspective incorporating the interaction between the state and society. Therefore, in addition to examining how the state distributes citizenship, it is also necessary to examine how social factions and groups competed to obtain citizenship.

Before applying the concept of citizenship to explain the Cultural Revolution, redefining the concept of citizenship that emerged with its development in Europe and the United States is necessary. In other words, a series of related questions must be posed, including: How should citizenship be defined in the Chinese context? How does this concept relate to the collective actions during the Cultural Revolution that is the subject of this book? Furthermore, why is it necessary to use the concept of citizenship to analyze collective action during the Cultural Revolution?

First, the definition of citizenship. In general, citizenship refers to status, on the one hand, and rights, on the other, regardless of the type of political regime. Citizenship is usually defined by the state, while rights are typically attached to it, and status determines the rights that can be entitled. Of course, rights also imply duties and obligations. That is to say, the status, rights, responsibilities, and duties inherent in citizenship form the basis of social governance and the social order. They also construct the core values of a nation as a civic virtue.[8] Since citizenship reflects the combinations of relations and types of governance in society, understanding citizenship in a given national setting requires understanding the social and political context in which citizenship operates, the system of power, and the political operation that forms citizenship. It is also necessary to understand the state ideology that legitimizes that system. In other words, understanding citizenship requires consideration of at least two dimensions. First, how does the state regulate citizenship, and on what basis is legitimacy supported? Second, what method and legitimacy do citizens use to compete for citizenship?

After founding the People's Republic of China (hereinafter referred to as the PRC) in 1949, China underwent a new democratic rule. From the mid-1950s onwards, the transition toward a socialist system began.

Creating a new type of state undoubtedly entails a reconstitution of citizenship , not to mention a new state that was created in a disruptive and revolutionary manner. Both the Common Program of the Chinese People's Political Consultative Conference issued in September 1949 and the Constitution of the People's Republic of China published in 1954 defined the new form of state as "a people's democratic state led by the working class and based on the alliance of workers and peasants."[9] Although Article 85 of the 1954 Constitution also provides that all citizens of the PRC are equal under the law, in practice, class and class struggle became the basis for the state's legitimacy to construct a socialist form of justice and fairness. Class thus became the root of citizenship construction of the new state after its founding. While all those who enjoyed the right of residence in the PRC were citizens, they were divided by political ideology into "the people" and "class enemies." As will be analyzed in subsequent chapters, the line between the people and class enemies was not fixed. Each time a political movement took place, depending on the situation, the state deemed those whose words and deeds were subversive to the regime as class enemies, and those who actively exposed and criticized class enemies were rewarded politically. This class struggle-based social justice formed the core value of citizenship in the Maoist era. Class antagonism and class struggle were embedded in the formation of loyalty to the political community and in developing citizenship qualities. This social fairness and justice built upon class struggle formed the core value of citizenship during the Mao Zedong era. Being a good citizen loyal to the party and country is motivated not only by positive factors, such as the pursuit of nobility and honor, but also by negative factors, such as avoiding punishment. The motivation to become a good citizen, whether to obtain political rewards or avoid punishment, formed the rational and interest factors behind factional behavior during the Cultural Revolution. The positive aspect of pursuing nobility and honor added fanaticism and idealistic irrationality to factional behavior at that time. As the analysis of examples from the Cultural Revolution in subsequent chapters of this book will present, both aspects were present in the collective behavior of the Cultural Revolution. However, the two factors have different dimensions of expression in different groups and different time periods. Whether it is the rational factor that carries the function of identity and rights or the idealistic characteristic that assumes the role of values, these elements contained in citizenship are a process that was formed and continuously reinforced through political indoctrination such as ritualized scenes and hero-building in various political movements. These were gradually internalized in the national identity formation process after the new state's founding.

It is important to note that the concept of citizenship is used in this book as a concept for analysis and not as a value because the concept can effectively explain collective behavior in the Cultural Revolution. That is to say, on the one hand, the concept can explain the interest and ideal elements of collective action in the Cultural Revolution, and on the other hand, as will be explained later in this book, the concept can also effectively explain the behavior of each of the two levels of the state and society and the impact of the interaction between the two on collective action. Although for most Chinese people in Maoist China, there was only class identity and not citizenship, the author believes that every Chinese person who lived in that era had a good understanding of various identities embedded in citizenship, the specific rights that different identities can enjoy (the specifics of citizenship will be discussed in Chapter 4) or the penalties that can be imposed on them, and the losses or gains that can result from a change of identity. Or one could say that the essence behind class identity is citizenship, as different classes are granted different forms of citizenship.

At the conceptual level, citizenship in the People's Republic of China during the Maoist era meant all rights, including the right to live, were centered on political and economic interests. In contrast to Europe and the United States, with a tradition of the rule of law, citizenship in China is not a legal concept but a political one. To illustrate, during the Maoist era in China, citizens were divided into two categories: the people and the enemies of the people. The people were further divided into government/party cadres, and the masses. The masses were then divided into different categories: activists, the general population, and those considered backward elements. These are by no means a legal distinction but a political one. The rights of citizenship differed according to political status. As noted earlier, Mann pointed out that under the authoritarian socialist system, the state did not provide civil and political rights to its citizens but more social rights when compared with other political systems. Here, however, Mann neglects to note that, in practice, social rights in capitalist and socialist countries are entirely different. Social rights in Europe and the United States are based on the legal premise that any member of the state is entitled to social rights, i.e., they are universal in nature. In contrast, social rights in socialist countries are only part of socialist-style justice and are unique. In essence, social rights are prizes granted by the state to those who perform socialist-style justice. In other words, social rights in the socialist state contain robust political implications and are a political good. This is the very essence of citizenship in socialist China.

From Marshall's research, we can learn that the citizenship in Europe and America was formed during the modernization process along with the differentiation, establishment, development, and evolution of various rights protected by the capitalist system. On the contrary, citizenship in Mao-era China was formed through the concentration of resources from society to the state, the establishment of the social, political, economic, and management system, and the infiltration of ideology after the establishment of the People's Republic of China. Based on these characteristics, citizenship in Mao-era China could not be divided into civil, political, and social rights, as in Europe and the United States. Conversely, integration and indivisibility were its main features, with the acquisition of political rights becoming fundamental in determining all other rights and benefits that could be obtained.

In addition, due to the different institutional spatial configurations of the state and society during the Mao Zedong era compared to Western countries, the distribution of citizenship within a specific institutional spatial configuration of the state and society was also markedly different. This will be discussed in detail in the next section.

Why is it necessary to use citizenship to explain collective action during the Cultural Revolution? The author believes that citizenship played the most significant role in the interaction between the state and society at the time. The interaction between the state and society during the Cultural Revolution was mainly realized through the distribution and competition of citizenship. Using citizenship to analyze collective action during the Cultural Revolution is valid in at least three respects.

First, the concept of citizenship can effectively explain many incomprehensible phenomena in factional behavior. For example, what were the factions fighting over during the Cultural Revolution? On the surface, they were fighting over which class would have more opportunities for education and employment, but why were people so fervent in pursuing specific benefits? The concept of citizenship can effectively explain this. Simply put, the interests sought by the movement's participants were what was included in citizenship within the political environment of China at the time. The citizenship in Mao Zedong's era China had specific interests, "revolutionary rights," and status rights that transcended specific interests. That is to say, the passionate participation of the movement participants was not only to pursue specific interests but to obtain higher-level citizenship through the pursuit of specific interests. As Gong Xiaoxia points out, "while active movement participation does not directly provide them with political and economic benefits, the revolutionary identity that people expect to acquire through participation can bring them perhaps the political future they hope for or the protection they need in rapidly changing political storms. Thus, participation indirectly provides people a pathway to state-controlled political and economic resources."[10] Movement participation is a tool for accessing state-controlled political and economic resources. Pursuing this revolutionary identity or revolutionary right can be more passionate than pursuing general interests. Moreover, precisely because citizenship implies an identity linked to rights, the competition for citizenship is not only a competition for "gains" but also a competition for how not to lose existing identity and rights. Since the state entirely controls the distribution of citizenship, changes in the structure of political opportunities during the Cultural Revolution brought about changes in the distribution of citizenship. It was a contest of wits for each participant to avoid losing their existing citizenship and even falling into the "untouchable" category in the rapidly changing environment. Generally speaking, aligning oneself with the most politically legitimate organization or state agent, or demonstrating through one's behavior that one is highly aligned with the ideology espoused by the state, proved to be an effective strategy to gain political rights before the Cultural Revolution. However, the organizations and individuals representing political legitimacy and the interpretation methods of ideology were constantly changing during the Cultural Revolution. Not only did this inject continued motivation into citizenship competition, but it also increased the complexity and variability of the competition. In this respect, it also explains why the Cultural Revolution developed to the point where it pushed the cult of Mao to the extreme. This is because Mao was the only "correct" factor representing the most stable political and ideological element of legitimacy in the Cultural Revolution. For this reason, the author believes that citizenship is a compelling concept that can be used to bridge the two aspects of the state and society and continuously and dynamically analyze collective behavior during the Cultural Revolution.

Second, related to the first point, citizenship is expected to explain the intensity of competition between factions and why competition rises to violence. In any country, citizenship is the most critical value in new nation-building. The author believes that the Cultural Revolution was not an accidental event but a part of the process of national reconstruction since the founding of the PRC in 1949. The process of national reconstruction is also a process of redefining the distribution of citizenship. To be a part of the new state, an individual must have citizenship. Therefore, pursuing citizenship is more important than any other specific interest. Moreover, precisely because the state distributes citizenship based on the fairness and justice of the socialist value system, the concept of citizenship combines the desirable acts of the value system with the actions of the individual. The concept of citizenship effectively bridges the gap between the value system's supposed actions and individuals' actual actions. In other words, "what I want to pursue" and "what I am pursuing" is the same as "what I should pursue." The identity of the two not only makes this pursuit passionate but, more importantly, makes it highly political and blurs the boundary between ideal pursuit and interest pursuit. Ultimately, the competition for citizenship becomes a political struggle between factions. Once it becomes a political struggle, violence becomes part of the competition because violence is simply the ultimate expression of political behavior.

Third, unlike other specific benefits, citizenship is an institutional concept. Because it is a political good, it is not something anyone can get at will. As will be explained in later chapters of this book, citizenship is distributed hierarchically by the state through various systems and policies such as the class origin system, the Household Registration System, the party cadres administrative rank system, the enterprise employment system, the enterprise ownership management system, and the policy of selecting activists. In this way, social groups are divided into two groups, "the people" and "the enemy," with the former can enjoy citizenship and the latter cannot. The content of citizenship varies within the social groups that can enjoy citizenship depending on their social status. State policy determines which groups are entitled to citizenship, which groups are not, what kind of citizenship a group is entitled to, what criteria the state uses to distribute citizenship, and the state agents responsible for its implementation.

The confrontation between "bloodline theory" and "class origin theory" in the Beijing Red Guard movement,[11] the antagonistic structure between the conservative faction centered on the "activists" and the rebel faction centered on the "backward elements" and ordinary workers in the Shanghai workers rebellion in late 1966,[12] and even the demands of the "National Red Workers' General Rebellion Group" (*quanguo hongse gongren zaofan zongtuan*, hereinafter referred to as "General All Reds"), which began to emerge in late 1966 as the rebellion of contract and part-time workers and soon developed into a nationwide rebellion alliance, etc.,[13] almost all of the factionalism and heretical ideas of the Cultural Revolution can be observed in the structure of the hierarchical distribution of citizenship by the state before the Cultural Revolution, and the competing demands for citizenship among members of society. Thus, the concept of citizenship provides the institutional context for factional competition on the one hand and organically links factional/mass behavior with that of the state (leaders/elites) on the other hand. It could be argued that a discussion of the distribution of and competition for citizenship is, in fact, a discussion of the state–society relations, and thus the institutional context of the relationship between the state agents and factions and the interaction between the two during the Cultural Revolution.

Therefore, studying the impact of citizenship on the Cultural Revolution requires examining the relationship between how the state distributes citizenship and factional competition for citizenship and how such a relationship influences collective behavior during the Cultural Revolution. Specifically, by using the concept of citizenship, we can answer the following related questions: How was socialist-style citizenship distributed in China during the Cultural Revolution? Why was citizenship allocated in different ways? How did the various forms of distribution affect the behavior of social groups? Why did members of society compete for citizenship? How did competing approaches affect the formation of social groups and their changes? Why did competition for citizenship escalate and even lead to violence? How did the interaction between citizenship allocators and competitors affect the Cultural Revolution and make it a social movement?

2 Institutional Spatial Configuration and Competition for Citizenship

2.1 *The State, Factions, and Movements*

The three concepts of state, faction, and movement are crucial to understanding the Cultural Revolution. The state and the factions were the two main actors during the Cultural Revolution, and the movement was both a product and a process of interaction between them. Based on these three concepts, the following issues must be considered when analyzing the Cultural Revolution.

First of all, the Cultural Revolution was a large-scale social movement. It lasted three years and contained several "sub-movements," i.e., small-scale social movements. This needs to receive more attention in the existing studies of the Cultural Revolution. The causes, development, and results of the sub-movements, how the previous sub-movement evolved into the next sub-movement, and the reasons for their evolution, etc., are a series of questions that need to be explored.

Second, how the sub-movements constituted the massive social movement of the Cultural Revolution, that is, the relationship between the parts and the whole, must be examined. Although the composition of participants and the goals of various sub-movements of the Cultural Revolution varied, factional conflict and collective violence were common features. So, what was the relationship between the sub-movements, factional conflicts, and collective violence? Here it is necessary to examine whether factional conflict and collective violence are the results of intrinsic factors of the sub-movement, whether they are the result of some external factors, or whether they result from the interaction between internal and external elements of the sub-movement.

Third, to explain the relationship between the sub-movement and the cultural revolution, it is necessary to analyze the relationship between the faction and the state as the main subjects of the movement. Usually, social movements do not necessarily follow the will of their initiators and participants but have their own laws of development. Studies of social movements also show that the interaction between the internal changes of the participating subjects and external factors is the main driving force in the development of the movement.[14] From this point of view, it is possible to discuss the various sub-movements of the Cultural Revolution and its main movement by analyzing the interaction between factions and the state. In this regard, we can discuss the occurrence, development, and results of the main movement of the Cultural Revolution by analyzing the interaction between factions and the state.

Here it is necessary to briefly explain and qualify the Cultural Revolution's sub-movements. Let's take the three years of the Cultural Revolution from May 1966 to April 1969 as the main movement. Anyone who has studied the Cultural Revolution can observe that it was very diverse in terms of its stages and geographical locations. It isn't easy to generalize it as a whole movement. This is precisely why the study of the Cultural Revolution attaches importance to the study of the local Cultural Revolution. As mentioned in the previous chapter, as a politically mobilized social movement, the collective behavior of factions was subject to the political goals of the Cultural Revolution. That is to say; factional rebellions had to be carried out within the framework of the political goals set by the state. From this point of view, the territoriality and diversity of the Cultural Revolution were subordinated to some common political goal of the state. The Cultural Revolution can be divided into at least three sub-movements according to their different political goals. The first sub-movement was from May to October 1966. Its primary political goal initiated by Mao Zedong was to revolt against the powers from bottom to top, to "smash all monsters and freaks," which resulted in the flourishing of the Red Guard movement. The second movement, from October to December of the same year, had the main political goal of "overthrowing the capitalist roaders within the party" and "purging the reactionary bourgeois line of Liu Shaoqi-Deng Xiaoping" from the central to local levels. The student organizations, known as the rebel faction, entered the political arena during this period. The third sub-movement, from January 1967 to April 1969, had as its primary political goal the establishment of the Three-In-One Regime. During this period, workers' organizations began to replace student organizations as the main force of the movement. Note that in the different sub-movement periods, the protagonists thrust into the political arena to bear heavy responsibilities were also different due to other political goals. This will be discussed in more detail in a later chapter. It should be noted that only the central government could set the cultural revolution's political goals, and from the beginning was never allowed to emerge in versions that were at odds with the state's political goals. Therefore, although there were many differences in the unfolding of the Cultural Revolution at the local level, these widely varying local versions of the Cultural Revolution cannot be called sub-movements. According to the author's earlier definition of sub-movement, these various regional versions of the Cultural Revolution are only different forms of expression of the three sub-movements. The three collective actions analyzed in Chapters 5–7 of this book represent one of these forms in each of these three sub-movements and certainly the most representative ones. Since the actors of these three collective actions were also the leading characters in each sub-movement, examining these three collective actions can best reflect the relationship between the sub-movements and the main movement in the Cultural Revolution.

Based on the above reasons, this book will analyze the interaction between factions and the state and analyze the occurrence, development, direction, and results of various sub-movements during the Cultural Revolution. Then, by analyzing the process of mutual transformation between various sub-movements, the author aims to explore the essence of the Cultural Revolution as a social movement. The author believes that this research method not only considers previous research's achievements but also helps to make up for its shortcomings. This is because this method does not simply analyze the internal or external causes of the Cultural Revolution as a social movement but analyzes it from internal and external factors.

2.2 *Institutional Spatial Configuration of the State and Society*

The social movements of any country are collective behaviors in which social groups compete among themselves or with the state to gain citizenship. As Mann points out, the relationship between the state and social groups can be seen in how citizenship is distributed. Understanding how citizenship is distributed and competed for is an understanding of the relationship between the state and society. Therefore, to understand the Cultural Revolution, it is first necessary to clarify the characteristics of the institutional spatial configuration of the state and society during the Cultural Revolution. In other words, understanding the institutional spatial arrangements that governed the relationship between the state and society during the Cultural Revolution is crucial to its study. The author will elaborate on this point later in the chapter. Here, we will first use comparative analysis to sort out the unique institutional spatial configurations of the state and society in Mao-era China and their relations under its constraints.

Generally speaking, in sociology, social space is divided into two parts: the private sphere and the public sphere. The private sphere is the space specific to the individual citizen and has no relation to the state or public institutions. The interaction between the state and society takes place in the public sphere. In countries with democratic institutions, social movements tend to move between two spheres, the public and the private. According to Hirschman, people begin to intervene in the public sphere when their demands are unmet or become frustrated in the private sphere, and social movements arise. Thus, social movements are collective actions that unfold in the public sphere. However, when the movement develops unexpected consequences or the participants' wishes are unmet, people withdraw from the public sphere and return to the private sphere.

The distribution of private and public spheres can be considered a product of historical development in Europe and the United States. Although the distribution of the two spheres in Western countries varies considerably with no uniform pattern, they have been developing in a more balanced manner than in East Asian countries. The private sphere is subject to numerous institutional safeguards, such as private property rights, autonomy, and other independently developed institutional factors. The state's role in Europe and the United States has varied from one period of history to another. Except for the anomalous period of Hitler's fascist rule, the state has rarely been dominant. Even in abnormal periods, the private sphere did not die out when the state intervened. Usually, there is a great deal of resistance to the state's involvement in the private sphere, characterized by gradualness. Similarly, the expansion of the public sphere is likewise gradual. Expanding the private sphere into the public sphere is also subject to resistance.[15] This has much to do with the gradual development of democratic politics. Since democratic politics is a process of individual participation in public politics, it occurs in the public sphere. Still, democratic politics is not something in which the entire population can participate from the beginning, and different social groups participate in politics at different times and in different ways. After the establishment of modern centralized politics, it was first the bourgeoisie that challenged the dominant aristocracy and the monarchical state polity for a share of political power. The bourgeoisie did this very successfully. Then the working class rose as an organized force and began challenging the dominant bourgeoisie, entering the state's political process through the ballot box. This is how democratic politics was established.[16]

The existence of private and public spheres, their relatively balanced state, and the gradual way of intervention greatly impact social movements, especially large-scale ones. Over the long course of history, the two spheres have been relatively institutionalized. In other words, the two are fairly independent, and certain institutions guarantee their independence. The existence of the private sphere constrains the public sphere, and the capacity and power of the state to mobilize society are limited. Except for a few state-mobilized social movements like fascism, most social movements in the West are not driven by the state but by society and result from individuals leaving the private sphere and entering the public sphere. Therefore, Western researchers define social movements as collective actions of social forces against the ruling class.

Moreover, it is precisely because of the existence of the private sphere that participants will withdraw from the public sphere and return to the private sphere once they become dissatisfied with the conduct of the movement and the change of its goals. This provides a vital outlet for social movements and can mitigate their radicalization. According to Hirschman's research, the movement of participants between the private and public spheres constitutes the cyclical change of social movements in the West.[17]

After the founding of the People's Republic of China in 1949, society became highly politicized. The complete transformation of socialist ownership to eliminate private property severely suppressed the private sphere, and the nation was politically shifted into the public sphere. At least until the end of the Cultural Revolution, it can be said that the private sphere no longer existed in China, and even if it did, no one was allowed to remain there. The fact that no private sphere existed is of great importance to the Cultural Revolution. We can explain this issue in at least four ways.

First, since there is no private sphere, inter-group competition for interests can only occur in the public sphere, i.e., the public sphere becomes the only place where inter-group interests compete.

Second, no private sphere significantly impacts the internal changes in the public sphere. That is to say; the public sphere is divided into several zones, each depending on varying political risks. This book refers to these areas as "sub-domains" of the public sphere. Given this, movement participants can choose different sub-domains according to their perceived interests and identity. Sub-domains became, in effect, the playing fields where sub-movements were carried out. The analysis of factional struggles during the Cultural Revolution in the next part of this book shows that the withdrawal of factions from the political arena during the Cultural Revolution did not mean a departure from the public sphere to the private sphere but rather a withdrawal from a politically risky public sub-domain to a less risky one.

Third, since there is no private sphere, there can be no option to withdraw from the public sphere, and therefore the intensification of social movements can be taken for granted. Once a social movement has begun, the participants have no other choice but to participate to the end. In the case of mobilized political participation, those participants are not allowed to withdraw from the movement, and if they do, they have to pay a high price. This is because the values necessary for the survival of the individual have to be realized in the public sphere. Therefore, the absence of a private sphere was the institutional context in which the Cultural Revolution intensified and generated violence.

Fourth, because there is no private sphere, public power is unchecked, and the state and its agents can mobilize society endlessly. This both intensifies the politicization of the public sphere and gives endless impetus to the escalation of movements.

The lack of a private sphere can be said to be a common feature of socialist countries and, in the case of China, is closely related to the process of the founding of the People's Republic. As mentioned earlier, the private and public spheres developed almost simultaneously in Europe and the United States, but China's development pattern differed. The communist revolution in China's modern history was a revolution to "strive for national survival." Before the revolution, China did not have a state in the contemporary sense. To build such a state, the political elite first had to build a party, and then the party would build a state. As such, the formation process was highly politicized. After the state was established, the political elites had to use the executive power of the state to achieve their goals. This has led to a "state dominion of society" type of state–society relationship in China.[18] Its developmental path has created a peculiarity in China's state–society relationship, thus forming a unique relationship between the private and public spheres.

To a large extent, the domination of society by the state means that the public sphere dominates the private sphere. As mentioned, the private sphere manifests more naturally and spontaneously in Western countries. That is, society can develop the private sphere relatively uninfluenced by the state. On the contrary, in Mao-era China, society could hardly rely on its own will and strength to develop the private sphere. This is because it was up to the state to decide whether to give society the right to develop the private sphere, whether it should be developed, and to what extent.

In addition, in Europe and America, citizenship is distributed between two areas: the public sphere and the private sphere. The citizenship necessary for personal survival and values are distributed in the private sphere, completely unrelated to the public sphere. The relatively independent private sphere not only protects the individual's right to survival but also plays a role in restraining the abuse of public power. In Mao's China, since there was no private sphere, all citizenship was concentrated in the public sphere, thus giving rise to a situation in which the state or agents of the state monopolized citizenship. What kind of social groups could enjoy citizenship and what type of citizenship they could enjoy were determined by the will of the state. Moreover, the state and its agents also determine how social groups were classified and to which group individuals belong. This further reinforces the state's monopoly on regulating and distributing the citizenship system. Additionally, since there is no private sphere, there is no citizenship in this sphere. Under such circumstances, competing in the public sphere to acquire citizenship became the only option for people of that era. If one could not obtain citizenship in the public sphere, one would lose the fundamental right to live in society. In other words, there was neither the possibility of leaving the public sphere nor was the act itself permitted. The fate of the people was to have no way out. This was the same for all except Mao Zedong himself, who was above the state system. Even Liu Shaoqi, a powerful state presidency holder, could not quit office and return to his hometown when criticized.

Moreover, the quest for citizenship in Western societies can be described as quantitative. In other words, after acquiring all the citizenship in the private sphere, it is a question of how to gain more citizenship in the public sphere. In contrast, the pursuit of citizenship in Maoist China, especially during the Cultural Revolution, was acquiring the right to live. Therefore, it can be considered a qualitative pursuit. The factional conflicts and brutal violence of the Cultural Revolution are, to some extent, understandable from the perspective of a collective action of competition for the right to exist.

2.3 *The State and Factions During the Cultural Revolution*

Having discussed in the previous section the institutional spatial configurations surrounding the distribution of, and competition for, citizenship, this section describes the analytical framework for the internal dynamics of the state and faction and the interplay between them. As noted earlier, the state and factions were significant players during the Cultural Revolution and are not abstract concepts, so some explanation is necessary.

The state is originally an association of administrative bodies and agents. Since their functions are different and they are in different locations of the organizational hierarchy, the scope of authority and the perception of interests of administrative agencies and state agents differed. They acted and reacted to the outside world differently. The distinction between the state and society in Maoist China is shown in Table 1. As part of society, factions usually consisted of people of a similar level of class, status, occupation, and education, but there are exceptions. Therefore, the analysis of the Cultural Revolution should be carried out in at least three ways.

Table 1 Conceptual diagram of the state and society in Mao-era China

State	**State Agency** **Policymaking Body:** • Central Government **Policy Implementation Agency:** • Provincial governments • Prefectural governments • County governments • Township governments	**State Agent** **Policy Decision Maker:** • Central Party and Military Cadres **Policy Implementers:** • Party and government officials of provincial governments and provincial administrative units • Party officials in prefectural governments and prefecture-administered units • Party and government officials in county governments and county-administered units • Party and government officials in township government and township-administered units
Society	Areas other than government agencies	Individuals, collectives, factions

The first is the internal changes within the state. This refers to the frequent changes in state policies, the division of interests among policymakers, power struggles, differences in understanding and methods of dealing with issues related to national development, and the impact of these factors on policy implementers, resulting in changes at the policy decision-making level and confusion at the policy implementation level.

The second is the internal changes within a faction. This includes whether the members of a faction have consistency in terms of social hierarchy, such as class, social strata, profession, and educational level, as well as whether there is consistency in their individual goals and the collective goals of the faction when participating in a movement.

The third is the continuous interaction between the state and factions. This includes how changes within the state affect changes within factions and how changes within factions, in turn, affect changes within the state. More specifically, it refers to the interaction process between the state and factions and how this interaction drives movements, leading to social chaos, intensifying movements, and violence.

The methodology for examining factional change is discussed first. As mentioned in the previous section, instability was the main feature of factional politics throughout the Cultural Revolution. At some times, a particular faction dominated the stage of the Cultural Revolution, while at other times, the leading role was taken by another faction. Just as remarkable as the turnover in the political arena was the precariousness of the factions, constantly dividing, regrouping, and regrouping again. A small faction can expand into a large faction overnight, or it can suffer a mass exodus of members in an instant. This uncertainty and fluidity of factional politics constituted the daily routine of the Cultural Revolution.

In reality, two levels are involved in analyzing changes within factions. One is the relationship between factions and their members, and the other is the relationship between factions. Here we can borrow Hirschman's concepts of "loyalty, voice, and exit" to analyze.[19] Specifically, members have three options concerning which faction they belong to. The first option is to maintain a high degree of identity with the faction and loyalty to the faction. The second option is to express dissatisfaction with the services provided by the faction and protest (voice) in the hope that the faction to which they belong will improve and thus provide competent services. If the faction accepts the member's opinion, improves, and provides satisfactory service, the member's loyalty to the faction will be restored. The third option is to withdraw (exit) and join another organization or form a brand-new organization that can meet their requirements if the members have negative feelings about the services the faction provides and do not see improvement through members' voices.

We can use the same understanding to explain inter-factional relations; there are two ways for any faction to survive, either trying to gain the loyalty of its members or improving it in response to their voice. If it fails to do either, the faction will have to face the crisis of its members quitting and the danger of the quitting members joining rival factions and coming back as a member of the opposing camp. Thus, it can be argued that factions had an invisible competition mechanism. The competition between factions during the Cultural Revolution pushed factions to radicalization. The reason is that only through ongoing radicalization could they gain a higher level of citizenship from the state or its agents. Unchecked competition and radicalization inevitably led to irrational collective action, including violence. It is important to emphasize here that the collective violence of the Cultural Revolution was not always the result of radicalization and competition for citizenship getting out of hand. In some cases, violence was a means of showing loyalty and thus attaining a higher level of citizenship. The author will elaborate on this point in Chapter 10.

So how should the relationship between the state and factions be interpreted? The relationship between the state and factions during the Cultural Revolution was highly complex. A faction that receives the support of a state agent today may not get it tomorrow. Conversely, a faction fully supporting a state agent today may become its opponent tomorrow. In explaining such changing state–society relations, the aforementioned Hirschman's concepts of loyalty, voice, and exit remain valid.

As mentioned, explaining collective action in the Cultural Revolution requires considering changes within the state. This is because, in most cases, factional differences follow how the state allocates citizenship. If citizenship allocation does not change, members' social identity is relatively stable, and competition rules are more easily institutionalized. If, on the other hand, there are frequent changes in the way the state distributes citizenship, there will be changes and divisions in social factions, reorganization, and so on. This is because only through reorganization can the most significant benefits be obtained from a new distribution method. However, the impetus for how the state changes its citizenship distribution comes from within the state, with contributing factors such as differences in perception of interest among leaders, policy direction differences, political environment perceptions, and internal power struggles.

Taking together the elements discussed above that lead to changes within factions and the state, it can be argued that changes in the relationship between factions and the state manifest themselves in the following six ways. The first is loyalty, i.e., factions are loyal to the state to the best of their ability when the state can satisfy the interests of social groups. The second is protest and voice for improvement when factions express their dissatisfaction with the citizenship allocated by the state or with the way the state distributes citizenship. The third is an exit when the state fails to satisfy the demands of a social group, and the faction chooses to withdraw. In China during the Cultural Revolution, state agents, especially at the level of policy decision-makers, were composed of splintered and plural political factions, so the "withdrawal" referred to here is not from the public sphere but from a risky sub-domains sector within the public sphere to a less risky one. It also implies withdrawal from a "patron–client" relationship with a political faction and includes the possibility of forming such relationships with other political factions. As in the case of the "dropout group" (*xiaoyao pai*), retreating to less politically risky areas is also possible. Fourth, sometimes changes in state policy can cause factions to change their goals. Fifth, factions may create new goals in response to changes in perceptions of the state's approach to the distribution of citizenship. Sixth, as the "heretical trend" of the Cultural Revolution demonstrated, the factional competition brought about the radicalization of social movements, leading to state policy changes. We can see how the state and factions interacted in these six areas, producing alternating sub-movements.

3 Institutionalization and Collective Violence

Why did inter-factional rivalries lead to collective violence? On this question, as shown in the section on the history of the Cultural Revolution in the introduction, Lucian Pye and Lynn White explain the causes of collective violence during the Cultural Revolution in terms of the role of the "resentment" mentality embedded in Chinese political culture and the institutional roots of "labeling" policy. As noted earlier, while explanations from the perspective of political psychology and institutions are essential for understanding collective violence during the Cultural Revolution, at least two issues have arguably not been fully explained in such studies. First, under what conditions did the political psychology of resentment translate into collective violence? Second, why did the policy elements that created social hierarchy surface precisely during the Cultural Revolution and generated large-scale collective violence? The author argues that factional competition during the Cultural Revolution resulted from the interaction between the state and society. That collective violence was primarily a result of the intensification of factional competition. As such, the causes of collective violence must be analyzed from the perspective of the interaction between the state and society.

As Max Weber's theory of the state explains, maintaining a stable social order and preventing conflicts between various interest groups from occurring is the most basic function of the state. The monopoly of violence is one of the critical ways the state maintains a stable social order.[20] Therefore, when discussing collective violence during the Cultural Revolution, it is essential to ask what role the state played. As we will see later, collective violence during the Cultural Revolution was closely related to the state. It can be considered a result of the state's political mobilization and internal institutional divisions. However, these do not necessarily lead to collective violence. For example, the competition for citizenship under the same institutional environment before the Cultural Revolution did not cause social chaos, so why did the competition for citizenship during the Cultural Revolution lead to collective violence? The author argues that in addition to the two institutional factors mentioned earlier, namely, the institutional spatial configuration of state and social integration and the distribution of and competition for citizenship, the low degree of institutionalization at all three levels of the state, society, and state–society interaction was also an important cause of large-scale collective violence. In general, collective violence arising from competition between interest groups in society is related to the degree of institutionalization between the state and society. A fourfold link between the degree of institutionalization of the state and the degree of institutionalization of society can be envisaged, as shown in Fig. 1.

First, in contexts where the state is highly institutionalized and society is relatively less institutionalized, spontaneous political participation in society is not supported or rewarded by the state. Therefore, the capacity of society for autonomy cannot quickly be developed. In environments

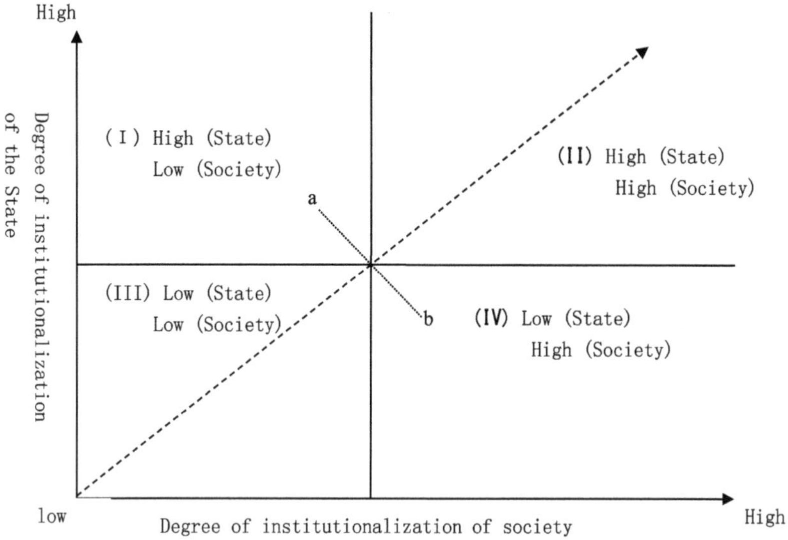

(Note) The arrows on the dotted line in the graph indicate the likelihood of collective violence occurring. The further to the upper right, the less likely violence is expected to occur. The likelihood of collective violence along line ab can be considered equal.

Fig. 1 The relationship between state-society institutionalization and collective violence

with relatively low levels of social institutionalization, there is a greater likelihood that mass political participation will cause disruption. But because of the high level of institutionalization of the state, it can generally be assumed that the state has the potential to control collective action in society and save it from escalation.

Second, where both the state and society are highly institutionalized, the state relies on laws and institutions to control society, and society has institutionalized channels to express its will. At this point, it can be argued that the competition around the distribution of citizenship and access to citizenship usually occurs in an institutionalized environment, and the resulting collective action is relatively unlikely to translate into violence.

Third, in countries such as the former Soviet Union and Mao-era China with low levels of state institutionalization and low levels of social institutionalization, where the state power is highly concentrated in the hands of the top leadership, collective violence is likely to arise when there is non-institutionalized political participation, as society does not have the capacity for autonomy to curb vicious competition among interest groups. At this point, if the state can maintain internal coherence, it can control the disorganization of mass behavior and provide stability and order to society. Once the state is divided within its institutions due to power struggles, and the state itself does not have a sound legal system to regulate its internal disputes, this will weaken the ability of the state to provide the basis for social order and make collective action ineffective. Even more dangerously, power struggles among top leaders and competing interests among social factions may take advantage of each other to further radicalize collective action.

Fourth, in cases where the state is relatively less institutionalized, and society is more institutionalized, competition for access to citizenship takes place in an environment where society functions to regulate sectarian interests so that the likelihood of collective violence can be considered relatively low. Even if there is a split within institutions of the state that affects the political participation of society, society instead of the state can maintain order and stability due to the high degree of institutionalization. In such cases, the power of social autonomy can somewhat prevent collective violence.

Of the four types mentioned above, the relatively more stable ones are those with a high degree of institutionalization of both the state and society. In contrast, a low degree of institutionalization of the state and society can be considered the least stable. Mao-era China can be positioned as the third least stable type of these four types. What criteria should be adopted in judging the degree of institutionalization of the state and society?

Max Weber divides the forms of state rule into three types. The first is the individual transcendental charismatic ruler, the second is the patriarchal and traditional hereditary type of rule, and the third is the rational-legal type of rule. Of course, there is also a mixture of the three types. Weber argues that these three types of rules are processes of political development and change. With the birth of a rational state premised on bureaucracy and law, the state's rule inevitably evolves from charismatic to traditional to rational-legal.[21] Weber's methodological analysis of rational-legal legitimacy emphasizes bureaucracy, law, and institutions. According to his theory, the modern state is institutionalized through governance by law, and the soundness of the rule of law can evaluate the degree of institutionalization of the state. Specifically, when examining the degree of institutionalization of the state, the following indicators are undoubtedly important: whether there is a sound bureaucratic system, whether the power relationships within the bureaucratic system, such as judicial, legislative, and administrative powers, are independent, and whether there are coordinated relationships based on laws or systems between various power relationships. The state's low institutionalization level becomes evident when we use these indicators to measure Maoist China.

First, the China of Mao's time was not a country governed by the rule of law but by the rule of man, in which the leader's charisma dominated. Second, a unified system of parliamentary and executive power was adopted in the relationship between the legislative-executive-judicial powers of the state.[22] In this system, judicial power was not independent, executive power was not neutral, and the legislative power was entirely controlled by the Communist Party. Third, no legal and institutional coordination existed between the legislative-executive-judicial powers and other powers.

The low degree of institutionalization of the state was the general situation in China during the Mao era. The situation deteriorated further during the Cultural Revolution when the degree of institutionalization of the state dropped to its lowest point. There are several reasons for this. First, building an institutionalized system of governance was at odds with the Cultural Revolution's goal of transforming bureaucracy through a mass movement. Second, the cult of Mao was at its height during the Cultural Revolution, which made the formation of laws and institutions even more impossible. Third, during the Cultural Revolution, all state institutions, including the party, were turned into instruments of mass mobilization rather than providers of social order. This is precisely why many local public security departments provided lists of black five families to mass organizations during the Cultural Revolution.

After defining the criteria for determining the level of institutionalization in a country and summarizing the level of institutionalization of China during the Mao Zedong era using these criteria, the next question is, what criteria should be used to assess the level of institutionalization in society? In the literature on civil society, the indicators of autonomous social organizations independent of the state and the existence of coordinated relations within these organizations and with other autonomous organizations, guaranteed by laws and institutions, are undoubtedly important. Moreover, this process, from its conception to forming internal mechanisms and establishing communication channels with other organizations, takes considerable time and does not happen overnight.

The results are clear when we use the above criteria to examine Chinese society under Mao. After the state's founding, long-standing civil organizations of all kinds were disbanded one after another, and society as a whole was mobilized into a highly politicized public sphere. Henceforth, there were no social organizations independent of the state. Although there were organizations such as work units, Communist Youth Leagues, and trade unions, these were mass organizations run by the party and the state and were fundamentally different from autonomous social organizations. While society appears to be highly organized, these organizations are instruments of state domination and control, and they do not have the function of advocating for self-rights of the society. Because members of society could not use work units, Communist Youth Leagues, and trade unions as instruments for the pursuit of citizenship, factions emerged during the Cultural Revolution and were used by society to pursue interests.

It should be emphasized that during this particular period of the Cultural Revolution, it appears that the masses enjoyed, to some extent, more freedom of association than before the Cultural Revolution and freely formed various mass organizations independent of the party and league organizations, but these were not autonomous social organizations. First, these organizations emerged from the party's political mobilization, and their organizational goals and legitimacy were subject to the state, so it is difficult to speak of autonomy. Second, there was no institutionalized mechanism of synergy within and between factions. Members of factions bond over temporary interests, but when external conditions change, the alliance immediately disintegrates from a collaborative relationship to an antagonistic one. Some factions tended to be antagonistic to other factions and even their parent organizations at the very beginning when they were established or separated from their parent organizations. Third, the formation of autonomy requires a long-term development process. Even without the reasons stated above, it would have been difficult for the factions to develop into autonomous organizations during the Cultural Revolution quickly.

For these reasons, the autonomy of factions during the Cultural Revolution was extremely low, and there was no institutionalized mechanism for cooperation and interaction between factions. As a result, when the control of the state over society was weakened, society could not resolve conflicts of interest on its own, and the competition among the masses was left in anarchy, which tended to radicalize and lead to large-scale violence.

Moreover, at the level of the interaction between the state and society, its institutionalization is equally low. In democracies, the interaction between the state and society is governed by laws and institutions. The state cannot mobilize society without limits, nor can society intervene in the state at will. But in the institutional, spatial configuration of state–society integration in Mao-era China, there were no explicit institutional and legal guarantees of state–society interaction. This is illustrated below in two ways.

First, the control of society by the state is not legalistic but political. This politicized method of domination is achieved by two primary means: (1) control of society through the emphasis on the ideology of class struggle and (2) control of society through the exclusive and hierarchical distribution of citizenship. The content of the Theory of Class Struggle is adjusted to the changing political goals and the necessity of political struggle, and its connotations change accordingly. The allocation of citizenship is also greatly influenced by factors such as national goals, policy changes, the intentions of national agents at the policymaking and implementation level, and power struggles between state agents, which makes it highly unstable. From this perspective, it can be said that in the institutional spatial configuration of state and society, the state controls society completely through ideology and organization. At the same time, its methods of governance are non-institutionalized.

Second, from the point of view of political participation, society did not have the mechanisms for political participation regulated by institutions and laws. Under such circumstances, the masses could only fight for their rights and interests by taking advantage of the opportunities offered by the political mobilization of the state. Thus, the political movement of the Cultural Revolution eventually became an arena in which the people fought for their citizenship.

The Cultural Revolution was a politically mobilized social movement in an environment of low institutionalization at all three levels of the state, society, and state–society interaction. In the case where there is a political struggle that affects the competition of interests between factions, the political participation of the masses generated by political mobilization is transformed into factional competition. Due to the lack of mechanisms for the state and society to maintain stable social order with self-discipline, factional competition is easily transformed from conflicts of interest to violence. This is because competition itself is a conflict of interest. As Tilly pointed out, most collective violent behavior originates fundamentally from non-violent collective behavior and is similar to non-violent collective behavior that occurs under similar circumstances and at the same time.[23]

4 Conscious Structures that Translate Loyalty into Benefits

The previously discussed analytical approach to the relationship between faction members and faction organizations and inter-factional relations is linked to another fundamental question in the study of social movements: why do rational participants bring about such irrational outcomes as violence? In other words, when we understand the Cultural Revolution as a political mobilization-type social movement revolving around the acquisition of citizenship, what is the relationship between the competition for citizenship and the fanatical and irrational behavior during the Cultural Revolution? Generally speaking, competition for citizenship does not necessarily lead to irrational behavior, and it is possible to conduct it within rational and institutionalized rules.

The author recognizes the idealistic elements in the behavior of the participants of the Cultural Revolution because citizenship not only contains elements of identity and rights but also carries the value system for forming civic virtue. Because, as stated, the distribution of citizenship is based on the socialist value system of fairness and justice, the boundaries between idealism and the pursuit of personal interests are blurred and easily substituted for each other. Nevertheless, the author disagrees with previous studies on the Cultural Revolution that overemphasize the irrational factors of the participants (e.g., fervent worship and loyalty to Mao Zedong) to explain the collective action of the Cultural Revolution. From the overall progress of the Cultural Revolution, it is undeniable that irrational behaviors such as fanatical leader worship and foolish loyalty were present. Still, the irrational characteristics of the participants cannot be judged from the whole to determine the motives and behavioral characteristics of individual participants. At the same time, idealism can result entirely from a rational choice. This book will proceed on the assumption that the participants are rational actors. Rational choice in the Cultural Revolution was conditioned by several factors, including the time background (the influence of Mao's thought, elements of traditional culture, institutional elements, socialist civic values, etc.). Despite external constraints, whether individuals or groups choose to participate in social movements, whether it is an active or passive choice, and whether their motivation is based on values or interests can be seen as the result of rational decisions in the political environment at the time. In other words, from the perspective of Rational Choice Theory, the collective behavior of the Cultural Revolution can also be seen as the result of factions' rational pursuit of interests or pursuit of the value of good socialist citizenship. Therefore, in understanding the factionalization and collective violence during the Cultural Revolution, in addition to the three institutional elements mentioned above, namely, the institutional spatial configuration of the state and society, the distribution and competition for citizenship, and institutionalization, beyond these three elements, we cannot ignore the role of the transformation of loyalty into interests, which was a common conscious construct of the time. Simply put, the Mao cult of personality was ostensibly a religiously irrational act, but there was a very rational element behind it. In other words, irrational behavior is just the result of rational choice. Here it is first necessary to analyze the relationship between the personal worship of Mao and irrational behavior to explain the conscious elements of rational choice that bring about irrational results.

Why worship Mao Zedong? This book argues that the worship of Mao is a manifestation of the worship of the state, the result of the integration of the worship of the individual leader and the state. Three factors explain this: the elements of tradition, the "party-state system," and the role played by Mao's philosophy of statehood in the founding of New China.

First is the element of tradition. It has been pointed out that there was no "state" in the modern sense of the word in dynastic China, and traditionally the emperor was the state and was regarded as the state. Therefore, only a perception of the emperor existed but not of the state.

Moreover, positioned as the "Son of Heaven," the emperor was the apex of the dynasty. This means that loyalty to the emperor embodies a universal moral standard. Unsurprisingly, this view of loyalty to the emperor and even the "emperor-equates-state" paradigm was transformed into a "state-equates-Mao Zedong" paradigm after the founding of New China. For some at the time, Mao was an emperor, a god. Moreover, it was the Chinese Communist Party, with Mao Zedong as its leader, that broke China out of the pain of the aggression of the Western powers for more than a century since the Opium War and restored the self-respect of the Chinese people. As such, the worship of Mao Zedong was even greater than that of the emperor, as seen in popular songs of the time.

Second, the influence of the party-state system. It is well-known that the concept of political parties and the state was introduced from Western Europe to China. However, the relationship between the state and political parties in China vastly differs from that of Western Europe. In Western society, the state came first, and then political parties. In other words, political parties emerged to transform the state from non-democratization to democratization. In contrast, in China, the party creates the state, and thus the "party-state" system is characteristic of China. In other words, political parties in Western Europe are within the state and are the instruments of the formation and implementation of state policy. In China, political parties are above the state and rule it directly. Simply stated, ruling Western European parties govern, while ruling Chinese parties rule.

We can trace the idea of the party creating the state back to Sun Yat-sen. Sun believed that the key to the success of the Russian Revolution was its "party-state system" (i.e., because it put the party above the state). Therefore, at the First National Congress of the Kuomintang in 1924, he mentioned: "We do not have a state to rule now; we can only say that we will build the state with the party and then rule it when it is built." Putting the party above the state and putting it in direct control of power was the core of Sun Yat-sen's "party-state theory." When Sun was alive, the party was the supreme authority in the relationship between the Kuomintang and the government, holding the power to appoint and dismiss government officials. The government implements the party's ideology, policies, and directives and is subject to the leadership and supervision of the party.[24]

The Chinese Communist Party, which believed in Leninism and was also influenced by Sun Yat-sen, had precisely the same understanding as the Kuomintang of "building the state by the party and ruling it by the party." The difference lies in that in the "party-state system" of the Kuomintang, the party is composed of elites while the people are placed outside of the political power; in contrast, the Communist Party is not composed of elites but consisting instead of numerous workers and peasants, and the people are incorporated into the political power and become the main class force to rely on.

The party-state doctrine, whereby the party rules the state, was confirmed and reinforced in the political system of the People's Republic of China after its establishment. As Kazuko Mori points out, since 1954, the Communist Party, the state, and the People's Liberation Army have formed a solid trinity. In this trinity, the party leads the other two in the form of the supreme power.[25] While the National People's Congress is ostensibly defined as the "supreme organ of state power," the supreme authority to rule the state belongs to the Communist Party of China.

Europe and the United States have often criticized the relationship between political parties and the state in China as unmodern and undemocratic. However, in light of the formation of the party-state system, it is natural that political parties should have absolute leadership over the state. This is because the party created the state and the army, so naturally, the party should have ownership and leadership. A small political elite created the party, so although both the Kuomintang and the Communist Party had many members, only a tiny group was central to the party's decision-making, and the top leader of this small group was given absolute power.

As discussed above, the relationship between the individual leader, the political party, and the state significantly impacts public perception. Thus, in the 1950s and 1960s, some people equated Mao Zedong, the Communist Party of China, and the People's Republic of China and confused the relationship between loyalty to Mao, loyalty to the Communist Party of China, and loyalty to the new China.

The third point is related to the role of Mao's philosophy of nation-building. New China can be considered to have been founded on Mao's founding philosophy. The formation of the relationship between the state and society in the early stage of the founding of the country, the economic decentralization during the Great Leap Forward (which will be discussed in detail in the next chapter), and the political decentralization during the Cultural Revolution can all be understood as the practice of Mao's founding philosophy. In particular, the Great Leap Forward and the Cultural Revolution can be considered Mao's personal attempt to rebuild the state. Since the new state was built under the guidance of Mao Zedong thought, it is difficult to distinguish between loyalty to the state and loyalty to Mao as a person.

As mentioned, the confusion between loyalty to the leader and loyalty to the state led the people of the time to turn their worship of Mao Zedong into the worship of the state. Why worship the state? Apart from the hope that socialist China brought to the Chinese people, the most important reason was that the institutional spatial configuration of state–society oneness determined the necessity of worshiping the state. Specifically, under the institutional spatial configuration of state–society integration, the state controls citizenship and distribution methods, and society completely depends on the state. On the one hand, the state provides society with almost everything it needs, from the necessities of life to housing, health care, education, and entertainment; on the other hand, the state extracts the maximum amount of loyalty from society. In such an institutional environment, expressing loyalty to the state and its agents becomes an essential method of acquiring citizenship for each member of society. In other words, when state institutions and policies personify the psychological element of state worship, the worship of the state becomes more than a mere emotional expression, and it can be translated into rational acts of interest.

As will be discussed more specifically in later chapters, the allocation of citizenship by the state is based on political criteria. Thus, political resources in citizenship are prioritized more than cultural resources. Loyalty to the party and the state is necessary to obtain political resources. The pattern of loyalty in exchange for resources was already in place before the Cultural Revolution, and the worship of Mao Zedong was pushed to its zenith by the central leadership during the Cultural Revolution. In such a political environment, "I" am most loyal to the Great Leader Mao than others. This kind of loyalty contest became essential for people to protect themselves and gain political and economic benefits. In this regard, the general public and central leaders like Zhou Enlai and Lin Biao were no exception. Against this backdrop, and with the "leftist before rightist" thinking prevalent during the Cultural Revolution, this competition to express loyalty pushed the worship of Mao Zedong to an extreme level of fanaticism rather than being driven by religious sentiment or blind impulse of the masses. From this perspective, it can be argued that there was actually rational judgment behind the frenzied worship of Mao Zedong. The extreme worship of Mao Zedong, the progress of the Cultural Revolution, the increasing degree of this extremism, the constant updating of its form, and the growing instrumentalization are the inevitable results of the competition for citizenship without exiting the institutional space. As already mentioned, of the many elements of the Cultural Revolution relevant to the acquisition of citizenship, especially the aspect of organizations and individuals that could represent ideological legitimacy, the only one that remained constant was Mao himself and Mao Zedong Thought. It was the only "correctness" that led to the escalation of the Mao cult.

Moreover, the formation of the psychological element of loyalty into interest is also related to the mass movement style of political participation. Mass movements were the only means of mass political participation in Mao-era China. For the state and top political leaders, political mobilization was a necessary means of implementing national goals or the personal will of the leaders. For members of society at large, a mass movement in the form of political mobilization was an opportunity to express loyalty to the party and the state or state agents and thus gain benefits. As the political movements since the founding of the state have shown, every political movement always brings about competition among members of society to express their loyalty or "revolutionary spirit." Of course, not all people wish to improve their political and economic position by expressing loyalty, and there is no shortage of sincere admirers. It is undeniable, however, that for many, the expression of loyalty is simply a way of hoping to save themselves in a changing political situation. In such an institutional spatial configuration of the state and society in which the possibility of withdrawal into the private sphere is eliminated, members of society do not have the possibility of relying on legal means other than state power to safeguard their rights and interests.

5 Concluding Remarks

The institutional spatial configuration of the State and Society, the system governing citizenship distribution, and the weak institutionalization of state–society interactions all mutually reinforced one another. These three institutional aspects, along with the underlying mindset that transformed loyalty into tangible benefits, fostered rational competition for citizenship among various social groups. Concurrently, this mindset introduced an irrational aspect to the otherwise rational competition. These four elements combined to create an environment ripe for factionalization and collective violence even before the onset of the Cultural Revolution. The subsequent eruption of violence during the Cultural Revolution will be explored in the chapters that follow.

Notes

1. T.H. Marshall, *Citizenship and Social Class* (Cambridge: Cambridge University Press, 1950); T.H. Marshall and Tom Bottomore, *Citizenship and Social Class* (Doubleday, Garden City: N.Y. 1965); Japanese version: T.H. Marshall, Tom Bottomore, translated by Nobuhiko Iwasaki, etc., *Citizenship and Social Class* (Legal Culture Society, 1993).
2. Reinhard Bendix, *Kings or People: Power and the Mandate to Rule* (Berkeley, CA: The University of California Press, 1978); Ralf Dahrendorf, *Class and Class Conflict in Industrial Society* (London: Routledge and Kegan Paul, 1959). Japanese version: translated by Kenichi Tominaga, *Class and Class Conflict in Industrial Society* (Daiyamundo, 1964); A. H. Halsey, "T. H. Marshall: Past and Present, 1893–1981," *Sociology*, vol. 18 (1984); Seymour Martin Lipset, "Tom Marshall—Man of Wisdom," *British Journal of Sociology*, vol. 24 (1973); David. Lockwood, "For T. H. Marshall", *Sociology*, vol. 8 (1974); B. S. Turner, *Citizenship and Capitalism: The Debate over Reformism* (London: Allen and Unwin, 1986).
3. Michael Mann, "Ruling Class Strategies and Citizenship," in Mann, *States, War, and Capitalism: Studies in Political Sociology* (Oxford: Blackwell Publishers, 1992), pp. 188–209.
4. Ibid. pp. 188–209.
5. Mann refers to the former Soviet system of government as an authoritarian-style socialist system, primarily to distinguish it from other forms of socialism.
6. Here, "infrastructural capabilities" is a specific concept of Mann's. He divides the capabilities of the modern state into two types, despotic capabilities, and infrastructural capabilities. The former refers to the ability of the ruling class of the state to act on its own without periodic, regular, and institutional consultation with the citizens, while the latter refers to the ability of the ruling class

of the state to effectively implement its political decisions in society through the penetration of civil society. See Michael Mann, "The Autonomous Power of the State: Its Origins, Mechanisms, and Results", in John A. Hall, ed., *State in History* (Oxford: Blackwell Publishers, 1986).

7. For a discussion of how the state's monopoly of political, economic, and social resources affected state–society relations in Mao-era China, see Andrew Walder, *Communist Neo-Traditionalism: Work and Authority in Chinese Industry* (Berkeley, CA: University of California Press, 1986); Cecilia Milwertz, "Control as Care: Interaction between Urban Women and Birth Planning Workers", in Kjeld Erik Brodsgaard and David Strand, eds. *Society, and National Identity* (Oxford: Clarendon Press, 1998), pp. 92–112.
8. For a related discussion, see Keith Faulks, *Citizenship* (London and New York: Routledge, 2000), pp. 1–28.
9. Lin Yunhui, Fan Shouxin and Zhang Gong, *Kaige xingjin de shiqi* (*The Age of Triumphantly Moving Forward*) (Henan People's Publishing House, 1989), p. 26, pp. 431–432.
10. Gong Xiaoxia, "The Origins and Development of Mass Violence in the Cultural Revolution" (wenge zhong qunzhong baoli xingwei de qiyuan he fazhan) op. cit. in *Journal of Social Sciences* (Cultural Revolution Special Issue) Shehui kexue xuebiao, wenge zhuanji, 1996, no. 7, p. 106.
11. Yang Lijun, "Social Origins of the Red Guard Movement: The Perspective of Class and Stratum", *Modern China,* no. 74, 2000, pp. 209–221.
12. Li Xun, *Dabengkui:shanghai gongren zaofanpai xingwangshi* (*The Great Collapse—A History of the Rise and Fall of the Workers' Rebellion in Shanghai*) (Taibei: Times Cultural Publishing Enterprises Ltd, 1996).
13. Tsuneto Yamamoto, "Grassroots Workers at the Early Stages of the Cultural Revolution: The Rebellion of Temporary and Contracted workers", *Asian Economy*, no. 2 (1981).
14. Doug McAdam, *The Political Process and the Development of Black Insurgency* (Chicago: University of Chicago Press, 1982); Herbert Kitschelt, "Political Opportunity Structures, and Political Protest: Anti-Nuclear Movements in Four Democracies," *British Journal*

of Political Science, vol. 16 (1986), pp. 57–85; Mary F. Katzenstein and Carole Mueller, *The Women's Movements of the United States and Western Europe: Consciousness, Political Opportunity, and Public Policy* (Philadelphia: Temple University Press, 1987); Sidney Tarrow, *Power in Movement: Social Movement, Collective Action and Politics* (New York: Cambridge University Press, 1994).

15. Albert O. Hirschman, *Shifting Involvements: Private Interest and Public Action* (Princeton, NJ: Princeton University Press, 1982). Japanese version: translated by Takeshi Sasaki et al. *The Phenomenology of Disappointment and Participation: Private Interest and Public Action* (Hosei University Press, 1988).
16. For example, Barrington Moore, Jr., *Social Origins of Dictatorship and Democracy* (Boston: Beacon, 1966); Japanese version: translated by Ryuji Miyazaki et al. *The Origins of Dictatorship and Democratic Politics in Society: The Lords and Peasants in the Formation Process of the Modern World* (Iwami Shoten, 1986).
17. Hirschman, op. cit., 1982.
18. For a discussion of state formation and state–society relations in China, see Yongnian Zheng, *Zhongguo minzu zhuyi de fuxing: minzu guojia xiang hechu qu* (*The Revival of Chinese Nationalism: To What End the Chinese Nation-State*) (Hong Kong: Sanlian Books, 1997).
19. Albert O. Hirschman, *Exit, Voice, and Loyalty: Responses to Decline in Firms, Organizations, and States* (Cambridge, MA: Harvard University Press, 1970); Japanese version: translated by Takayuki Miura, *The Theoretical Construction of Organizational Society: Exit, Voice, and Loyalty* (Minerva Books, 1975).
20. Max Weber, translated by Lin Rongyuan, *Economy and Society*, volume 2, (The Commercial Press, 1997), pp. 719–735.
21. Ibid., pp. 263–490, 720.
22. Kazuko Mori, op., cit., pp. 105.
23. Charles Tilly, *From Mobilization to Revolution* (Englewood Cliffs, NJ: Prentice-Hall, 1978); Japanese version: translated by Horie Fukashi, *The Theory of Political Change, (*Ashi shobo, 1984), pp. 218.
24. Yongnian Zheng, op., cit. (1997), pp. 149–154.
25. Kazuko Mori, op., cit. (1993).

CHAPTER 3

State-Building and the Reorganization of Institutional Spatial Configurations Before the Cultural Revolution

As mentioned in the previous chapters, the author sees the Cultural Revolution from the perspective of the interaction between the state and society since the founding of the People's Republic of China. Suppose the Cultural Revolution is understood from this perspective. In that case, it is necessary to sort out the process of the formation and development of state–society relations since the country's founding, that is, to explain the gradual expansion of the public sphere and the gradual shrinking of the private sphere. Furthermore, the author also looks at the Cultural Revolution as a state-building stage, like the socialist transformation and the Great Leap Forward in the country's early years. As a result of the state's distribution of citizenship, it is also necessary to examine the Cultural Revolution in the context of state-building. This chapter, therefore, focuses on three issues: first, the process of restructuring the institutional spatial configuration of the state and society; second, the impact of state-building on the institutional spatial configuration of the state and society; and third, the positioning of the Cultural Revolution in state-building.

After founding the People's Republic of China in 1949, the most critical issues facing the Chinese Communist Party were the consolidation of power and the integration of society. These two subjects needed to be pursued simultaneously and without delay. Strengthening the regime and

Y. Lijun, *Contest for Citizenship and Collective Violence during China's Cultural Revolution*, IPP Studies in the Frontiers of China's Public Policy, https://doi.org/10.1007/978-981-97-0906-9_3

integrating society simultaneously would inevitably lead to an excessive expansion of state power because this process requires the maximum use of the state's capacity for integration. Usually, in a state of revolutionary regime change, the re-establishment of social order is often not the result of society's development; the state inevitably plays a leading role in its reconstruction. In addition to these common conditions, the People's Republic of China has unique features. Since the early years of the state were limited by social conditions such as the paucity of resources and the low level of modernization, it was all the more necessary to construct an integrated state and social system to draw resources from society, the institution of the state and society integration is also related to the concept of the socialist state, which is a characteristic common to socialist states. In terms of institution-building, the formation of the system of state and social integration is achieved mainly through creating economic and political systems. In turn, creating ideology provides values with legitimacy and legality for the state and society integration. The institution-building and ideological propaganda complement each other and together form the institutional basis of socialist citizenship and its legitimacy, as well as provide and legitimize the code of conduct for the citizens of the new socialist state. Here, the author examines the encroachment of state power into society in the early years of the state's founding through three aspects: the formation of the economic system, political organization, and the advancement of ideology.

1 Penetration of State Power into Society at the Economic Level

The nine years between 1949 and 1958 can be described as a period in which the state formed and consolidated the structure of a highly controlled society. In terms of the economic system, the penetration of state power into society took different forms in the countryside and the cities. The countryside underwent land reform, agriculture collectivization, and public ownership transformation. While in the cities, it was accompanied by the socialist transformation of capitalist industry and commerce.

1.1 *Reorganization of the Institutional Spatial Configuration of the State and Society in the Countryside*

The land reform between 1949 and 1952 confiscated the surplus land of landlords and rich peasants and distributed the land equally to the entire agricultural population on a township basis.[1] The land reform changed the landholding situation in the countryside and made it possible for "the land to the tiller," but land ownership still belonged to the individual. Next, the collectivization of agriculture was promoted. First, Mutual Aid Groups were formed and then gradually developed into Primary Agricultural Cooperatives and Advanced Agricultural Cooperatives. These developments show that agricultural collectivization is the process of changing land ownership from private to collective ownership.

The so-called Mutual Aid Groups are collective production and mutual labor organizations based on rural self-employment. The land is privately owned by the farmer, who has the right to decide on the cultivation and management of crops, etc.[2] The Primary Agricultural Cooperative is an agricultural organization that recognizes the private ownership of land and where farmers contribute their land to be used in a unified manner. The farmer owns the land but cedes the land use rights to the cooperative, which manages it.[3] In short, the Primary Agricultural Cooperatives' separation of ownership and land use is a characteristic.

The Advanced Agricultural Cooperatives is the transition from private ownership of land and other means of production to the public ownership of cooperatives to establish collective ownership. The basic organization in rural areas during the Advanced Agricultural Cooperative period is still the production team, but the distribution of remuneration has changed. It is no longer based on land remuneration but on the amount of labor performed as labor remuneration. This transition reflects a shift toward collective ownership and establishing socialist economic principles in rural areas.[4]

Advanced Agricultural Cooperatives had achieved collective ownership of land and other production materials. This change in ownership from private to collective ownership changed the institutional spatial configuration of the state and society. State power became more involved in society than before, and the private sphere was significantly reduced compared to the period of the Primary Agricultural Cooperatives. This is confirmed by the statistics on the development of agricultural co-operatives: in 1950, only 10.7 percent of the peasant households in the country participated in Mutual Aid Groups; in 1957, 97.5 percent did so.[5] The percentage of peasant households participating in cooperatives increased considerably with the reform of land collectivization and the reorganization of the Advanced Agricultural Cooperatives. Still, the decision to join the Advanced Agricultural Cooperatives was not institutionalized regulation but left to the individual's will. In other words, individuals were free to join or not. While the government used various methods, such as political propaganda, inducement of benefits, and administrative constraints to reward farmers for joining and discourage them from leaving,[6] it could not force them.

The system of state control of society in the countryside was finally completed during the period of the People's Commune.[7] The People's Commune introduced the new proposition of changing ownership from collective to national ownership.[8] They introduced two systems of integration of government administration and economic management (Political and social unity) and "production team-based three-tier ownership," which had not existed during agricultural collectivization. The integration of government administration and economic management system is an organizational form in which production, administration, and political functions are combined to achieve effective governance of the countryside. In particular, the Advanced Agricultural Cooperative, responsible for production, and the "township government," the lowest local state administration, are united.[9] On the one hand, the state's direct management of production activities was achieved. On the other hand, the "production team-based three-tier ownership system" expanded management scope and enhanced the public nature of production activities. This included transferring production management functions, such as the calculation of labor points, from production teams and production brigades to larger organizations with higher administrative levels, such as the people's communes, thus, injecting more public functions into work and life. In other words, the establishment of the people's commune system[10] brought about a qualitative change in the relationship between the state and society in the countryside, as the peasants' production activities and daily life were integrated into the state's administrative system, and the public nature of work and life was gradually increased. Rural society was gradually nationalized and politicized, and the private sphere was made public through the change in ownership, a shift toward the socialist economic system.

1.2 State–Society Interactions in the Reorganization of Institutional Spatial Configuration in Rural Areas

As mentioned earlier, at the end of the 1950s, with the essential completion of the transformation of ownership, the state encroached directly into rural society by taking complete control of the economic resources of the countryside and achieved direct control over all corners of agrarian society. It must be noted here that the state played a decisive role in advancing state power in society, but we should also note society's active participation. In other words, the socialist transformation of the 1950s must be understood in the context of the interaction between the state and society. Specifically, policymakers, policy implementers, and peasants each played a role in the phased changes of Mutual Aid Groups, Primary Agricultural Cooperatives, and Advanced Agricultural Cooperatives.

First, in expanding from the public to the private sphere, in addition to the policy decisions of the central government and its vigorous promotion of policies, local cadres, as policy implementers, played an active role. For example, between 1952 and 1953, grassroots cadres in the countryside, either to retain their cadre status or to gain the appreciation of their superiors, were overly aggressive in promoting the collectivization of agriculture, forcing peasants to join Primary Agricultural Cooperatives, and engaging in radicalization. In March and October 1953, the central government issued two directives to restrain the process of cooperitization in the countryside.[11] The same situation occurred again in 1954. Why were grassroots cadres so active in promoting the collectivization of agriculture then? The author argues that in addition to their identification with the goals of state policy, the expansion of the public sphere also led to the expansion of state agents' power at the policy implementation level. From the perspective of competition for citizenship, there is also a need and a drive among state agents at the policy implementation level to compete for loyalty to the state or higher state agents. This dynamic makes grassroots cadres actively work to advance state policies, bringing about rash progress in policy implementation. This was one of the characteristics of Chinese politics during Mao's time and continues to influence local politics in China even today. It is worth noting that because the state agents at the level of policy implementation have the most direct relationship with the advancement of state policies when there is tension between the state and society, the public's anger is most likely to be directed at the state agents at the level of policy implementation. This was why the masses attacked many local and grassroots cadres during the Cultural Revolution.

Second, in addition to the role played by state agents, many peasants actively participated and promoted agricultural collectivization through their efforts. For example, in 1955, out of 3711 cooperatives in the Xinyang District of Henan, 1577 were organized by higher-level Work Teams, 1573 by township branches, and 561 by peasants on their initiative.[12] Although there were also examples of passive resistance from peasants in 1954 due to some grassroots cadres forcing them to join Primary Agricultural Cooperatives, on the whole, it is clear that in response to the state's call, peasants spontaneously cooperated with state policy. The reasons for the peasants' active participation in the process of agricultural collectivization shared with the motives of the grassroots cadres' behavior, which, in addition to their identification with the policy of agricultural collectivization, also included elements of actively striving to be good citizens of the country.

From the above discussion, the achievement of agricultural collectivization resulted from the interaction between the state and society. Many researchers have emphasized factors such as Mao Zedong's changing ideas about the founding of the country and national policies to explain why the speed of socialist transformation accelerated. However, in reality, this also includes the active promotion of state agents as policy implementers, as well as the spontaneous cooperation and competition of citizens in order to gain more citizenship from the new country.

Moreover, to understand the interaction between the state and society in the formation of state–society relations, in addition to the role played by state agents and farmers, it is essential to focus on the interaction between state policies and market rules. The author uses the Unified Purchasing and Marketing decision-making process as an example.

The agrarian reform in the countryside in the early 1950s brought about "land for the farmer" and completely dismantled the power of the landowning class in the countryside. This enabled the state to use agricultural reserves in industrial development. Still, it must also be noted that the land reform also destroyed the food supply system between urban and rural areas, which had previously been maintained mainly by the landlords, resulting in a shortage of food supply in the cities. To solve this problem, the state had to establish a new food supply system between urban and rural areas, and in 1953 decided to implement a policy of "unified purchase and marketing" of agricultural products.[13]

Implementing the "unified purchase and marketing" policy brought about a series of changes in China then, especially its significant impact on the organization and construction of grassroots cadres. The unified purchase depended on the cadre system at the grassroots level and meant the compulsory purchase of surplus grain from peasants at low prices to meet the grain distribution needs of cities. The unified marketing policy implies developing a food distribution system in the cities. To successfully promote these two policies, it was necessary to establish a system of cadres from the central to the grassroots level, a Household Registration System to control the movement of people between urban and rural areas, and a grain rationing system. The establishment of the unified purchasing and marketing system was thus closely linked to building state power at the grassroots level in urban and rural areas. The two could be said to have taken place simultaneously. The results of implementing the Unified Purchasing and Marketing Policy regarding the institutional change in the new China were enormous. Three of the most relevant ones are illustrated below.

First, the division of the countryside from the cities prevented the movement of people between urban and rural areas.

Second, it contributed to the development of rural cooperatives. That is, implementing the unified purchase policy strengthened the party's power at the grassroots level in the countryside, thus facilitating the transformation of the organization of agricultural production from a subsistence farming economy to a collective agricultural economy run by village cadres.

Third, the absorption of economic activity into the bureaucracy reinforced the bureaucracy and linked economic activity to politics, so the market, the central element of economic activity, began to shrink.

In thinking about changes in the institutional spatial configuration of state–society in the countryside, special attention must be paid to the role played by the unified purchasing and marketing system. The establishment of this system led to the dismantling of the traditional Chinese system of free trade between urban and rural areas and its replacement by a state-led planned economic system.

1.3 Reorganization of the Institutional Spatial Configuration of the State and Society in Cities

The encroachment of state power into urban areas can be illustrated in three main ways: support for the development of state-owned enterprises, the socialist transformation of capitalist industry and commerce, and the establishment of the work unit (*Danwei*) system.

Policies to Support the Development of State Enterprises

In the early years of the country's founding, the government actively supported and encouraged the development of state-run enterprises. State enterprises in the early years of the country consisted of three main components: (1) industrial and commercial enterprises established by the Communist Party in the base areas, (2) capitalist industrial and commercial enterprises confiscated from the founding of the People's Republic, and (3) foreign enterprises nationalized during the Korean War.[14] In the first half of the 1950s, especially between 1949 and 1953, state enterprises assumed the dual functions of production and political organization.[15] In other words, state enterprises were both the archetypal socialist industry and commerce and an essential channel for advancing state power into society. The state preached equality by implementing a glamorous low-wage-high-benefit-high-social-security policy within state-owned enterprises to gain social support for the new government.[16] In addition, the state implemented various preferential policies to support the development of state-owned enterprises. For example, taxation policies provided differential treatment between state ownership and private enterprises (public–private distinction), including a provision that exempted state enterprises from 135 commodity taxes. In contrast, private enterprises producing similar goods were not exempt from taxation.[17] Such preferential policies for state enterprises were similar to the preferential policies for foreign enterprises adopted after reform and opening to attract foreign investment.

Socialist Transformation of Capitalist Industry and Commerce

In the early years of the country's founding, the government supported the development of state-run enterprises while carrying out the gradual socialist transformation of capitalist industry and commerce.[18] In December 1953, the Central Committee of the CCP put forward "the gradual socialist industrialization of the country and the gradual socialist transformation of agriculture, handicrafts, and capitalist industry and commerce by the state," which was the general party line in the transitional period. The government adopted a policy of confiscation of bureaucratic capital. Still, it adopted a "public–private partnership—fixed shares and fixed interest" policy for national capital based on the principle of "use, restriction, and transformation of capitalist industry and commerce."

Although the socialist transformation of capitalist industry and commerce officially began in 1953, the Five Anti-Corruptions Campaign launched in February 1951 was the first step in controlling capitalist industry and commerce by political means. Its results profoundly impacted the structural changes in the economy. Specifically, the campaign aimed to combat taxation, embezzlement of national resources, cutting corners and skimping on materials, and theft of economic intelligence. However, according to the government's findings, only 10–15 percent of the national capitalists were found not to have violated the law.[19] In other words, the Five Anti-Corruptions Campaign targeted 85–90 percent of private enterprises.

According to the research of Jin Guantao and Liu Qingfeng, the Five Anti-Corruptions Campaign, which mainly targeted private enterprises, had three effects on the economic structure of the time. First, by making private enterprises pay for the taxes they had evaded, the state increased the "state shares" of private enterprises. As a result, some of the private enterprises became joint state-private ventures. Second, the state sought to infiltrate the power of private enterprises through the Five Anti-Corruptions Campaign. In other words, by establishing party groups within the enterprises and training activists, the state brought private enterprises under the party's leadership almost as much as state-run enterprises. Third, under the guise of preventing improprieties by private enterprises, the state monopolized the sale of raw materials and goods.[20]

It is easy to see here that the Five Anti-Corruptions Campaign was a political approach to urban economic reform. It can also be described as a process of state power encroaching into private enterprises and economic markets. The involvement of state power in private enterprises and economic markets had a significant impact on the reorganization of the institutional spatial configuration of state and society, especially as the penetration of state power into the market directly brought about the market's shrinking. Because prosperity and development of capitalist industry and commerce are inextricably linked to the market based on the rules of free competition, the market's shrinking has dealt a massive blow to capitalist industry and commerce. Specifically, the state monopolized the sale of raw materials and goods. As a result, from 1954 onwards, many private enterprises could not get enough raw materials to maintain normal production activities.[21] Under such circumstances, the only way for private enterprises to survive at that time was to enter into joint state-private partnerships. Because of these factors, the socialist transformation of capitalist industry and commerce advanced rapidly from the end of 1955 to the first half of 1956. On 1 January 1956, private industrialists and business owners in Beijing applied for a joint state-private partnership for the entire industrial and commercial sector in Beijing. Ten days later, a joint state-private partnership for the whole industry was achieved. This action in Beijing greatly influenced other cities. By the end of January 1956, private businesses in Shanghai, Tianjin, Guangzhou, Wuhan, Xi'an, Chongqing, Shenyang, and other large cities and more than 50 medium-sized cities announced whole-industry joint state-private partnerships.

From information published in 1960, it can be confirmed that the proportion of socialist industry and commerce to capitalist industry and commerce in the country's gross industrial and commercial product has gradually increased. After the Five Anti-Corruptions Campaign in 1952, the proportion of capitalist industry and private commerce dropped sharply to 17.1 and 57.2 percent. After implementing the Unified Purchasing and Marketing Policy in the cities in 1954, their shares fell further to 5.3 percent and 25.6 percent. By 1956, capitalist industry accounted for less than one percent of the country's industrial output, and private commerce accounted for only 2.7 percent.[22] In other words, the socialist transformation of capitalist industry and commerce was completed in 1956.

The pace of socialist transformation since the mid-1950s has been somewhat excessive and was established in 1953 at the same time as the nation's First Five-Year Plan. It was envisaged that in addition to the primary industries that had already been nationalized, joint state-private partnerships and full nationalization in industry and commerce through paid purchases would be achieved by 1962; Agriculture would be collectivized in stages by 1967. Many researchers have placed the phenomenon of earlier-than-expected completion of socialist transformation within the framework of state policy, arguing that the speed was the result of a vigorous state policy of socialist transformation or that it was the result of a change in state policy. But it is clear from the preceding analysis that the socialist transformation of capitalist industry and commerce in the cities, like the collectivization of the countryside, can be said to have resulted from the interaction between the state and society. In particular, the intervention of state power in the market led directly to the shrinking of the market and the private sphere, whereby competition that should have taken place in the private sphere had been transferred to the public sphere. The change in the institutional spatial configuration of the state–society certainly brought about a change in the methods of competition. Showing loyalty to the state and becoming a business desired by the state also became part of the competition, which became how the business survived. Statistics show that in 1955, the state-private partnership and the "private enterprises working on state orders" (*jiagong dinghuo*) accounted for almost the same share of the country's industrial output. Still, by 1956 nearly all enterprises were state-private.[23] In addition to the impetus of state policy, this phenomenal rate cannot be ignored, considering the effects of inter-enterprise competition, the competition within local government, and the interaction between state policy and the market.

Establishment of the "Work Unit" System

The complete incursion of state power into urban society is closely related to the "work unit" (*Danwei*) system. The work unit system was formed during the development of state enterprises and the socialist transformation of capitalist industry and commerce. It can be said that the work unit system results from a series of institutional formations carried out by the state in urban areas, resulting from the interaction between the state and society. In other words, the socialist transformation of capitalist industry and commerce increased the proportion of stake ownership, thereby increasing the state's financial capacity. The result was the ability of the state to provide social benefits such as housing and health care for urban workers.[24] The formation and development of the institutional welfare system further strengthened the workers' sense of belonging to the state and their integration into state enterprises. The work unit system results from expanding space within the system and shrinking space outside the system.

As an intermediary between the state and the individual, the work unit is far-reaching in creating state-social relations. We can illustrate this in five ways.

First, the work unit assumed the essential function of the state in unifying society. Thanks to the work unit's existence, the various social organizations and groups in urban society have been brought effortlessly under the state's rule since its founding.[25]

Second, as shown in Fig. 1, the work unit functions as a means of social infiltration by party and government power. The work units have internal structures corresponding to the party organization, for example, the party committee office, the propaganda department, the organization department, and the Youth League committee that corresponds between party and government organizations. The most important function of these bodies is to convey the party line, establish guidelines and policies, and control the thinking of party members, league members, and the general public within the work unit. That is to say, the work unit plays an important role in the integrated system of party and government. In this system, the party's lines, principles, and policies are conveyed from the central government to the grassroots through party and government organizations at all levels, and the results of policy implementation and social reactions are also reported to the central government by party and government organizations at all levels. In this way, the state's control over society can reach individuals directly.

The original institutional design of the state was to establish a system of transmission from the central to the local level and vice versa. However, in reality, this information conveyance system does not operate through a single circular loop of the central party and government organizations → local party and government organizations → grassroots party and government organizations ...... → individuals → ...... → grassroots party and government organizations → local party and government organizations → central party and government organizations. Because of the absence of any other communication channels or the monopoly of the party and government organizations on upward and downward communication channels, several circular loops with a chain structure from the central to the grassroots level are formed: central party and government organizations ⇔ local party and government organizations ⇔ grassroots party and government organizations ⇔ individuals. In other words, a relatively closed loop is formed between the higher-level party

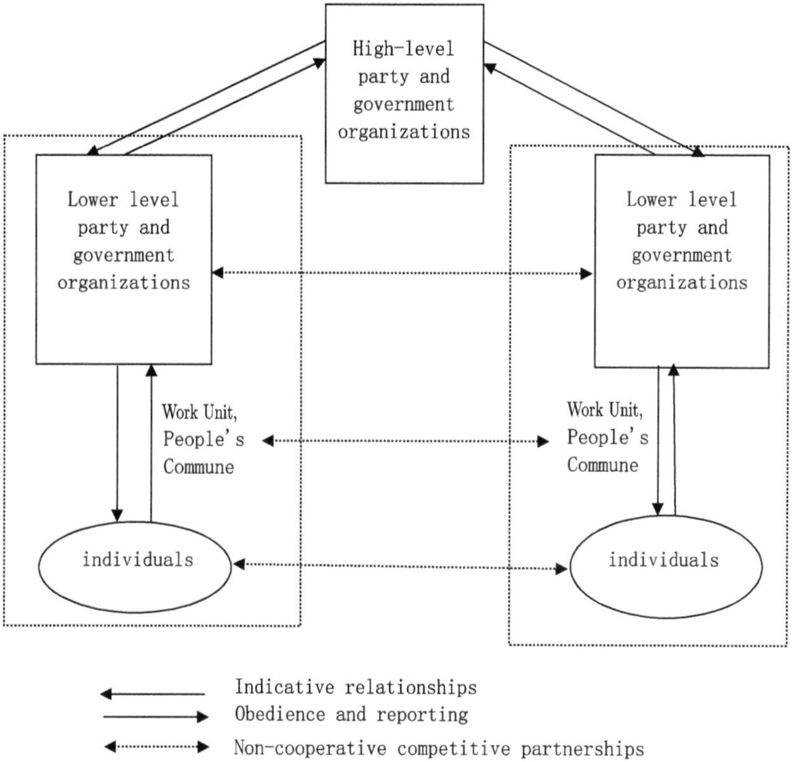

Fig. 1 Diagram of the relationship between the work unit—people's commune and the party-government organizations

and government organizations and the lower-level party and government organizations under their jurisdiction. In this case, except for the central party and government organizations, party and government organizations each assume two roles at all levels. One is to obey the higher party and political organizations and express loyalty to them. The other is to protect the organizations under their jurisdiction and subordinates and draw loyalty from them. But the party and government cadres are appointed by higher authorities. Authority is not derived from below but delegated from above, so obedience to orders is considered much more important than protecting subordinates. Moreover, it is a requirement of institutional and organizational life that subordinates obey their superiors.

Because of the existence of this chain structure, it can be said that the design of state rule over society is essentially vertical from the top down. The system guaranteed almost no cooperative relationship between the same level of party-state organizations, work units, and people's commune. At the same time, the Communist Party is organized in a pyramidal structure; the more organizations at the same level, the lower the hierarchy is. The resources needed by party-government organizations at all levels are allocated by the higher-level organizations that govern them, making it extremely easy for competition to arise between institutions at the same level. For example, competition between party-government organizations at the same level, competition between work units, competition between workshops and individuals within a factory work unit, etc. The lack of institutional and legal constraints on these rivalries made them highly susceptible to chaos. Before the Cultural Revolution, these competitions were subject to party discipline, and no significant disruptions occurred. Still, party discipline during the Cultural Revolution, which had governed competition, lost effectiveness, and competition fell into anarchy.

Third, when considered from the standpoint of the distribution of citizenship, the work unit is the only organization that enjoys the right to redistribute resources apart from the state. The state does not have a direct relationship with the individual but invests resources in the work unit, which distributes them to the individual. Specifically, the work unit provides lifelong employment, housing, health care, education for children, and work, among other socialist-style benefits. As detailed in the next chapter of this book, these socialist benefits are included in the new Chinese citizenship. The allocation of citizenship by the state through the work unit created a dependency relationship between the state and the work unit and between the work unit and the individual, which had two consequences. One is competition between work units for citizenship, and the other is the excessive power of state agents within work units, which created a so-called "patron–client" relationship between state agents and workers in the work units regarding obedience, reward, and punishment.

Fourth, since the work unit is the only organization authorized by the state to allocate citizenship, the penalties imposed on individuals by the state and its agents are of great importance and can determine their right to live. Although the Constitution provides for various rights of citizens, in reality, it does not work, and it is common for the internal rules of the work unit to override the law.

Fifth, the work unit is the only way for an individual to function in society, and an individual must obtain their identity from the work unit.

From the above five characteristics, it can be seen that the work unit is an essential setting for the state to control society. In addition to the work unit system, the residential committees and street committees, which are the grassroots power structures in the place of residence, also have the function of controlling the population directly or indirectly.

1.4 Establishment of the Household Registration System

The Household Registration System is the most crucial system of state control over society. According to the provisions of the Common Program, Chinese citizens enjoyed the freedom of residence and movement from 1949 to 1957. In 1958 the Standing Committee of the National People's Congress issued the Regulations of the People's Republic of China on Household Registration, according to which citizens of the People's Republic of China could not move freely from or to their place of residence without the permission of the public security authorities. There are many reasons for such a Household Registration System, such as maintaining social security, identifying undesirable elements, etc. Still, the most direct reason is for the state to control the unchecked movement of rural populations into the cities.[26]

The state allocates resources to society through the Household Registration System. Without a household registration, individuals cannot obtain food, clothing, and housing but also find employment, pursue higher education, and marry. The Household Registration System has been highly effective in helping the state anchor individuals in their place of origin and employment and in controlling the overpopulation of cities during industrialization. However, the system has blocked urban and rural cohesion and widened the gap between rural and urban areas and large cities and small towns.

In addition, the Household Registration System integrated the welfare system (such as the grain rationing system with employment and played an essential function in urban management.[27] These welfare systems can be said to be auxiliary systems derived from the Household Registration System.

2 Penetration of State Power into Society at the Organizational Level

Here the author illustrates the involvement of the public sphere in the private sphere at the organizational level in terms of the Communist Party organization and the creation of the governed mass groups.

2.1 The Building of the Communist Party Organization

The People's Republic of China has a much larger bureaucracy than traditional China and the Kuomintang period. Regarding the cadre organization's structure, the PRC's bureaucracy penetrated the township governments. However, at the de facto social control level, the reach of Communist leadership extended all the way to work units in cities and the villages in the countryside.[28] How, then, was the Communist Party able to achieve the infiltration of power to the end of society? Three significant ways can be cited as follows.

The first is related to the structure of the modern Chinese political system. According to Kazuko Mori, the two most striking features of the modern Chinese political system are the strong trinity between the party, the state, and the military and the highly monolithic top-down approach known as democratic centralism.[29] Thus "the leadership of the Communist Party reached all aspects of (1) the military, (2) the government and state power, (3) the Political Consultative Conference and democratic parties, and (4) mass organizations such as trade unions/Youth Communist League/Women's Federation."[30] The party's leadership of the state was institutionally ensured by the party organizations within state institutions and by counterpart ministries set up within party organs at all levels. Both began formation after the state's founding and were fixed in 1955.[31]

Second, the Communist regime had two main organizational platforms: work units in the cities and people's communes in the countryside.

Third, the Communists had a unique method of building grassroots power. In traditional China and under the Kuomintang, the only people selected as bureaucrats were elites with knowledge and economic means. Moreover, the bureaucracy penetrated society vertically, i.e., the central government administered the localities by sending officials to those localities. In contrast, the Communist Party built the bureaucracy both vertically and horizontally. It carried out vertical infiltration from central to local power, devoting itself to building capacity at the grassroots level and absorbing people and resources horizontally. It can be said that the latter preceded the former in some periods. This can be illustrated by the example of the Communist Party's grassroots power-building in the countryside. This began as early as the 1930s during the base-area period, when the Communists entered a village and immediately carried out land reform, selecting from among the local peasants talented people who would cooperate with the Communist regime and training them to become members of the Communist Party. The Communist Party's grassroots power in the countryside was built up by absorbing grassroots forces into the organization. It can be said that land reform was the process of building grassroots power in the countryside by the Communist Party. Once the peasants became part of the system, it indeed became more straightforward for the Communist Party to infiltrate rural society, no longer from outside, but to expand its organization within the countryside. The training of grassroots cadres in the cities was essentially the same. As the collectivization of the countryside and the socialist transformation in the cities proceeded, the state strengthened its ability to unify society. At the same time, the number and network of Communist Party grassroots cadres expanded.[32]

Fourth, the party managed the grassroots in urban areas by establishing neighborhood committees and successfully permeated its system of power to the micro level of the city. According to MacFarquhar's research, the formation of neighborhood committees in urban areas was almost complete by 1954.[33]

2.2 *Managed Mass Organizations*

Developing the managed mass organizations and expanding the party's influence over them is also critical when considering the state's penetration into society. The main managed mass organizations were the Trade Unions, Youth Leagues, Women's Federation, Young Pioneers, Cultural Union, Overseas Chinese Association, etc. What these groups have in common is that their members share a common identity. In 1952, there were 865,000 members of the New Democratic Youth League and 10,023,000 trade union members, which increased to 2255 and 17,467,000, respectively, in 1957.[34] These groups have no autonomy in the process of political participation and are literally managed mass organizations. But the role of these groups was significant in connecting individuals with the pipeline of party policy. As Mao Zedong noted in

1955, for the first time in thousands of years, such a large number of people were included in various organizations, thus transforming a "fragmented" China into a "unified China."[35]

Concerning the managed mass groups, Satoshi Amako explained the following. "By placing party members and activists in the leadership of these organizations, the Communist Party was able to reach the masses directly so that party policy could penetrate the masses, thereby gaining their support and mobilizing them, and at the same time taking in and reflecting their opinions and sentiments. The people's groups → The Communist Party is the conduit through which the opinions and sentiments of the masses can be realized and reflected in policy. More importantly, these mass organizations played a crucial role in serving as a platform for the party to attract and enlist new members."[36]

3 Ideological Saturation and the Creating of National Identity

After the establishment of the People's Republic of China, in addition to rebuilding the political and economic systems discussed earlier, it was more important to build the legitimacy of the regime and the political parties and the national identity of the new country's citizens. The ruling party needed to justify to the people within the political body why the party and its political leaders had the legitimacy to exercise the state's leadership in place of the previous government, political parties, and leaders. In other words, the new government needed to have an ideology that justified the regime's legitimacy and use public opinion, education, and other propaganda tools and a series of political activities to root this ideology in the hearts and minds of the people. Thereby, a new national identity was created, replacing the old political legitimacy with a new one. Building regime legitimacy and national identity were essential for the new China under the Communist Party. This is because the influence brought about by this regime change can be considered unprecedented in both breadth and depth. From a historical perspective, regime change during the dynastic period was more limited to the dynasty or the scholarly class. It did not involve the general public, nor did it usually lead to large-scale changes in the social structure or economic system. However, the establishment of the People's Republic of China was different. The change of political power meant a change of political party and a comprehensive transformation of political structure, social structure (here, mainly referring to the political changes in social structure), economic structure, and even diplomatic relations. The former Kuomintang government became a "reactionary government" and a "lackey of imperialism."[37] Once in the upper stratum of society, the powerful and elite lost their political, social, and even economic status. The low-income groups once struggling in the lower stratum of society were hailed as the country's masters. The allied United States, which had assisted the Nationalist government during the war against Japan, became American imperialism in the new nation's diplomatic relations. All these changes and reversals required a set of ideologies to build their legitimacy and thus gain widespread acceptance. So what were the main elements of this ideology? By what methods was it formed in the population? What impact did these methods of constructing ideology have on collective action in the Cultural Revolution?

3.1 The Ideology of the New State

The exploration of the ideological content of the new post-state is an interesting and ambitious subject, and here only the discursive systems of the state, party, class, and class struggle that are directly relevant to this book are explored. In the author's view, these concepts were central to the construction of the ideology and value system of the new state, which directly influenced the distribution and competition for citizenship in the new socialist state.

First, the new state orientation and class in the context of history and world development. Socialist China is based on the Marxist political theory that combines the wisdom of Marx, Engels, and Stalin. According to this theory, there are stages of human development leading through primitive society, slave society, feudal society, capitalist society, socialist society, and finally, communist society. The socialist system, based on public ownership and a planned economy, is considered superior to the capitalist system because it eliminates exploitation and private ownership. Thus, it is bound to overcome and replace the capitalist system. A socialist society, with the continuous improvement of the productive forces, will transition to a communist society, which is the ultimate system of human development. Communist society will be a fair and just human society that eliminates the three significant differences (class, urban and rural, intellectual and physical labor) and distributes them according to needs. This ideology of socialist superiority plays a central role in the state-building discourse of socialist China. First, it places the socialist system above the capitalist system and points to its uniqueness as a path to the communist system. Second, it justifies and legitimizes the theory of class division and class struggle. That is, the class struggle against the landlords, rich peasants, and capitalists is to help carry out ideological reformation and eventually eliminate class and exploitation and achieve the ambitious goal of communism. As Mao Zedong said in *On the People's Democratic Dictatorship*, "The elimination of classes, the elimination of state power, the elimination of the Party, all mankind will have to take this path; the question is only time and conditions." From this perspective, the regime in the new socialist China is founded on class, its source of legitimacy has a class dimension, and the socialist values of fairness and justice are also built on class. However, as will be detailed in Chapter 4 of this book, the central element in the distribution of citizenship in the socialist state is precisely class. This not only determines how the competition for citizenship is waged using class struggle, but it also brings nobility to the competition for citizenship, imbuing the individual's pursuit of self-interest with a romanticism of the pursuit of merit and the achievement of national goals. Considered on the psychological level of individual heroism, pursuing the public good and lofty goals is often more impassioned. This may explain why idealistic elements can be seen in several participants of the time. However, it is essential to point out here that not only is the idealistic pursuit of the public good and rational competition for citizenship not contradictory, but the two are superimposed, which is precisely the characteristic of collective behavior during the Cultural Revolution. Even those who had apparent personal interests at the time of their participation in the movement were often moved by their moral actions when they jumped on the criticism platform to criticize the "capitalist roaders within the party" regime amidst the highly politicized use of language and passionate political scenes.

Second, the new national position and political party in the history of China's anti-imperialism and anti-feudalism. In *On New Democracy* (*Xin minzhuzhuyi lun*), Mao Zedong defined the revolution led by the Chinese Communist Party not only as part of the "world socialist revolution" but also as an essential part of the anti-imperialist and anti-feudal movement in China since the Opium War. Taking the May Fourth Movement as the boundary, he divided China's century-long anti-imperialist and anti-feudal movement into two periods: the old democratic revolution led by the bourgeoisie and petty bourgeoisie and the new democratic revolution led by the proletariat. He argued that the period of the old democratic revolution, which experienced the failure of the reform movement to learn from the West and was eventually led by Sun Yat-sen's bourgeois-democratic revolutionaries who launched the Xinhai Revolution to overthrow the Manchu dynasty and establish a democratic republic, could not assume the anti-imperialist and anti-feudal aims because of the class fragility of its bourgeois revolution. The task of anti-imperialism and anti-feudalism for a hundred years could only be accomplished by the new democratic revolution led by the Chinese Communist Party established after the May Fourth Movement.[38] Therefore, in the Common Program of the Chinese People's Political Consultative Conference promulgated in September 1949, the founding of New China was defined as "the victory of the Chinese people, after more than a hundred years of heroic struggle, in the people's revolution against imperialism, feudalism, and bureaucratic capitalism under the leadership of the Chinese Communist Party." All these are concentrated and condensed in two sentences: "The People's Republic of China is founded! The Chinese people have stood up from now on!" (Mao Zedong's proclamation from Tiananmen Square on the day of the founding of the country) and the words "Without the Communist Party, there would be no new China" (a song composed in 1943 and widely sung after the founding of the country).

3.2 *Ideological Infiltration and its Impact on the Cultural Revolution*

The construction of the identity and ideology of the new state after the founding of the country was generally achieved through political campaigns, various political rituals such as national parades, school education, the building of political discourse, symbols (images of people's heroes and people's monuments), etc. Only the most representative political movements are used here to explain how national identity and ideology permeated society in these two ways and how they influenced collective behavior during the Cultural Revolution.

As many researchers have pointed out, top-down, nationally mobilized political movements were the norm in Chinese politics before the Reform and Opening-Up period. Political movements played a vital role in building the new national identity and ideological saturation. According to Su Yang's statistics, after the state's founding and up to 1969, the

CCP launched 48 large and small movements,[39] covering every group in society, involving all political, economic, social, and cultural areas. At the state level, the main role of political movements is to solve a specific public policy issue and a process of political mobilization and social transformation. More importantly, the state successfully infiltrated national identity and ideology into every corner of society in this process. For example, the land reform movement solved the problem of equal land rights in the countryside, eliminated the political and economic status of the landlord class, and created the peasants' sense of identity with the party and the state in the political movement. It successfully introduced rural society to concepts of class and class struggle unfamiliar to the peasants of the time. The anti-American and assist North Korea movement achieved not only the goal of nationwide material supply but also fostered a hatred for American imperialism and building awareness that "to defend the motherland is to protect the hometown" by drumming up support for the army with the image of the People's Volunteer Army "majestically crossing the Yalu River."[40] However, the long and continuous political campaigns and mobilization created a culture of political campaigns with the rituals of political movements, the ways of struggle, the language of politics, etc., as examples. On a social level, the participants in a political movement over a continuous period condensed a set of participation strategies and developed methods used in the movement for self-preservation or access to avenues of upward mobility (e.g., the patron–client relationship, competing to become activists, etc.). These would influence collective behavior in the Cultural Revolution.

4 The "Great Leap Forward" as Economic Decentralization

4.1 *The Idea of Power Decentralization*

As discussed earlier, from the founding of the People's Republic until the mid-1950s, the state undertook a series of social transformations and institutional reconstruction. State power continued to saturate society through a socialist transformation in the countryside and the cities, establishing party organizations and managed mass groups from the central to the grassroots level. At the same time, the ideology of national identity and class struggle of the new state was instilled into every corner of society through various political movements and political mobilization. As a result, the public sphere was expanded, and the private sphere was gradually compressed. Eventually, all spheres of society were publicized and highly politicized, thus forming the state-social relations of state and society integration. Establishing new state–society relations marks a revolutionary milestone in maximizing the integration of social resources and achieving modernization through social integration. However, it also gives rise to various issues, as elaborated below.

First, the privileging of bureaucrats began to appear along with establishing the new bureaucracy. This is not a problem unique to China but to other socialist countries such as the Soviet Union. For example, in the 1960s, Milovan Djilas proposed a "new class" by analyzing state socialism. Djilas pointed out that, unlike previous revolutions, the communist revolution, carried out to eliminate classes, gave birth to a new single class in full possession of power, what he calls the "absolute rule of a new class: the communist oligarchy."[41] In the 1980s, Mikhail S. Voslensky analyzed the privileged class in the former Soviet Union based on Djilas's theory, exposing the truth that the ruling class in the Soviet Union used power for personal gain, pointing out that there was a ruling class and a ruled class and the antagonism between the two in the former Soviet society.[42] In other words, the emergence of a new privileged class is a problem in all socialist countries.

Second, the highly centralized system reduced social activism. For example, in 1956, the director of an average state enterprise had financial autonomy of only 200–500 yuan. The director of a company joint state-private venture had much less financial independence than in a state enterprise.[43] The total lack of autonomy of the head of the enterprise concerning production and business activities means that the enterprise can only engage in production activities according to the instructions and orders of its superiors, which hinders the economic dynamism and autonomous development of society.

Third, the total control of society by the state led to social resistance. For example, there were frequent incidents in the countryside where peasants opposed the Unified Purchasing and Marketing Policy and attacked cadres.[44] There were also strikes by urban workers due to the shortage of daily necessities and resistance to state policy by intellectuals dissatisfied with their political treatment.[45]

Fourth, the state's monopoly on citizenship increased the individual's dependence on the state. Under the Mao Zedong era, almost all the necessities of life for urban residents could be met within the work unit. In addition, the work unit provided social security such as a pension and medical insurance. From nurseries to pensions, the work unit took care of its employees "from the cradle to the grave," so to speak. In addition, a letter of introduction from the work unit is required to purchase airline or train tickets and accommodation for business trips. Although the work unit guaranteed all aspects of the employee's food, clothing, housing, and transportation, the individual's entire life was tied up in the work unit. The individual's dependence on the organization became extensive and comprehensive.

To solve the above problems, Mao began to think about power decentralization. In his speech, *On the Ten Relations,* published on April 25, 1956, Mao Zedong discussed the relationship between the state, production units, and individual producers, saying, "The state and the factory, the state and the workers, the factory and the workers, the state and the cooperatives, the state and the peasants, the cooperatives and the peasants, all must be taken into account and cannot be neglected. Focusing on only one side is detrimental to socialism and the dictatorship of the proletariat." "Concentrating everything at the central or provincial level without giving factories any power, flexibility, or benefits may not be appropriate."[46] Stating further, "I am afraid it is not right to centralize everything in the central government or the provinces and cities, without giving the factories a little power, a little room to maneuver, a little interest."[47] In the same document, Mao elaborates that local power should be expanded based on ensuring unified leadership from the central government: "local power should be expanded on the premise of consolidating unified central government leadership, giving more independence to localities and allowing them to do more things. This is better for building a strong socialist state."[48]

Under the theoretical guidance of *On the Ten Relations*, the Eighth National Congress of the Chinese Communist Party was held in September 1956. At the Congress, Liu Shaoqi and Zhou Enlai made important speeches on the relationship between the central and local levels and put forward seven principles on central and local administration.[49] The main idea was to consolidate major powers at the central level and decentralize minor powers at the local level (centralization of major powers and decentralization of minor powers).

In October 1957, the enlarged session of the Third Plenary Session of the Eighth Central Committee of the CCP issued the Provisions on the Reform of the Industrial Management System (Draft), Provisions on the Reform of the Commercial Management System (Draft), and the Provisions on the Reform of the Financial Management System (Draft). These draft provisions stipulated that to adjust the relationship between the central and local governments and between the state and enterprises, the power to control industry, commerce, and finance will be decentralized in the following areas. First, to shift the jurisdiction of some central government-owned enterprises from the central government to the provinces, cities, and autonomous regions. Second, to expand the financial powers of localities and enterprises. Third, to expand the local power of distribution of goods. Fourth, to expand the planned economic management power at the local level and enterprises. Fifth, the prices of agricultural and industrial products should be managed hierarchically; that is to say, the provinces, cities, and autonomous regions should have the right to set their own prices considering the state's uniform pricing. Sixth, to expand the power of the local level and in enterprises to manage personnel.[50]

4.2 *Characteristics of Economic Decentralization During the Great Leap Forward Period*

The main objective of economic decentralization during the Great Leap Forward was to curb the expansion of the bureaucracy. However, the following three economic decentralization features make achieving this objective extremely difficult.

First, the decentralization of power during the Great Leap Forward period was not a decentralization of power from the state to society but a decentralization of power within government agencies, from the central to local governments. This decentralization was limited to the policymaking authority of the central and local governments and their jurisdiction over the enterprise. It was an internal adjustment of power within government agencies. There is an essential difference between this and the decentralization that grants autonomy to enterprises. The former is a decentralization of power within the state, while the latter is a decentralization of power from the state to society. Decentralization from the state to society is decentralization from the public to the private sphere, which brings about institutional changes in the spatial configuration of the state and social relations in the form of a reduction in the public sphere.

Decentralization from the central to the local government, on the other hand, is only a reorganization of power within the public sphere.

Second, the decentralization of power during the Great Leap Forward was not political but was limited to economic decentralization, which is very different from the political decentralization of power during the Cultural Revolution, which will be discussed later. The reason why Mao did not engage in political decentralization at the same time during the Great Leap Forward may be related to the fact that the problems he faced at that time were mainly economic in nature, or it may be due to his fear that decentralization in the political field at the same time would lead to social chaos, etc. This is just speculation, and no historical information on this issue is available. But what needs to be emphasized here is that economic decentralization without political decentralization directly led to the phenomenon of solving economic problems through political means seen throughout the country. For example, on June 2, 1958, the State Council of the Central Committee of the CCP issued a circular to local governments, requiring nine central ministries and commissions to complete all the procedures for decentralizing the more than 880 enterprises and institutions under their jurisdiction by June 15. In other words, the time given by the central government for administrative decentralization was only two weeks.[51] Not only does the central government treat economic decentralization as a political task to be completed, but local governments also treat economic decentralization as a political task to be carried out. For example, after the jurisdiction of the central ministries was delegated to the local level, they were decentralized from the provinces to the cities, counties, and communes. Such a practice undermines the original cooperative relationship between enterprises and reduces their productive capacity for reasons such as insufficient local management capacity.[52] In addition, economic decentralization and political centralization brought about competition among local governments to show their loyalty to the central government, which led to the Great Leap Forward period when the "wind of exaggeration" and "satellite launching" were prevalent.

Third, the decentralization of power during the Great Leap Forward period left much to be desired regarding policy decisions. While the policy specified what powers were to be delegated to the local level, it was almost silent on the implementation phase of how they should be delegated.

5 The Cultural Revolution as Political Decentralization

A detailed analysis of the Cultural Revolution will begin in Chapter 5, and the parts relevant to this chapter will be briefly discussed here. Both the Cultural Revolution and the Great Leap Forward were part of the nation-building process in the People's Republic of China. If the Great Leap Forward is seen as economic decentralization, then the Cultural Revolution can be seen as political decentralization. Why was it necessary to launch the Cultural Revolution to carry out political decentralization? This question can be understood from the following four perspectives.

First, the Cultural Revolution, like the Great Leap Forward, aimed to prevent the overstaffing of bureaucracies and the privileging of cadres. But the decentralization of the economy during the Great Leap Forward led to the delegation of power to the local level, resulting in excessive financial power for local governments and financial difficulties for the central government. For example, before 1957, about 9300 enterprises and public institutions were under the central government's jurisdiction. Still, after decentralization, the number decreased to 1200, with 88 percent transferred to local governments. During the First Five-Year Plan, the proportion of industrial production value of centrally managed industrial enterprises, which accounted for 39.7 percent of the total industrial production value in the country, rapidly declined to 13.8 percent in 1958 after the delegation of power. Additionally, the significant decentralization of financial power from the central to local governments resulted in a decrease in the proportion of financial resources at the disposal of the central government from 75 percent during the First Five-Year Plan to 50 percent, while local governments experienced a rapid increase in the proportion of financial resources at their disposal from 25 to 50 percent.[53] By the 1960s, to rectify the mistakes of economic decentralization during the Great Leap Forward, the state began to recover economic power from local governments and transfer it back to the central government. As a result, problems such as the central fiscal deficit and excessive economic power of local governments were alleviated. However, compared to the First Five-Year Plan period, the bureaucratic system of the 1960s was more centralized, and the institutions were more cumbersome.

Second, like the Great Leap Forward, the Cultural Revolution can be seen as an experiment in Mao's ideas about the state and social construction. As Satoshi Amako points out, "the state had a low status in

Mao's thinking,"[54] and "for Mao, the state was ultimately a means and a tool."[55] The society Mao wanted to build was a "people's commune in which workers, peasants, merchants, academics and soldiers were all one, and the township (government) and the community (cooperative) were all one."[56] Building such a society was Mao's consistent governance policy after the country's founding. As mentioned earlier, to solve the problem of over-bloated bureaucratic organization and bureaucratization of state cadres, Mao first introduced economic decentralization within the government agencies. Still, it did not achieve the expected results. Mao began considering a political approach, seeking support outside the bureaucracy to solve it. In this sense, the Cultural Revolution can be seen as Mao's experiment in "democracy" at the political level, using the power of the masses.

Third, because of the divisions within the central leadership over the philosophy and methods of state-building, Mao sought to seek support from society. As many researchers have pointed out, there was a disagreement between Mao and Liu Shaoqi over the philosophy of state-building, with Liu advocating specialization and technicalization and Mao actively promoting collectivization. Because of their divergent views, Mao could not implement his state-building program in a way that completely ignored the presence and views of Liu Shaoqi, an agent of the state, in the existing administrative system. So to seek social support from outside the bureaucracy, it was necessary to go beyond the bureaucracy to delegate political power to the masses and use their power to carry out his will.

Some researchers argue that Mao Zedong brought down Liu Shaoqi to regain lost power. But this book argues that Mao Zedong caused Liu Shaoqi to step down to push for bureaucratic reform. Because Liu Shaoqi was the most representative figure in actively promoting the Soviet model and defending the existing bureaucracy from the state's early days, it can be argued that Liu's presence was an obstacle to Mao's practice of his ideas.

As will be discussed in detail in later chapters, there were social problems in China before the Cultural Revolution, such as the privileging of cadres, the discontent of intellectuals, the low level of trust between cadres and the masses, and the demonstrations and marches of peasants and workers for their rights. The policies implemented by the state before the Cultural Revolution provided the institutional background for the formation and development of social factions during the Cultural Revolution. In this sense, the mobilization orders from Mao Zedong to rebel triggered members of society to form factions to express their grievances and realize their interests.

NOTES

1. Under the "Land Reform Law of the People's Republic of China," landowners' land was confiscated, but the remaining land of the rich peasants was not nominally confiscated but expropriated. In fact, there is no difference at all between confiscation and expropriation in economic terms. Once the land was expropriated, the holder lost ownership of the land forever. In a political sense, however, the political rights conferred on the confiscated are quite different from those conferred on the expropriated.
2. Zhang Letian, *A Farewell to the Ideal—A Study of the People's Commune System* (Oriental Publishing Center, 1998), p. 62 (*gaobie lixiang:renmin gongshe zhidu yanjiu,* dongfang chuban zhongxin).
3. The size of the cooperative was about 10 households on average at the beginning of its establishment, expanding to 26.7 households in 1955 and even more to 51.1 households in 1956, Hideo Yamamoto, (*China's Rural Revolution*, Toyo Keizai Shinbun, 1975), p. 220.
4. The size of "advanced cooperatives" expanded even more than that of "primary cooperatives," reaching an average of 246.4 households per commune in 1956 (Hideo Yamamoto, op. cit., pp. 222–223).
5. National Bureau of Statistics, ed., *The Great Decade: Statistics on the Economic and Cultural Construction Achievements of the People's Republic of China* (People's Publishing House, 1960) (guojia tongjiju bian, *weida de shinian: zhonghua renmin gongheguo jingji he wenhua jianshe chengjiu de tongji*, renmin chubanshe).
6. Zhang Letian, op. cit., pp. 60–61.
7. In 1958, an enlarged meeting of the Political Bureau of the Central Committee of the CCP was held in Beidaihe, and the General Assembly issued the Resolution of the Central Committee of the CCP on Certain Issues Concerning the Establishment of Rural People's Communes. Since then, the people's commune system has been officially implemented.

8. Masao Fukushima, *A Study of the People's Commune* (Ochanomizu Shobo, 1960), p. 252.
9. *Resolution of the Central Committee of the CCP on the Establishment of People's Communes in the Countryside.*
10. It was in February 1962 that the Central Committee of the CCP formally adopted the *Regulations on the Work of People's Communes (Final Draft)*, which provided for a three-tier ownership system based on production teams.
11. For example, in 1954, a section of grassroots cadres forced peasants to join primary cooperatives in disregard of the principle that peasants have the right to join and leave the cooperatives freely. As a result, there were frequent incidents of peasants destroying farming tools and slaughtering livestock. Once the means of production, such as farm tools and draft animals, were reduced, it would take at least two years to restore them. Therefore, in January 1955, the central government issued an urgent *Circular on Rectifying and Strengthening Agricultural Production Cooperatives* and an *Urgent Instruction on the Protection of Draft Animals.* Jin Guantao and Liu Qingfeng, *The Transformation of Chinese Society (1840–1956): The Fate of its Ultrastable structure in Modern (kaifang zhong de bianqian)* (Chinese University of Hong Kong Press, 1993), p. 426.
12. "The Work of Rural Party Organizations Should Be Centered on Cooperativeness", an editorial in *People's Daily*, November 9, 1955.
13. Before the founding of the People's Republic of China, the food supply system between urban and rural areas was primarily maintained by landlords. At the beginning of the founding of the People's Republic, the state tried to maintain the food supply system by buying grain from farmers while collecting agricultural taxes from the countryside. However, most peasants who had been poor for a long time were not willing to sell their surplus grain to the state but kept it for family consumption and storage, resulting in a shortage of urban food supply. Jin Guantao and Liu Qingfeng, op. cit., pp. 419–428.
14. Yang Xiaomin and Zhou Yihu, *China's Work Unit System* (China Economic Press, 1999), p. 99 (*Zhongguo danwei zhidu,* zhongguo jingji chubanshe).
15. Id., p. 101.
16. Id.

17. The Editorial Committee of the Contemporary China series, ed., *Economic System Reform in Contemporary China* (China Social Science Press, 1984), p. 44 (*dangdai zhongguo* congshu bianji weiyuanhui bian, *dangdai zhongguo de jingji tizhi gaige,* zhongguo shehuikexue chubanshe).
18. Mao Zedong, *Selected Writings of Mao Zedong*, volume 2 (People's Publishing House, 1986), p. 70 (*Mao Zedong zhuzuo xuandu, xiace,* renmin chubanshe).
19. The Editorial Committee of the Contemporary China series, ed., *Contemporary Chinese Economy* (dangdai zhongguo jingji) (China Social Science Press, 1987), p. 122.
20. Jin Guantao and Liu Qingfeng, op. cit., p. 430.
21. He Jun, "On All-industry Joint State-Private Venture—The Role of Implementing All-industry Joint State-Private Venture from the Situation of Seven Industries in Private Industry in Shanghai," *Study*, March, 1956 (lun qunmian heying: cong shanghai siying gongye de qige hangye de qingkuang kan shixing quanmian heying de zuoyong, *xuexi*).
22. *The Great Decade: Statistics on the Economic and Cultural Construction Achievements of the People's Republic of China* op. cit.
23. Ibid.
24. For the growth of the state's financial capacity and the increase in the state's capacity to allocate housing and provide benefits such as health care to urban workers, refer to the previously published book, *The Great Decade: Statistics on the Economic and Cultural Construction Achievements of the People's Republic of China.*
25. All organizations in the city, such as enterprises, shops, hospitals, schools, associations, government agencies, etc., are called "work units." The work units can be divided into "enterprise units," "institution units" (including scientific, educational, cultural, and medical institutions), and "administrative units," three categories.
26. Yang Xiaomin and Zhou Yihu, op. cit., pp. 53–54.
27. Id., p. 106.
28. Jin Guantao and Liu Qingfeng, op. cit., p. 413.
29. Kazuko Mori, *Reading Modern Chinese Politics* (Yamakawa Press), pp. 29–31.
30. Id., p. 39.
31. Id., pp. 29–67.

32. Gilbert Rozman, *China's Modernization* (Jiangsu People's Press, 1988), p. 353 (Chinese version: zhongguo de xiandaihua) (English version: Gilbert Rozman, *The Modernization of China* (New York: Free Press, 1981)).
33. Roderick MacFarquhar and John K. Fairbank, eds., *The Cambridge History of the People's Republic of China* Part 1 (China Social Science Press, 1992), p. 96 (English version: Roderick MacFarquhar and John K. Fairbank, eds., *The Cambridge History of China*, volume 14: *The People's Republic, Part 1: The Emergence of Revolutionary China, 1949–1965* (New York: Cambridge University Press, 1987)).
34. National Bureau of Statistics, ed., *China Statistical Yearbook 1991* (China Statistics Press, 1991), pp. 304, 780.
35. Roderick MacFarquhar and John K. Fairbank, eds., p. 96.
36. Satoshi Amako, *China: Changing of a Socialist Power*, p. 4.
37. Mao Zedong, "The Chinese People Stood Up," in *Selected Works of Mao Zedong*, volume 5 (People's Publishing House, 1977), pp. 3–7 (Zhongguo renmin zhanqilai le, *maozedong xuanji,* diwujuan*).*
38. Mao Zedong, On the New Democracy (January 1940) (xinminzhu zhuyi lun), *Selected Works of Mao Zedong*, volume 2 (People's Publishing House, 1991), pp. 662–711.
39. Su Yang, translated by Song Xi, *Collective Killing in Rural China during the Cultural Revolution* (Hong Kong: Chinese University of Hong Kong Press, 2017), pp. 170–171.
40. Lyrics of the *Battle Hymn of the Chinese People's Volunteers*, composed in 1950.
41. Milovan Djilas, *The New Class: An Analysis of the Communist System* (New York: Praeger, 1957); (Japanese version: translated by Rinjiro Harako, *The New Class: An analysis of the Communist System* (Jiji Press, 1957)).
42. Michael S. Voslensky, translated by Boku Sakyma and Mithuyuki Funato, *Nomenklatura: The Red Aristocracy of the Soviet Union (Nomenklatura: die herrschende Klasse der Sowjetunion in Geschichte und Gegenwart)* (Chuokoron-sha, 1981).
43. The Editorial Committee of the Contemporary China series, ed, op. cit. *Economic System Reform in Contemporary China*, p. 52.
44. Office of Socialist Thought Education, Committee of the CCP, Renmin University of China, ed., *Selected Reference Materials*

on Socialist Thought Education, 3rd–4th series (Renmin University of China Press, 1957), p. 174 (zhongguo renmindaxue weiyuanhui shehuizhuyi sixiang jiaoyu bangongshi bian,*shehui zhuyi sixiang jiaoyu cankao ziliao xuan*, disan-siji, zhongguorenmindaxue chubanshe).

45. Mitsuyoshi Himeida et al. *Modern Chinese History*, volume 2 (Tokyo University Press, 1984), pp. 583–584.
46. Mao Zedong, op. cit. in *Selected Writings of Mao Zedong* (part 2), p. 729.
47. Id., p. 727.
48. Id., p. 729.
49. Liu Shaoqi, *Political Report at the Eighth National Congress of the CCP*, single volume (People's Publishing House, 1956), pp. 32–33 (*zai zhongguogongchandang di8jie quanguodaibiao dahui shang de zhengzhi baogao*); Zhou Enlai, *Report on the Proposal for the Second Five-Year Plan for the Development of the National Economy*, single volume (*guanyu fazhan guomin jingji de di'erge wunian jihua de jianyi de baogao*) (People's Publishing House, 1956), p. 61.
50. The Editorial Committee of the Contemporary China series, ed, op. cit. *Economic System Reform in Contemporary China*, pp. 57–59.
51. Id., p. 71.
52. Id., p. 71.
53. Id., p. 155.
54. Satoshi Amako, op. cit. *China: Changing of a Socialist Power*, p. 77.
55. Id., p. 78.
56. Id., p. 79.

CHAPTER 4

The System of Citizenship Distribution Before the Cultural Revolution

This chapter discusses the change process in public and private spheres, and the impact state-building has had on institutional space since the state's founding. This provides a macro-institutional context for the allocation of citizenship. In this chapter, the author will focus on the following series of questions: How is the system of citizenship allocation constructed by the state in the institutional configuration of state–society integration? How does this citizenship distribution system determine how citizenship is contested? Furthermore, how is the distribution of and competition for citizenship relevant to collective behavior during the Cultural Revolution?

1 The Social Context of Faction Formation—Citizenship and the Citizenship Distribution System

In terms of the institutional spatial configuration of the state and society, pre-Cultural Revolution China was characterized by a highly centralized system of power in which the public sphere took precedence over the private sphere, resulting in complete domination by the state over society. The state's method of citizenship distribution in such an institutional spatial configuration led to extreme state domination and subdivision,

Y. Lijun, *Contest for Citizenship and Collective Violence during China's Cultural Revolution*, IPP Studies in the Frontiers of China's Public Policy, https://doi.org/10.1007/978-981-97-0906-9_4

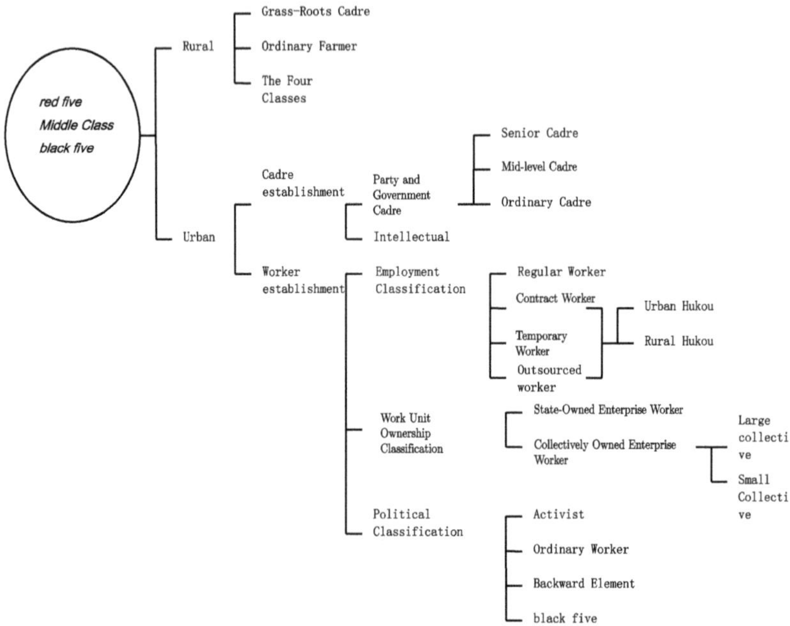

Fig. 1 Social stratification in China before the cultural revolution

eventually resulting in a multi-layered and rigid hierarchy of social stratification, as illustrated in Fig. 1. As we shall explore further, this subdivided social order was formed by the state's system of citizenship distribution, making Fig. 1 an illustration of the pre-Cultural Revolution system of citizenship distribution.

1.1 Class Differences and Class Struggle

In pre-Cultural Revolution China, a person's political status was determined by their social class of origin, which formed the foundation of the hierarchical structure and shaped their entire life. Those who belonged to social groups with a "bad" origin, known as the "black five" (landlords, rich peasants, counter-revolutionaries, bad elements, and rightists), were considered class enemies, deprived of citizenship, and subjected to continuous criticism and ideological re-education throughout their lives. From the country's earliest days until the abolition of the class system in January 1979, there was no single day without political oppression of people with bad origins. The severity of their persecution only increased during heightened class struggles and political movements. Even during the Cultural Revolution, when society was mobilized as the "Four Major Democracies" (speaking out freely, airing views fully, writing big-character posters, and holding great debates), this rule remained in effect, and political repression was even more severe. A review of documents issued by the Central Committee of the CCP during the Cultural Revolution reveals frequent expressions such as "resolutely forbidding the landlord, the rich peasant, the counterrevolutionary, the bad element, and the rightist to speak and act indiscriminately,"[1] appearing in various documents.

Those who are excluded from the circle of the people are designated as the state's legally recognized enemies. As Bergson noted, the state uses dictatorial ceremonies toward these manufactured enemies to sanctify its goals and win the people's loyalty.[2] Similarly, the people express their loyalty to the state by fiercely criticizing these enemies and striving to gain more citizenship rights than others. In other words, for those competing for citizenship, enemies are an essential political resource, and possessing more resources is a crucial factor in determining victory or defeat. Hatred toward class enemies is equivalent to love for the party. Thus, considering the logic that deeper hatred toward enemies correlates with stronger feelings for the party, the collective killing of the five black categories during the Cultural Revolution becomes understandable.

In addition to the class differences between the people and the enemy, class distinctions were also emphasized within the people's camp. This was driven by the need for socialist ideology and political pragmatism, focusing on the "wavering" and "two-faced" nature of the middle classes and the imposition of ideological transformation.[3] The political function of this was twofold: to strike a balance between workers-peasants and the old middle class, mainly intellectuals, through the decentralized distribution of political and economic resources and to exclude intellectuals from the power structure to prevent their rise as a political force and legitimize such exclusion.[4] In China, where modernization was at a very low level, this method effectively utilized political and economic resources, enabling the Communist Party to gain the loyalty of the worker-peasants while using the talents of intellectuals to modernize and legitimize its monopoly on state power. But this approach also made the distinctions between intellectual and physical work more prominent, which caused previously hidden conflicts to become more obvious.

On the other hand, the worker-peasants, who were at the bottom of the political and economic hierarchy before the founding of the state, were granted the status of masters of the state and experienced the joy of "turning over" and the satisfaction of being in charge, as described in many literary works and songs. However, after the mid-1950s, the bureaucracy became structured, and the bureaucrats, supposed to represent the people's interests, gradually separated from the people's class. As their sense of privilege grew, the worker-peasants began to realize they were treated poorly compared to the party and government cadres who wielded power and the middle class who enjoyed better wages.

It was not the black five class or the "wavering" middle class, enemies of the people, who first challenged the Communist regime through collective protests, but rather the worker-peasants, who were the "masters" of the country. In the six months from September 1956 to the spring of 1957, more than ten workers' strikes involving about ten thousand participants occurred in cities,[5] and there were peasant riots in the countryside as early as 1954.[6] However, these protests had little impact on the Communist regime, which viewed the old intellectual and business elites as the most powerful rivals at the time. To Mao Zedong and his colleagues, the protests by a few individuals in society represented "internal contradictions of the people." Thus, there was no need to worry about them.[7] Mao judged that the bureaucratic style of the party cadres was the cause of the worker-peasant protests and called on intellectuals and members of the democratic parties to assist in the rectification campaign.[8] However, even Mao could not disregard the voices of social minorities, including the intelligentsia. Consequently, the original aim of the Hundred Flowers and Hundred Schools of Thought campaign, which was to "rectify the wind" in the party, shifted to the Anti-Rightist campaign, targeting intellectuals and members of democratic parties. The Anti-Rightist movement ended with Mao using the phrase "luring snakes out of holes" to disguise his disappointment with the intellectuals and convince his colleagues of his judgment and strategic planning abilities. Nevertheless, the documented facts lead to the following conclusions.[9]

Like the Cultural Revolution, the Hundred Flowers and Hundred Schools of Thought movement was Mao's attempt to promote socialist democracy. Mao's vision of socialist democracy is expressed in the concept of "democratic centralism,"[10] which entails that in a socialist state, "the people enjoy broad democracy and freedom but must also be subject to socialist discipline."[11] Additionally, cadres are expected to follow the "mass line," meaning they should "come from the masses to the masses" and work in partnership with them.[12] At the political system level, the construction and enhancement of the power function of the People's Congress was also emphasized in the concept of Democratic centralism.[13] However, in practice, the supervisory role of the People's Congress had not been utilized, and democratic centralism was limited to the "Four Major Democracies." Both the Hundred Flowers and Hundred Schools of Thought Movement and the Cultural Revolution aimed to prevent the bureaucratization of party cadres and the establishment of a bureaucracy by giving the people the right of supervision through the "Four Major Democracies." This was intended to create a democratic and centralized socialist state. However, from a conceptual and practical standpoint, there was no institutional guarantee of either democracy or centralization of power, and no relationship or institutional link exists between democracy and centralization. Furthermore, in the institutional spatial configuration of the time, social group relations were dominated by tensions over competition for political resources. As a result, the movement to build a socialist democratic state ultimately became an arena for social groups to compete for access to citizenship.

Some people may doubt there was competition for citizenship during the Hundred Flowers and Hundred Schools of Thought and Anti-Rightist movements. Still, in the author's opinion, it was precisely because of the competition for citizenship within the party and government cadres, between party and government cadres and intellectuals, and among intellectuals that the movement expanded to the point of producing 550,000 rightists. During this period, the competition for citizenship manifested in distinct ways in the early stage (A Hundred Flowers and a Hundred Schools of Thought) and the later stage (Anti-Rightist movement). Intellectuals and members of democratic parties boldly advocated for their right to participate in politics, this was particularly evident in the early stage. In essence, the statements made by intellectuals and democratic party members during this period, which included their criticism of the One Party Dictatorship (dangtianxia), their proposal for the establishment of a political design institute (zhengzhi shejiyuan), and their call for a multi-party electoral system,[14] proved instrumental in shaping the national political system. However, these statements also clearly reflected the fervent desire of intellectuals and democratic parties to be granted the right to participate in politics and share political power with the Communist Party.

As the political climate evolved during the later period, the competition for citizenship took on increasingly intricate and convoluted forms. In fact, the conditions during this period bore a striking resemblance to those of the Cultural Revolution and therefore merit closer examination. In July of 1957, Mao Zedong famously characterized the conflict between the rightists and the people as "the contradiction between the enemy and us, an irreconcilable confrontation between life and death."[15] On the 28th of the same month, the *People's Daily* published an editorial entitled, The Anti-Rightist Struggle is a Major Test for Communists, which advocated that "Communists must thoroughly criticize the rightists inside and outside the party and resolutely oppose sentimentalism."[16] The editorial made it abundantly clear that the Anti-Rightist movement was cast as a struggle between "us" (the people) and the "enemy," with this conflict permeating every level of the party. Resolving this contradiction hinged on engaging in class struggle. Thus, their response was a matter of political survival for the party cadres in power tasked with judging the rightists. In other words, while uncovering additional rightists might not necessarily earn them the favor of their superiors, failing to do so could result in accusations of showing leniency toward their class enemies. In such a charged atmosphere, the mantra of "preferring more to less" and "preferring the left to the right" became the guiding principles for these cadres' self-preservation.

Furthermore, the fate of the political movement hung in the balance, as did the adoption of a favorable stance toward it. Consequently, the competition among cadres intensified sharply, resulting in a surge of rightists. There even arose the absurd scenario of a cadre placing their name on the rightist list simply to meet their superior's rightist quota. In response, some intellectuals and party members resorted to false accusations and other means of self-preservation. Despite the CCP Central Committee's publication of Criteria for Determining Rightists on 15 October 1957,[17] which provided detailed guidelines for identifying rightists, the Anti-Rightist movement represented a life-and-death struggle for both party cadres and intellectuals. This struggle was conducted within an institutional framework from which there seemed to be no escape, leaving most individuals competing fiercely for survival. Conversely, a minority viewed the Anti-Rightist movement as an opportunity to eliminate dissidents and secure a political future.

The preceding analysis makes it clear that the policy of class division and intensified class struggle played a critical role in the state's management of the social situation. The division of classes furnished the theoretical basis for rationalizing the redistribution of political and economic resources that occurred following the establishment of the state. The emphasis on class struggle became essential for consolidating the regime's stability. Individuals or groups whose speech or actions threatened the regime's stability could be designated as enemies, and a dignified mass movement could be launched to crush them. However, the division of classes also gave rise to numerous adverse effects, such as excessively accentuating the differences between classes and reinforcing the sense of class antagonism, thus exacerbating social division. More significantly, the decentralized allocation of political and economic resources among classes deviated from the state's goal of building an egalitarian society, leading to relative deprivation across all classes.

The demarcation of classes and the prioritization of class struggle can be likened to a double-edged sword with inherent risks. As the collective behavior of the Cultural Revolution attests, following a succession of political movements, society gradually recognized that one's class origin could serve as a form of political currency and that the tenets of class struggle could be wielded as a means of marginalizing dissidents. The cadre children expanded the class policy into the bloodline theory. They tried to use it to exclude academically accomplished middle-class children from the path to promotion, thereby cementing their status as successors.[18] In addition, following the publication of Mao Zedong's big-character poster, *Bombard the Headquarters*, it became possible to take the lives of disgruntled party and government cadres without assuming any responsibility for these deaths as long as they were labeled as "capitalist roaders within the party." For party and government officials at all levels, their authority could be readily usurped by portraying them as "having carried out the capitalist reactionary line." As the author explains in subsequent chapters, during the Cultural Revolution, the Theory of Class Struggle was subject to arbitrary interpretations by various social groups following their own interests, fully exploiting it to achieve their aims. However, when class struggle, which was initially a tool of social domination, was employed by these groups to seize power, the only recourse for maintaining social stability was to rely on the military.

1.2 *Urban–Rural Differences*

The division of social groups resulting from the division of classes was an essential factor in the formation of factions during the Cultural Revolution. Still, it was not the only factor, for there were other class divisions besides class differences. These divisions complemented and greatly influenced the differentiation and restructuring of factions during the Cultural Revolution.

As shown in Fig. 1, the implementation of the hukou system served as a barrier between China's rural and urban societies, resulting in people being divided into two social groups, rural hukou holders and urban hukou holders, with significant differences in interests between the two.[19] The transition from a rural to an urban hukou was considerably difficult.

Various political and economic factors further stratified the urban–rural divide generated by the Household Registration System. Turning our attention first to rural society, following the implementation of rural land reform, economic disparities were no longer a factor in social stratification, making political factors the dominant determinants of social stratification in the countryside. By the significance of power and class origin in social reorganization after the country's founding, peasants were classified into a hierarchy consisting of grassroots cadres, general peasants,[20] and "four types of elements (landlord, rich peasant, counter-revolutionary, and bad element)."

Social stratification in urban areas is more intricate than in rural areas, characterized by multiple structures. According to the State Ministry of Labor and Personnel management system, the urban population was bifurcated into two establishments: cadres and workers (*bianzhi*).[21] The term "cadre" here does not denote specific managerial positions like a president or factory manager, but rather a distinction recorded by *Dang'an*[22] to distinguish social status. Before the Reform and Opening-Up period, regular urban employees were classified into two types of *Dang'an* records: cadres and workers.[23] Disparities existed in archival documents *(Dang'an), and* significant differences in wages, housing conditions, medical benefits, and pensions. Transitioning from worker to cadre was as challenging as changing one's household registration.[24] Broadly speaking, cadre status comprises two categories: those appointed by the state who hold positions in the state administrative system and those who graduated from formal, state-established secondary and tertiary institutions and were likewise assigned work by the state.[25]

Furthermore, there exist even closer and more detailed distinctions within the cadre and worker establishments. Within the cadre establishment, there are two broad categories: managerial positions in state bodies (e.g., party and government cadres) and specialists in various fields (e.g., intellectuals and technicians).[26] State, party, and government cadres are classified into 30 levels based on the administrative hierarchy, with levels 13 and above comprising senior cadres (i.e., divisional, regional, administrative division-bureau levels), levels 14 to 17 are middle-level cadres (i.e., regiment, county, section chief levels), and levels 18 and below comprising general cadres.[27] The senior cadres above level 13 are further divided into three groups: President-Premier from level 1 to level 3, Minister-Provincial from levels 4 to 7, and Secretary-Director-District Chief from levels 8 to 13.[28] Different levels enjoy different political and economic entitlements (Table 1).

The above forms the core of the hierarchical structure in urban areas. The position of specialists in each field in the hierarchical structure is shown in Table 1. It is worth noting, however, that the chart is based primarily on salary packages, i.e., it compares the economic status of cadres and intellectuals and does not represent a relevant comparison of the political status between the two.

The hierarchy of social classes within the worker establishment is more complex than that of peasants and cadres, owing to various factors. Depending on their form of employment, the working class can be divided into four distinct categories: regular, contract, temporary, and outsourced workers.[29] While ordinary workers are entitled to numerous benefits within the organization, this is not the case for the other three types of workers. Furthermore, urban and rural hukou holders face different prospects even among contract, temporary, and outsourced workers. Urban hukou holders can transition to regular workers, whereas it is rare for rural hukou holders to do so.

In addition to the differences mentioned above, before reform and opening, all resources necessary for the survival of workers could only be obtained from the work unit. As a result, the work unit was the sole domain where workers could attain social status. Consequently, the administrative categorization of the work unit had a profound impact on the social status of the individual worker, whether the work unit was state-owned or collectively owned or a centrally administered enterprise or a locally administered enterprise. These variances determined the divergent social statuses among workers. There exists a notable contrast between

Table 1 Comparison of levels between party, government and military cadres and intellectuals

<table>
<tr><th></th><th colspan="4">Party Government Military Cadres</th><th colspan="3">Intellectuals</th></tr>
<tr><th>Level</th><th>Military</th><th>Central Government</th><th>Local Government</th><th>Party</th><th>Technical</th><th>Humanities</th><th>Science</th></tr>
<tr><td rowspan="7">High-Level Cadre (Level 13 and above)</td><td></td><td>Chairman</td><td></td><td>Standing Committee of the Political Bureau</td><td rowspan="4"></td><td rowspan="4"></td><td rowspan="4"></td></tr>
<tr><td>Military Commission</td><td></td><td></td><td></td></tr>
<tr><td>Commander of Three Military Branches</td><td></td><td></td><td></td></tr>
<tr><td rowspan="2">Military Level</td><td rowspan="2">Ministerial Level</td><td rowspan="2">Provincial Level</td><td rowspan="2">Provincial Party Committee</td></tr>
<tr><td rowspan="2">Senior Engineer</td><td rowspan="2">Professor</td><td rowspan="2">Researcher</td></tr>
<tr><td rowspan="2">Division Level</td><td rowspan="2">Director Level</td><td rowspan="2">Regional (Municipal) Level</td><td rowspan="2">Regional Party Committee</td></tr>
<tr><td rowspan="2">Engineer</td><td rowspan="2">Lecturer</td><td rowspan="2">Assistant Researcher</td></tr>
<tr><td rowspan="2">Mid-Level Cadre (Level 14-17)</td><td rowspan="2">Regiment Level</td><td rowspan="2">Division Level</td><td rowspan="2">County Level</td><td rowspan="2">County Party Committee</td></tr>
<tr><td rowspan="2">Assistant Engineer</td><td rowspan="2">Assistant Lecturer</td><td rowspan="2">Intern Researcher</td></tr>
<tr><td rowspan="4">Ordinary Cadre (Level 18-24)</td><td rowspan="2">Brigade Level</td><td rowspan="2">Section Level</td><td rowspan="2">Commune Level</td><td rowspan="2">General Party Branch</td></tr>
<tr><td rowspan="2">Technician</td><td rowspan="2"></td><td rowspan="2"></td></tr>
<tr><td rowspan="2">Company Level</td><td rowspan="2">Clerk Level</td><td rowspan="2">County/Commune Group Level</td><td rowspan="2">Party Branch</td></tr>
<tr><td></td><td></td><td></td></tr>
</table>

Source Kazuko Mori, "The issue of 'Cadres' in Chinese Politics," edited by Hankichi Goto, *The Political Structure of Modern China*, Japan Institute of International Studies, 1982, 169 pages; Gu Jieshan et al.eds., *Analysis of Social Interest Groups in Contemporary China*, Heilongjiang Education Publishing House, 1995, 310 pages, etc. The author completes this table by referring to the above information

employees of state-owned work units, who enjoy lifetime employment and extensive health coverage provided by the big labor insurance (dalaobao), which also extends to their family members, and those in collectively owned work units, whose small labor insurance benefits (xiaolaobao) are limited to themselves only. Furthermore, the employment system for workers in large collectives, which fall under provincial, regional, and municipal jurisdictions, offers a similar level of job security to that of a lifetime system. On the other hand, those in small collectives governed by street offices received lower wages and lacked any form of social insurance coverage.[30] Notably, trade unions could only be established in state-owned work units.[31] Their presence significantly affected workers' benefits, as they provided various employee benefits.

Regarding political status within the work unit, the social groups of workers can be classified into four categories: activists, ordinary workers, backward elements, and black five workers.[32] Among them, urban "grass-roots cadres" who hold non-disengaged roles, such as production team leaders and union leaders, are considered workers and mainly fall under the category of activists.[33]

Considering the above, it can be seen that the establishment of political, economic, and social management and bureaucratic state institutions led to various social hierarchies. The social groups that were created by social hierarchies shared a common social background during the Cultural Revolution, which enabled them to comprehend the objectives of their actions and the elements at stake, laying the groundwork for collective behavior during that period. Meanwhile, the existence of multiple hierarchies also led to complex and diverse perspectives of the collective understanding of social groups. Such a situation was in favor of readjustment and reorganization of the collective understanding of social groups in accordance to changing political objectives and circumstances. Consequently, factions could divide and reorganize themselves at any given time during the Cultural Revolution, making it one of the primary reasons behind their existence.

Disparities in class, social stratification, occupation, income, and welfare, and the resulting inequality structures exist in any country. However, it is essential to highlight that pre-reform and opening of China had a social status system that was difficult to change through individual efforts because it was determined by the state and regarded as an innate and unalterable status. Figure 1 illustrates that the identity system during the Maoist era can be categorized into two types. The first type is inherited from one's parents and established before birth, such as the class origin system and Household Registration System. The state assigns the second type upon entry into society, and it is typically challenging to change through individual efforts afterward. Examples of this include the cadre and worker status and the level of work unit ownership.

The state played a significant role in determining the innate nature of both identity systems. It can be contended that the state created social groups and a closed hierarchical society through a series of institutions and policies, including the implementation of class divisions and the Hukou system. This closed hierarchical society restricted upward social mobility, and the state maintained control over the limited opportunities. The state's monopoly on social mobility and the competition among social groups for opportunities to move upward profoundly impacted collective behavior during the Cultural Revolution. The following points make it evident.

1.3 Opportunity Distribution Gaps in Social Mobility

Social mobility can clearly indicate how access to resources that determine the social status of individuals and groups is distributed.[34] Here, the author briefly elaborates on the correlation between upward social mobility and the distribution system of citizenship in China before the reform and opening era.

In pre-reform and opening of China, what were the main methods members of society achieved upward mobility in society? As mentioned earlier, in China at the time, all resources were monopolized by the state, and society did not have any resources that enabled upward mobility. Consequently, significant upward mobility in society could only rely on alterations in state policy. As a result, upward social mobility cannot be discussed separately from state policy. The connection between state policies and upward mobility can theoretically be categorized into three primary types in a state and social system without a private sphere. The first type involves mobility achieved through collective acts of protest that pressure the state to modify its hierarchical structure and policies. The second type occurs when the state proactively adapts its policies to offer opportunities for upward mobility to specific segments of society. The third type pertains to upward mobility under unchanging policy conditions, which examines how members of society can attain upward mobility without any adjustments in state policy. This involves identifying various mobility pathways, determining which social groups are granted the right to follow these paths, and exploring the coping strategies those excluded adopt. In this context, we will analyze the primary methods of upward social mobility at that time, concentrating on the following three distinct types.

First, since the hierarchical structure of pre-Cultural Revolution China was created by implementing state policies. Therefore, the most effective way to achieve upward mobility was through collective protests aimed at changing state policies. However, in pre-Cultural Revolution China, the conditions did not exist for creating a large-scale social action that could bring about changes in state policy. Although pre-Cultural Revolution China was an unequal and hierarchical society with a wide range of interests between social groups, social dissatisfaction did not reach an intolerable level. Social conflicts were not so deep as to trigger large-scale protests. Thus, achieving upward mobility through collective protests ultimately amounts to a detached assumption.[35] For example, while there were dozens of strikes and marches in urban and rural areas in the mid-1950s, they did not have the energy to trigger large-scale social movements and did not impact the hierarchical structure. Moreover, as the Anti-Rightist movement demonstrated, the state had sufficient capacity to deal with the protests of social groups, and the Communist Party's rule had complete legitimacy. Under such conditions, it is inconceivable that a nationwide social action challenging state power would have occurred without the political mobilization of the state during the Cultural Revolution. It can be argued that the decline of the state's capacity for social cohesion during the Cultural Revolution was due to the weakening of its capacity resulting from divisions within state institutions. This point will be discussed in detail later in analyzing collective behavior during the Cultural Revolution.

Second, the hierarchical structure in pre-Cultural Revolution China was caused by state policies, so changes in state policies played a crucial role in significant upward mobility. As many researchers have pointed out, in the history of the People's Republic of China, policy changes occurred in a cycle of steadiness and radicalism. Influenced by domestic political conditions and international situations, state policies had to be adjusted as national goals changed. In addition, policies were subject to fluctuations because different groups with influence over policymaking held political power.

Since the state had a monopoly on the distribution of citizenship, policy adjustments could bring about changes in the distribution of citizenship, affecting upward mobility in society. For example, the labor shortage during the Great Leap Forward led to the employment of some 20 million peasants in cities.[36] But in the early 1960s, the setbacks of the Great Leap Forward led the government to force these peasants back to the countryside to work as farmers to avoid the collapse of the national economy.

In addition, policy adjustments brought upward mobility for some social groups while simultaneously causing downward mobility for others.

For example, Liu Shaoqi's "two labor systems" of using more temporary workers and fewer regular workers in 1964 brought employment opportunities to some urbanites. The policy also allowed some sent to the countryside in the 1960s to return to the city.[37] However, some workers who should have been employed as regular workers after the probationary period could not do so and were instead employed as temporary or contract workers.[38]

Adjustments in relevant policies can bring about social mobility. Thus, waiting for policy changes, although a highly passive response for those who expect to improve the status quo, was one of the possible ways to implement it in pre-Cultural Revolution China. If the expectation is understood as hope, having future expectations may be seen as a way to vent their frustration with the status quo. The state has sometimes taken the initiative to adjust its policies to alleviate social discontent. For example, to soothe peasant dissatisfaction with the people's commune and the food crisis, the government changed its rural policy twice, in November 1958 and February 1962, to allow peasants to keep their household plots (*ziliudi*).[39] However, while using policy adjustments to balance the distribution of benefits among social groups may have temporarily healed malcontent, such adjustments were made based on maintaining a hierarchical structure. Therefore, society remained inherently stratified. As a result, the cycle of radicalism and steadiness inevitably brought about widespread social unrest.

The third is the question of how people achieve upward mobility in society in a constant institutional environment. For young people, going to university, joining the People's Liberation Army, and marrying someone of higher status are the main channels of upward mobility. Especially for the youth whose lives were defined by an innate identity system such as class origin and the Household Registration System, these channels were the only way to change their social status. Still, only a few could attend university, join the army, or marry upward. For older people, upward social mobility can be equated with political advancements, such as being selected as an activist, joining the party, or entering the bureaucracy. However, access to political advancement was limited and unequal. Here, we will briefly examine the distribution of opportunities for upward social mobility in four areas: education, military service, marriage, and political advancement.

Path to Higher Education

Equality of opportunity in education was important in pre-Cultural Revolution China as schooling was one of the main channels of upward mobility. Opening the doors of education to workers and peasants was the goal of education set by the People's Republic of China in the early years of its founding.[40] In reality, children of ordinary workers/ peasants of red five origins were not in a position to receive tutoring from parents, and household chores made securing time for studies difficult. Poor economic conditions and the educational disparity between urban and rural areas meant that only a few could attend university.[41] However, the ability of children from non-red five families to attend university depended not only on their effort but was also largely determined by state policy. After the state's founding, the university admission criteria included merit, class origin, and political performance. Of these, merit and class origin played a pivotal role and were greatly influenced by the political and economic situation of the time. For example, during the First Five-Year Plan from 1953 to 1957 and the adjustment period from 1960 to 1962, merit became a relatively more critical selection criterion than class origin to prioritize economic development. Children from middle-class families and some from black five families were admitted to university during this period. However, class origin was most important during the Great Leap Forward period from 1958 to 1960 and the Four Clean-ups campaign in 1963 to the Cultural Revolution. Children from black five families were wholly excluded from universities.[42] The emphasis on class line meant educational opportunities were favored by cadre children and those of middle-class origin. Before the Cultural Revolution, the competition for higher education was between the cadre and middle-class children. This competition for interests that existed before the Cultural Revolution combined perfectly with the emphasis on class struggle during the Cultural Revolution to shape factional behavior. Chapters 5 and 6 will analyze these reasons in depth.

Path to Military Service

For young people during the Cultural Revolution, joining the People's Liberation Army was as important a pathway to upward political mobility as taking school entrance exams. Likewise, joining the army had the same strict restrictions as taking the entrance examinations. It was just that the criteria for joining the military were more explicit than those for advancing to higher education. They were limited to those of red five class origins who had not made political mistakes. But there was also an order of precedence in this criterion, with Communist Party members and members of the Communist Youth League enjoying absolute priority. Although before the Cultural Revolution, joining the PLA was a supreme honor, and only a strictly screened minority of young men were able to obtain the opportunity, it is worth paying particular attention to the fact that it was not always possible to achieve upward social mobility simply by joining the PLA. According to Hiroko Maeda's research, "in principle, even if a person with a rural household registration joins the PLA, they can only return to their place of origin for employment after demobilization or retirement from military service, and it was usually difficult to find employment in cities. Urban hukou can only be obtained if they are promoted above a certain rank within the military."[43] In addition, if a person with an urban hukou cannot be promoted to a certain rank after joining the army, obtaining a cadre commission after demobilization or retirement was difficult, even if employment in the city was possible. For example, Wang Hongwen, a member of the Gang of Four, joined the Chinese People's Volunteer Army between 1951 and 1956. Although he joined the party in the army, he was only an ordinary soldier, so when he was demobilized, he could only work as a typical worker in the 17th Cotton Factory in Shanghai.[44] Overall, the path to changing status by joining the PLA was narrow, whether one had an urban hukou or not. Many Cultural Revolution sources show that demobilized soldiers were active in the factional struggles of the Cultural Revolution everywhere. Their organizational skills gained through military training and various talents were valued and used in factional struggles during the Cultural Revolution. Wang Hongwen's rebellion, analyzed in Chapter 7 of this book, is a classic and compelling example. Due to his dissatisfaction with his treatment in the factory after demobilization and resentment over being compelled to take off his shirt in public for stealing factory cloth, Wang was the first to rebel after the start of the Cultural Revolution. He eventually became a leading figure in the Central Committee of the CCP.[45] In late 1966, former soldiers dissatisfied with their treatment after demobilization organized the National Rebellion Union of Demobilized Soldiers, which was soon banned as a reactionary organization.[46]

Path to Marriage

Because of the significant disparities in living standards, employment, and educational opportunities between large and small and medium-sized cities, marriage was a means for young people to seek inter-urban migration and change from rural to urban hukou. However, according to the Regulations of the People's Republic of China on Registration of Hukou, published on 9 January 1958, "a rural person who marries an employee-worker in an urban area cannot obtain an urban hukou even if they marry, except for those who have difficulty living independently due to long-term convalescence and have no relatives in the rural area to take care of them."[47] In addition, inter-urban migration requires one of three hard-to-obtain approvals: permission for employment, job transfer, and city relocation.[48]

In addition, there is a relationship between marriage and class origin. The impact of class origin on marriage is both positive and negative. Specifically, suppose an ordinary worker-peasant of red five class origin marries into a family of revolutionary cadres, revolutionary soldiers, or revolutionary martyrs. In that case, the former will likely have improved access to political and economic resources. On the flip side, if a person of red five origin marries a person of black five class, the political evaluation of the former in school and workplace will be influenced by the spouse's class origin, and the marriage will affect their future. In short, in pre-reform and opening China, class origin was a vital political resource that played a decisive role in promoting life in society. Marriage to a person of high-class status was one of the ways to obtain this resource.

However, in pre-reform and opening China, marriage choice was not entirely personal for some people. For example, according to the Provisions on Marriage of Commissioned Officers issued by the General Political Department of the People's Liberation Army in 1955, the spouses of cadres of the People's Liberation Army above the regiment level and those engaged in confidential work must be members of the Communist Party or the Communist Youth League. In addition, the regulations prohibit military cadres from marrying immediate relatives of counter-revolutionaries and children of unreformed families of landlords, rich peasants, and capitalists.[49] Such restrictions on marriage did not only apply to the military but also to cadres above a certain rank in the public security organs and other departments engaged in confidential work. Even Zhang Chunqiao, who was at the top of the Central Committee during the Cultural Revolution, divorced his wife Wen Jing, who was very close to him so that his career would not be affected by his wife, who betrayed the communist revolution and collaborated with the Japanese in 1943.[50] Marriage could be both a resource and a burden on both sides.

Path to Political Activism

In addition to attending school, joining the army, getting married, joining the Communist Party-Communist Youth League, and being elected as an activist were the main ways to seek political advancement and achieve upward mobility. In simple terms, the political hierarchy in China before the Cultural Revolution was a pyramidal structure consisting of party bureaucrats → activists → ordinary people, → class enemies.[51] In this structure, joining the Party and being elected as an activist was a step toward a career. But the criteria for membership and selection as activists were highly ambiguous, and the personal values of the cadres in charge heavily influenced the selection process.[52] According to Hua Linshan's research, the number of activists was set at fifteen percent after the country's founding. Still, it was unclear what kind of people went through what process to become activists. Generally speaking, activists were first selected by higher cadres from the general population and then nurtured through exceptional guidance and training.[53]

Here it is necessary to briefly explain the activist selection process. In the first stage, a political survey of the entire population under the jurisdiction of a state agent is carried out. In the second stage, the survey results are categorized in percentage terms as progressive, intermediate, and backward. In the third stage, those selected as progressive are given special training in party policy and other areas. In the fourth stage, those judged to be qualified are selected as activists.[54]

It is clear from this process that the state agents have absolute power in selecting activists. Therefore, the behavioral characteristics of state agents directly involved in the determination process and the methods used by those seeking political advancement during the competition directly impacted the selection results of activists in the process of distributing citizenship.

First, let's look at the methods used by those trying to become activists. Since there are no objective criteria for being progressive, there are only two methods of competition for those who want to become activists. One is to try to be more in line with the party's requirements than others, a simple method of competition. The other is the unscrupulous method of gaining a competitive advantage by excluding competitors using denunciation or false accusations, establishing good relations, and showing loyalty to state agents who hold selection authority.

On the other hand, for state agents who hold power to select to expand their power base and successfully implement central policies, and complete tasks assigned by higher-level leadership, they also need to obtain the support and cooperation of lower-level personnel. This is the "patron–client relationship" within the unit studied by Andrew G. Walder.[55] As many studies have demonstrated, the patron–client relationship within the work unit led to divisions among workers. It was the cause of many factional struggles among workers during the Cultural Revolution. Furthermore, state agents involved in implementing policies, including the selection of activists in the distribution of citizenship before the Cultural Revolution, faced shocks and persecution due to their direct involvement in excluding individuals from the patron–client relationship during the Cultural Revolution.[56]

The behavioral traits exhibited by state agents at the policy implementation level during the distribution of citizenship, including the selection of activists, are a complex matter. To shed light on this issue, we will examine their position within the bureaucratic system.

When considering the role of state agents at the policy implementation level within the bureaucratic system, assuming that they lack personal interests and are solely motivated by advancing national policies is overly idealistic. In reality, it is difficult to avoid the possibility that subjective value judgments and the pursuit of personal interests in implementing state policies did not influence state agents at the policy implementation level. There are three reasons for this.

First, state agents' personal preferences and values at the policy implementation level influence their understanding of national policies.

Second, although state agents at the level of policy implementation cannot change national policies, they have some degree of control over their performance. For instance, the outcomes and social impacts of different measures vary widely, whether one strives to promote a policy or adopts the approach of "one set of rules for face-to-face interactions and another set of rules behind the scenes" or "policies from above, countermeasures from below."

Third, at the policy implementation level, state agents play a "top-down" role in the system of state political operations, which determines the characteristics of their behavior—that is, when there is a contradiction between the state's policies and social demands, state agents are highly likely to sacrifice the latter and prioritize the state's commands.[57] There are three reasons for this. First, power does not come from the people but is conferred by the state and superior party and government organizations. Second, loyalty to the party is the most important criterion for evaluating cadres. Third, there is also competition among state agents. Although there are cases of state agents at the policy implementation level making insincere or perfunctory expressions of support for state policies to protect social interests, those who dare to do so usually pay a heavy price for their behavior once they are investigated.[58] This is because, in most cases, cadres who do not implement state policies and party orders are punished by the Party Central Committee as being against the party and obstructing national construction.

2 Citizenship Distribution and Collective Behavior During the Cultural Revolution

From the analysis in the previous section, it is clear that the hierarchical structure created by the pre-Cultural Revolution system of citizenship distribution had at least three effects on the occurrence and development of collective behavior during the Cultural Revolution.

First, the multiple collective perceptions of members of society before the Cultural Revolution laid the foundation for factional mobility during its course. As we can see from the previous analysis, factions emerged after the Cultural Revolution began. Still, the state created the basis for their formation (social groups with different interests, class divisions, class antagonisms, etc.) before the Cultural Revolution. The state distributed economic and political benefits to various social groups, creating multiple hierarchical structures, resulting in each person having a sense of belonging to multiple social groups simultaneously. This sense of belonging to several groups allowed multiple choices and combinations. That is to say, when the goal of the Cultural Revolution changes, or factions cannot fully meet the interests of their members, or members' value perceptions of their affiliated factions change, or members are excluded from their affiliated factions or their values, participants are prone to adjusting their sense of belonging to other factions or forming new factions in a short period.

However, the state was decisive in determining the sense of belonging according to which social groups would form factions. In other words, the goals of the state and the changing power structure of the central leadership during the Cultural Revolution affected the stability of factions and prevented them from developing their capacity for autonomy.

Second, as will be detailed in subsequent chapters of this book, the hierarchical structure produced by the pre-Cultural Revolution citizenship distribution system and competition methods, as well as the widespread social discontent they created, made Mao Zedong's call for change through rebellion possible, and received a widespread response in society. Pre-Cultural Revolution Chinese society was hierarchical, and each interest group in society was part of its own hierarchical structure, maintaining the overall hierarchy together with other groups. This kind of hierarchy not only existed between the so-called "enemies" and "us" but also among the people due to the implementation of the citizenship distribution system. Social groups designated as "enemies" did not have the right to enjoy citizenship, which is understandable. According to the Communist Party's Theory of Class Struggle, discrimination against "enemies" is a legitimate revolutionary action. However, the hierarchical structure within the people is in contradiction with the socialist ideal of equality.

The discrepancy between social ideology and reality was a significant factor in the Cultural Revolution, manifested through Mao Zedong's ideology and the collective behavior of the masses during that period. Specifically, establishing an egalitarian society is both a Chinese social ideology and a political ideal of Mao Zedong. However, since the late 1950s, and especially since the 1960s, a hierarchical social structure and a privileged bureaucratic system gradually emerged, greatly contrasting with Mao's political ideals. As mentioned earlier, since the late 1950s, Mao Zedong emphasized that the detachment of bureaucrats from the masses was the root cause of internal contradictions among the people. Therefore, Mao's idea of the Cultural Revolution was to mobilize social groups outside the bureaucratic system and use their power to transform it to achieve the equality he pursued.

Mao Zedong's vision of the Cultural Revolution in 1966 can be summarized by the phrases "no construction without destruction,"[59] and "create order out of chaos."[60] His idea of "construction" can be found in his 1966 May 7th Directive, which states that "the People's Liberation Army should become a big school. ...." The understanding of "destruction" can be seen in his letter to Jiang Qing on 8 July 1966.[61] In the letter, Mao identified the targets of the Cultural Revolution's struggle and the movement's goals: "create great disorder under heaven to achieve great order under heaven. After seven or eight years, do it again. Let the demons and monsters come out and Cultural Revolution Committee have a big fight. The current task is to overthrow the rightists throughout the party and the country (although it is impossible to do so completely)." Mao also made statements about the struggle between the right and the left.

> (Ex. omitted) If there is a rightist coup against the Communists in China, I conclude that they too will have no peace and will probably be short-lived, for all revolutionaries representing the interests of more than ninety percent of the people will not tolerate it. At that time, the rightists may use my words to gain power for a while, while the leftists will certainly use some of my other words to organize and bring down the rightists. This Cultural Revolution is a serious exercise. In some regions (such as Beijing), the deeply rooted order was overturned overnight. The complex system collapsed instantly in some institutions, such as Peking University and Tsinghua University. Wherever the rightists were more arrogant, their failure was more disastrous, and the leftists became more enthusiastic. This is a nationwide exercise in which the left, the right, and the wavering centrists will all get their respective lessons.[62]

At that time, only Jiang Qing, Zhou Enlai, and Wang Renzhong read this letter, but the rest of the leaders of the Central Committee did not have the chance to see it. From the May 7th Directive and Mao Zedong's letter, it can be inferred that for Mao Zedong, the Cultural Revolution was an exercise in using the struggle between the left and the right to overthrow the right, to establish an integrated system of party, state, and society combining military, academic, industrial, and civilian elements.

On the other hand, at the social level, the discontent generated by the divergence between ideas and reality had spread among people long before the Cultural Revolution. Social groups with a lower status in the hierarchical structure were likely to rise up for social transformation once they were given the opportunity. From this point of view, Mao's idea of the Cultural Revolution was consistent with the aspirations of the deprived social groups, which was the most important reason for their alliance with Mao. Still, the goals of the two were very different. Specifically, Mao and many participants in the Cultural Revolution were united in their dissatisfaction with hierarchical divisions. However, from the two documents mentioned above, we can learn that Mao did not intend to destroy the bureaucracy but to transform it through a mass movement with the power of the people to achieve a system in which the government and the people are one. In other words, what Mao sought was not the creation of the state system in the general sense but the transformation of the human mind and the construction of the ideological sphere. However, for the movement participants, gaining more citizenship and achieving upward movement in society were the main objectives of participating in the activity. Of course, in addition to their interests, the ultimate goal of participants who rose up in rebellion inspired by Mao's concept of equality was to change what they perceived as unequal or irrational structures, not just ideological transformation. This difference in goals dictated that Mao could not form long-term alliances with various social groups. Thus, during the Cultural Revolution, Mao supported the different factions that best achieved his political goals. First, Mao supported the Red Guards, who advocated the bloodline theory, then the rebel Red Guards, and in 1967 when the Cultural Revolution spread throughout the country, Mao chose to support the workers' rebel group in Shanghai. The reasons for the trade-offs and their background will be detailed in subsequent parts of this book.

Third, from the analysis of upward social mobility, we can see that in the pre-Cultural Revolution multi-level society, opportunities for upward mobility were extremely scarce. The state and even state agents had absolute decision-making power in the distribution of opportunities. From this perspective, it can be said that even before the Cultural Revolution, there was already a tense relationship between the state/state agents and society members regarding the allocation and competition for social resources. Because state agents, especially those at the policy implementation level, directly participated in the distribution of citizenship, this led to attacks on many cadres during the Cultural Revolution by people dissatisfied with the citizenship allocation.

Fourth, from the previous analysis of social structure, it can be seen that competition among different social groups for citizenship existed even before the Cultural Revolution. Still, it did not lead to large-scale social conflicts and collective violence as in the Cultural Revolution. This is mainly because the state played an important role in maintaining social stability. Before the Cultural Revolution, competition for citizenship was pursued by different social groups within the framework of state-defined institutions, which were maintained not by law but by the personal charisma of Mao Zedong and other top leaders. Despite social dissatisfaction, dissenters did not have the opportunity to express their grievances through protests because the institutional framework was relatively stable. However, with the Cultural Revolution, political differences and power struggles among the top leadership emerged, causing political turmoil that shook the foundation of state rule and weakened the state's ability to control and manage inter-group competition. Furthermore, the expansion of the Cultural Revolution's impact led to the loss of legitimacy for party organizations at all levels to manage and control society, directly weakening the state's ability to control social movements and resulting in societal disorder. In such a context, inter-group competition for interests could only be pursued in a state of anarchy.

Notes

1. "Circular of the Central Committee of the CCP transmitting a circular on factories and mines requiring military control" (Zhonggong zhongyang zhuanfa dui xuyao shixing junshiguanzhi de changkuang de bugao) (March 16, 1967); "Several provisions of the Central Committee of the CCP on the propaganda of newspapers in provinces, cities and autonomous regions" (Zhonggong zhongyang guanyu gesheng shi zizhiqu baozhi xuanchuan wenti de jixiang guiding) (March 16, 1967); "Circular of the Central Committee of the CCP" Zhonggong zhongyang tongzhi) (March 17, 1967); "Circular of the Central Committee of the CCP transmitting the March 18 bulletin of the Military Control Committee of the Beijing Public Security Bureau" (Zhonggong zhongyang zhuanfa Beijingshi gonganju junshiguanzhi weiyuanhui 3yue 18ri de bugao) (March 18, 1967); "Circular of the Central Committee of the CCP on Several Provisions on the Handling of Red Guards' Confiscated Materials in the Cultural Revolution Movement" (Zhonggong zhongyang guanyu zai wenhuadagemingzhong chuli hongweibing chaojia wuzhi de jixiang guiding) (March 20, 1967), etc., edited by The Party History, Party Construction and Political Work Teaching and Research Department of the National Defense University of the People's Liberation Army (Zhongguo renminjiefangjun guofeng daxue dangshi dangjian zhenggong jiaoyanshi), *Research Materials on the Cultural Revolution* (part 1) *(Wenhuadageming yanjiu zhiliao* (shang), (The National Defense University of the People's Liberation Army, 1998), pp. 365–370.
2. Albert James Bergesen, op. cit., 1978, pp. 19–29.
3. In his "Analysis of the Classes in Chinese Society", published on December 1, 1925, Mao summarized the revolutionary attitude of

the middle class as, "That wavering middle class whose right wing may be our enemy and whose left wing may be our friend. We must always beware of them and not let them disturb our front". Mao, Zedong. "Analysis of the Classes in Chinese Society". In *Selected Works of Mao Zedong*, vol. 1, no. 9. Beijing: People's Publishing House, (1996). Moreover, in his "On New Democracy", published in April 1940, Mao discussed "the possibility of participating in the revolution on the one hand, and the compromise with the revolutionary enemy on the other, which is the Chinese bourgeoisie's 'one body, two mandates,' This is the two-faced nature of the Chinese bourgeoisie". Mao, Zedong. "On New Democracy". In *Selected Writings of Mao Zedong*, vol. 1, 360. Beijing: People's Publishing House, (1986).

4. For a study of the decentralized distribution of political and economic property of workers, peasants, intellectuals and cadres, see the previous paper by Yang Lijun", The Social Origins of the Red Guard Movement: The Perspective of Class and Stratum", *Contemporary China*, The Japanese Association of Contemporary China Studies, no. 74, September 2000, pp. 209–221.
5. Jin Fu, *China in the Decade Before the Cultural Revolution* Communist Party (History Press, 1998), p. 11. (*wenge qian shinian de zhongguo*, zhonggong dangshi chubanshe).
6. Edited by the Office of Socialist Thought Education of the Committee of the Chinese People's University, *Editorial of Reference Materials on Socialist Thought Education*, Series III-IV, (published by the People's University, November 1957), p. 174.(*zhongguo renmin daxue weiyuanhui shehuizhuyi sixiang jiaoyu bangongshi bian, shehuizhuyi sixiangjiaoyu cankao ziliao bianji,* di3-4ji, renmindaxue chubanshe).
7. Jin Fu, op. cit., p.12.
8. Mao Zedong, "On the Issue of Correct Handling of the Internal Contradictions of the People", *Selected Writings of Mao Zedong*, vol. 2, (People's Publishing House, 1986), p. 791.
9. Jin Fu, op. cit., pp. 4–52.
10. Mao's conceptions of democratic centralism are scattered in the following articles of his: "Conversations with the British Journalist Berendt"(he yingguojizhe beilante de tanhua)(*Selected Works of Mao Zedong, vol. 2*, People's Publishing House, 1991), p. 383; "On Coalition Government" (lun lianhezhengfu)(*Selected Works of*

Mao Zedong, vol. 3, People's Publishing House, 1991), p. 1057; "On the Correct Handling of the Internal Contradictions of the People"(guanyu zhengquechuli renminneibu maodun de wenti) (*Selected Works of Mao Zedong,* vol. 5, People's Publishing House, 1987), p. 368; "Speech at the Enlarged Central Working Conference"(zai kuodade Zhongyang gongzuo huiyishang de jianghua) (*Selected Works of Mao Zedong, part. 2,* (People's Publishing House, 1986), p. 816.

11. Mao Zedong, op. cit., "on the correct handling of the internal contradictions of the people", p. 368.
12. Mao Zedong, op. cit., "Speech at the Enlarged Central Working Conference", p. 816.
13. Mao Zedong, op. cit., "On Coalition Government", p. 1057.
14. Office of Socialist Thought Education of the Committee of the Chinese People's University, op. cit., *Editorial of Reference Materials on Socialist Thought Education,* Series III-IV, pp. 1–362.
15. Jin Fu, op. cit., p. 37.
16. Id. 37.
17. Id. 44.
18. Yang Lijun, op. cit. "The Social Origins of the Red Guard Movement", pp. 209–221.
19. For more information on China's household registration policy, please refer to "China's Household Registration Transfer Policy" by Hiroko Maeda, *Asian Economy*, May 1996.
20. According to a study by Kazuko Mori, cadres in China can be classified as "state cadres" (not engaged in production cadres) and "grassroots cadres" (engaged in production cadres). "The "grassroots cadres" are not clear in the cities, but in the countryside they include a large body of production brigade level (secretary of the brigade party branch—fellow committee member, captain), production brigade level (secretary of the Party Committee—deputy secretary), director of the women's federation, militia organization commander, etc., see, Kazuko Mori, "The issues of 'Cadres' in Chinese Politics", Hankichi Goto, ed., *The Political Structure of Modern China* (Japan Institute of International Studies, 1982).
21. Li Peilin, ed., *Report on Class Stratification in China's New Era* (Liaoning People's Publishing House, 1995), p. 68 (*zhongguo xinshiqi jiejijieceng baogao).*

22. A permanent dossier or archival system that records the "performance and attitudes" of citizens in China during the Mao Zedong era. it has been an important part of the government's efforts to maintain control of its people.
23. Li Peilin, ed., op., cit., *Report on Class Stratification in China's New Era*, p. 68.
24. Id. 68.
25. Id. 68.
26. Kazuko Mori divided China's state cadres into three broad categories: management cadres in various institutions, specialists in the production sector, and experts in the fields of culture, education and medicine. (Kazuko Mori, op. cit., "The issues of 'cadres' in Chinese politics", p. 166).
27. Id. 170.
28. Id. 170.
29. Party History Research Office of the Central Committee of the CCP, *Chronology of the Events in the Party History of the Party of China* (People's Publishing House, 1981), p. 141 (zhonggongzhongyang dangshi yanjiushi, *zhonggongdangshinianbiao*, renmin chubanshe).
30. Li Xun, "The 'Economist Wind' in Shanghai during the Cultural Revolution", Luo Jinyi et al., eds., *Beyond the Catastrophe—A Re-discussion of the Cultural Revolution* (Taiwan) (Feng Yun Forum Press, 1997), pp. 30–32 (wengezhong fasheng zai shanghai de 'jingjizhuyifeng', Luo Jinyi deng bian, *haojie zhiwai:zailun wenhuadageming*, Taiwan fengyunluntan chubanshe).
31. Shanghai Federation of Trade Unions, "Report on the request of workers in collectively owned factory enterprises to organize trade unions", 30 August 1966. (shanghuishi zonggonghui, "guanyu jitisuoyouzhi gongchang qiye zhigong yaoqiu zuzhi gonghui de qingshi baogao).
32. The distinction between "backward elements" and "black five" workers is very ambiguous. Usually the "black five" class and the "backward element" workers are classified differently, but sometimes they are uniformly treated as "backward elements".
33. Hua Linshan, "The origin of Mass Rivalry Factions during the Cultural Revolution", Liu Qingfeng, ed., op. cit., *The Cultural Revolution: Facts and Analysis*, p. 196.

34. In sociology, the term "social mobility" refers to the movement of individuals and groups between different socio-economic positions. In this context, "vertical movement" means moving up or down the socio-economic scale. People who increase in property and income and rise in status move "upwards", while the opposite is called "downwards". In contrast to "vertical movement", there is also "horizontal movement". Horizontal movement refers to geographical movement between residential areas and cities, as well as between local societies. In this book, the term "social mobility" refers mainly to "upward mobility".
35. Although small-scale demonstrations aimed at improving economic status had taken place before the Cultural Revolution, I am not aware of any large-scale protests. The "rightist" views on politics expressed in 1956 in the "Hundred Flowers, Hundred Schools of Thought" may be defined as a large-scale speech protest. However, this was not a spontaneous movement, but rather the result of a state-mobilized "rectification" campaign.
36. Hiroko Maeda, "The Household Management System and Population Movement in the People's Republic of China", *Asia Economics*, February. 1993, p. 23.
37. Li Xun, op. cit., "The 'Economist Wind' in Shanghai during the Cultural Revolution", pp. 36–37.
38. Tsuneto Yamamoto, "The Policy Trinity of Labor, Education, and Deportation in 1960 and Its Destruction", in Mitsuyuki Kagami, ed. *Frustration in Contemporary China,* p. 155.
39. Satoshi Amako, op. cit., *China: Changing of a Socialist Power*, pp. 91–92.
40. Atsuyoshi Nishijima, "The Present State of Chinese Education", *China Studies Monthly*, February 1964, p. 16.
41. Stanley Rosen, op. cit., 1982,; Hong Yung Lee, op. cit., 1988, p. 81.
42. Zhong Weiguang, "The History of the Birth of the Red Guard Group in Qinghua High School", *Beijing Spring*, October 1996, pp. 7–11("qinghua fuzhong hongweibing dansheng shishi", *beijingzhichun)*; Zheng Yi, "Qinghua High School, the Red Guards and Me", *Beijing Spring*, November 1996, pp. 27–29 ("qinghua fuzhong, hongweibing yu wo", *Beijing zhi chun)*; Zhang Chengzhi, translated by Shinji Kojima, *The Era of the Red Guards, (*Iwami Shinsho, 1992), pp. 20–26.

43. Hiroko Maeda, op., cit., "China's Policy on Transfer of Family Registration", p. 75.
44. Li Xun, op. cit. "The 'Economist Wind' in Shanghai during the Cultural Revolution", pp. 49–50.
45. Li Xun, The Great Collapse - A History of the Rise and Fall of the Workers' Rebellion in Shanghai (Taipei: Times Publishing, 1996), pp. 49–52.
46. "Comrades Zhou Enlai and Li Xiannian's Speech on Receiving Cadres of the Director and Above of the Department of Finance and Trade" (February 18, 1967), Congress of the Red Guards of the Capital Universities and Colleges, the Political and Legal Commune of the Beijing Academy of Political Science and Law, eds, *Reference Materials on the Great Proletarian Cultural Revolution* 1967, No. 29, p. 456. ("zhouenlai lixiannian tongzhi jiejian caimaokou sijuzhang yishang ganbu de jianghua", shoududazhuanyuanxiao hongweibing daibiao dahui, Beijing zhengfaxueyuan zhengfa gongshe bian, *youguan wuchanjieji wenhuadageming de cankao ziliao).*
47. Zhang Qingwu, ed., *General Knowledge of Household Registration* (Law Press, 1983), pp. 63–67.(*hukou dengji cangshi,* falvchubanshe).
48. Hiroko Maeda, op. cit., "The 'Household' Management System and Population Movement in the People's Republic of China", p. 25.
49. Cadre Department of the General Political Department and Military System Research Department of the Academy of Military Sciences, eds., *Outline of the Cadre System of the People's Liberation Army* (Military Science Press, 1988), pp. 343–344. (zongzhengzhibu ganbubu, junshikexueyuan junzhiyanjiubu bian, *renminjiefangjun ganbuzhidu gaiyao,* junshi kexue chubanshe).
50. Xu Jingxian, *Ten Years a Dream* (Hong Kong: Times International Press, 2005), pp. 389–409. (*shinian yimeng,* shidaiguoji chuban).
51. Hua Linshan, op. cit., pp. 194–197.
52. Id. pp. 194–197.
53. Id. p. 196.
54. Id. p. 197.
55. Andrew G. Walder, op. cit., *Communist Neo-Traditionalism,* 1986.

56. Andrew G. Walder, "The Chinese Cultural Revolution in the Factories: Party-State Structures and Patterns of Conflict", Elizabeth J. Perry, ed., *Putting Class In Its Place: Worker Identities In East Asia* (Berkeley: University of California Institute of East Asian Studies, (Berkeley: University of California Institute of East Asian Studies, 1996); Lynn T. White, op. cit., *Policies of Chaos: The Organization Causes of Violence in China's Cultural Revolution;* Hua Linshan, op. cit., "The origin of Mass Rivalry Factions during the Cultural Revolution".
57. For example, during the period from 1954 to the winter of 1955, the cooperative movement faced resistance from the peasants. In order to accomplish the tasks assigned to them by the higher authorities, some policy implementers ignored the principle that individuals were free to join and withdraw from their cooperatives and imposed economic penalties in the form of confiscation of private property and agricultural loans on individual peasants and mutual aid groups that did not participate in the cooperative movement, or labeled them as "bad elements following the capitalist road" and imposed political punishment. In addition, during the "Great Leap Forward" period, certain rural grassroots cadres, out of loyalty to the Party and the state, forced peasants to provide large amounts of food in the name of "striving for a bumper agricultural harvest". (Qiang Yuanjian and Lin Bangguang, "An Experimental Discussion of the 1955 Party Debate on Agricultural Cooperativization", *Party History Research*, January 1981, p. 11) ("shilun 1955 nian dangnei guanyu nongye hezuohua wenti de zhenglun", *dangshi yanjiu)*.
58. The results of Zhang Letian's investigation of the rural areas in the Haining area of Zhejiang Province show that from 1962 to 1963, a production team in the Haining area, after obtaining the tacit approval of the production chief, ignored the state's production plan and planted large quantities of grain crops in areas where mulberry trees should have been planted. As a result, it was unable to complete the task of raising mulberry assigned by the state (Zhang Le Tian, op.cit., pp. 93–94). In this example, the grassroots cadres prioritize farmers' interests over the state interests. However, it is worth noting that this was a case that occurred at an extraordinary time. That is, in the early 1960s, China had just experienced a nationwide food famine, and for the grassroots

cadres, ensuring that no one starved to death in their areas of jurisdiction was the state's primary requirement for cadres. From this point of view, on the surface, grassroots cadres acted in the interests of society over the interests of the state, but one cannot completely deny that there was still a side of petty calculations behind them. Of course, it cannot be asserted that there are no grassroots cadres who act purely to protect the interests of society, somehow at the expense of the interests of the State.

59. The phrase "no construction without destruction" appears in Mao Zedong's 1940 essay "On New Democracy" (op. cit., *Selected Writings of Mao Zedong*, p. 384). In this article, when discussing the relationship between the imperialist-semi-feudalist culture and the culture of New Democracy, Mao uses such expressions as "no construction without destruction", "no flow without obstruction", and "no progress without struggle". These expressions were used again in the "May 16th Notification" of 1966 and were given the following new interpretation: without destruction there can be no construction. Destruction, is to criticize, is to revolutionize". "The word 'destruction' is in the head and the word 'construction' is in it".
60. Wang Nianyi, op. cit., 1988, pp. 6–7.
61. Id. pp. 2–9.
62. Minoru Takeuchi, (compilation), "Mao Zedong's Cultural Revolution", Gendai Hyouron Sha, 1974, pp. 268–269.

CHAPTER 5

The Rise and Fall of the "Old Red Guards"

The previous chapter discussed the pre-Cultural Revolution social structure and the contextual factors implied in the distribution of opportunities for social mobility that contributed to the antagonistic system of the Cultural Revolution. The state's role in forming social groups and structuring the competition for citizenship before the Cultural Revolution was clarified. The correlation between the multiple hierarchical structures before the Cultural Revolution and the collective action during the Cultural Revolution was analyzed. Then, through what political structures of opportunities did the various social groups formed before the Cultural Revolution participate in the movement? What role did the state play in the formation, fragmentation, and reorganization of factions, and how did this influence the evolution of the Cultural Revolution? These are the central questions to be explored in Chapters 5, 6 and 7.

In this book, the author argues that the first three years of the Cultural Revolution consisted of three distinct movements, and the process of the Cultural Revolution was the transfer of power between these three sub-movements. Each of these sub-movements had different actors and targets. The "Old Red Guards" (*laohongweibing*), composed of Beijing middle school students; the "Tsinghua University Jinggangshan Corps" (*jinggangshan bingtuan*), composed of university students; and the

Y. Lijun, *Contest for Citizenship and Collective Violence during China's Cultural Revolution*, IPP Studies in the Frontiers of China's Public Policy, https://doi.org/10.1007/978-981-97-0906-9_5

"Shanghai Workers Revolutionary Rebel General Headquarters" (gongzongsi) organized by rebel workers in Shanghai, with each becoming the protagonist of their respective movement. We will focus on these factions in Chapters 5, 6 and 7, examining their formation, differentiation, and restructuring processes in the interactions between the state and factions and among factions. Furthermore, the book attempts to clarify three key points: first, the processes and reasons for faction formation, differentiation, and decline; second, the driving forces behind each movement; and third, the reasons for the transfer of power from one movement to the next.

Before diving into the discussion, it is essential to understand the timeline of the three sub-movements of the Cultural Revolution described in this book. The first three years of the Cultural Revolution can be divided into three sub-movement periods according to the different movement actors. The first sub-movement of the Cultural Revolution took place from the end of May to October 1966 and can also be described as the political mobilization period. Mobilising the people against "all the monsters and freaks" (hengsao yiqie niuguisheshen) was the main political goal in this period. Red Guard organizations centered around the children of cadres in Beijing's junior and senior high schools were the first to rise up in rebellion, providing Mao Zedong with a bottom-up approach to launching the movement, thus being selected as the protagonists of the Cultural Revolution. In tandem, it allowed the cadre children to pursue collective interests. However, as their pursuit of interests deviated from Mao Zedong's overall intention in launching the Cultural Revolution, the political resources granted to them were eventually withdrawn.

The second sub-movement of the Cultural Revolution can be divided from October to December. The goal of the Cultural Revolution in this period was "to oppose against and overthrow the capitalist roaders within the party." Criticizing and exposing the "reactionary line" between Liu Shaoqi and Deng Xiaoping became the national public goal of the Cultural Revolution. Against this background, the university rebel groups, with the spirit of doubting everything and overturning everything, became the main actors of the movement. This became an opportunity for those non-vested interest groups in the pre-Cultural Revolution period to legitimately pursue their collective interests.

The third sub-movement of the Cultural Revolution took place from January 1967 to April 1969. Regime reconstruction was the purpose of the Cultural Revolution during this period. Seeking to curb the spread of economism and restore order in production, the "Shanghai Workers Revolutionary Rebel General Headquarters" became the main actor of the third movement. During this period, non-vested interest groups among workers had the opportunity to pursue power and interest. This intensified the competition for interests between factions in many cities and regions, leading to large-scale fighting. However, in Shanghai, Zhang Chunqiao used violence to counter violence and factions to control factions to complete the re-establishment of power, thus realizing Mao's vision of "revolution and production" in parallel. In the process, the "Shanghai Workers Revolutionary Rebel General Headquarters" also completed its quasi-governmental transformation.

Strictly speaking, two further sub-movements can be divided with 1968 as the boundary in the third sub-movement of the Cultural Revolution. Since 1968, especially after the summer of 1968, the nationwide movement's focus had completely shifted toward rebuilding order. The protagonists of this period's movement were no longer social organizations but had returned to party, government, and military organizations at all levels. However, the violent conflicts triggered by the rebuilding of order during this period are not the focus of this book and will not be discussed here.

In this chapter, our analysis will center on the first period of the movement. The Red Guards, the cadre children in Beijing's major high schools, were well-known as the first organized group to enter the political arena of the Cultural Revolution. After October 1966, they took the name "Old Red Guards" to distinguish themselves from other groups of the same name and emphasize their legitimacy.[1] Why did these promising high school students rebel? To understand this phenomenon, we must examine the motives and actions of the "Old Red Guards" in the context of the interaction between the state and society. It can be argued that the emergence of the Red Guards resulted from a combination of two elements: the socialization of the political struggle between central leaders and the politicization of the competing interests of young people. This question can be examined from two aspects: first, how the power struggle of the Central Committee influenced the competition among social groups; second, how various social groups used the opportunities provided by the power struggle of the central leadership to compete for interests among themselves. Examining this interaction from these two perspectives is essential for understanding the formation, development, and decline of other factions during the Cultural Revolution and the early Red Guards organization.

1 The Socialization of Political Struggle at the Central Leadership Level

1.1 Criticism of Wu Han and the "Beijing Gang"

Before the Cultural Revolution, the Party Central Committee was divided into many policy aspects. Among them, those directly related to the Cultural Revolution can be summarized as the confrontation between the radical faction represented by Mao Zedong and the moderate faction represented by Liu Shaoqi.[2] The former advocated the continuing revolution and firmly intended to criticize the existing system, while the latter valued economic development and the stability of the current system, emphasizing formalization and technicalization. The opposition between the two was already quite deep before the Cultural Revolution.

At that time, Mao had taken a back seat to the second line of the high command, and the operational power of the party and government was held by the first line, centered on Liu Shaoqi.[3] From the late 1950s onwards, Mao Zedong's dissatisfaction with Soviet-style socialism and the bureaucratic system gradually surfaced. He believed that the existing Soviet-style bureaucratic system was overly focused on administration and organization and had evolved into a new system dominated by bureaucratic forces and professional technical groups with a tendency to produce a "new class." In the 1960s, Mao was deeply concerned about the social contradictions brought about by the failure of the Great Leap Forward and the situation in Soviet Eastern Europe marked by the Hungarian Incident in 1956. Mao began to regard as "revisionistic" the series of adjustment period policies implemented by Liu Shaoqi and other front-line leaders of the central government, which prioritized the pursuit of economic efficiency. Against this background, Mao once again put forward the "Theory of Continuing Revolution," aiming to transform the bureaucracy through class struggle. From late 1964 onwards, Mao began to emphasize the existence of capitalist roaders within the party.

As J.R. Townsend points out, in the early 1960s, Mao's personal appeal was as strong as ever for ordinary Chinese. Still, within the top leadership, his authority and the importance of his voice were weakening.[4] Of particular importance was the fact that stepping back from direct supervision of the party and government and relinquishing his position as president of the state had left him relatively isolated in the administrative process. By losing the executive body within the bureaucracy, Mao lost the "legs" necessary to get orders through. In the record of his meeting with Edgar Snow in 1970, Mao stated the following about his political position in the central leadership: "At that time, I had no power in the party, no power in propaganda, no power in the various provinces, no power in the various localities, such as the Beijing Municipal Party Committee."[5] As such, the Great Proletarian Cultural Revolution envisaged by Mao Zedong was likely to achieve the following three goals through a bottom-up political mobilization by organizing social forces outside the bureaucracy. First, the destruction of Liu Shaoqi and Deng Xiaoping and their influence on the leadership. Second, the replacement of the central leadership, which Liu Shaoqi and Deng Xiaoping heavily influenced. Third, to establish a new basic line to replace the Liu–Deng line and to build new power institutions to promote this new line.[6] The arts and culture became the initial breakthrough point to achieving these goals.

On November 10, 1965, Shanghai's *Wen Hui Bao* published Yao Wenyuan's op-ed Comment on the new historical play Hai Rui Dismissed from Office. The gist of the article was to accuse Wu Han of celebrating the integrity of Hai Rui (an official dismissed from office in the Ming Dynasty for advising the emperor). Still, Wu Han was regarded as insinuating Peng Dehuai's criticism of Mao Zedong. Designated as the leading body of the literary and artistic circles, the Cultural Revolution Group of Five, consisting of Peng Zhen, Lu Dingyi, Zhou Yang, Wu Lengxi, and Kang Sheng, limited its criticism of Wu Han to academic arguments to avoid politicization. However, Mao Zedong classified it as a political struggle, believing that Wu Han's phenomenon represented a revisionist tendency and emphasizing the need to fight it back as a profound political and class struggle.[7] Taking this as an opportunity, the opposition between the radical faction represented by Mao Zedong and the moderate faction represented by Liu Deng began to heat up.

As Amako Satoshi notes, confronting the bureaucracy represented by Liu–Deng was difficult despite Mao's supreme authority over the general public. The failure of the Great Leap Forward and the economic recovery brought about in the aftermath of the adjustment policies increased the prestige of Liu Shaoqi–Deng Xiaoping in the Party Central Committee. The central party and government organs, including the Party Central Committee, the Central Secretariat, the Propaganda Department, the

Organization Department, the Rural Work Department, and the Beijing Municipal Committee, were all under the strong influence of Liu and Deng.[8] In contrast, the minority actively supporting Mao Zedong among the current members and executive members of the Central Politburo were few.[9]

According to Harding's research, Mao Zedong's power base during the Cultural Revolution was mainly supported by three groups. The first group was the young intellectuals centered around Jiang Qing, whose main members included Mao's wife, Jiang Qing, Mao's political secretary Chen Boda, Kang Sheng, and other confidants in the central government, as well as Guan Feng, Qi Benyu, Lin Jie from the Philosophy and Social Sciences Department of the Chinese Academy of Sciences, and Zhang Chunqiao and Yao Wenyuan from the Shanghai Municipal Party Committee Propaganda Department at the local level. The second group was the PLA, led by Lin Biao. The third group was various social forces mobilized during the Cultural Revolution.[10] From the end of 1965 to the beginning of 1966, except for the first group, the alliance between Mao Zedong and the other two groups had not yet formed.

Since Mao Zedong and other radical factions were clearly in the minority in the central leadership, it was difficult to use normal party administrative means to remove Liu Shaoqi, Deng Xiaoping, and others from the central leadership and establish a new basic line to replace theirs. In this context, Mao Zedong had three response methods available. The first was to ensure the support of the military. The second was to monopolize the propaganda power and ideologically judge the opponents while occupying the ideological high ground. The third was to publicly publicize the central leadership's political struggles and even socialize them to seek societal support. From the launch of the Cultural Revolution from the end of 1965 to May 1966, Mao Zedong simultaneously used the above three methods.

In ensuring the support of the military, Mao Zedong took advantage of the power struggle within the leadership of the PLA at the time, namely the opposition between Defense Minister Lin Biao and Chief of Staff Luo Ruiqing, and supported Lin Biao in driving Luo Ruiqing out of power, in exchange for the support of the People's Liberation Army under Lin Biao's command.[11] The landmark event was the Army Forum on Literature and Art, called by Jiang Qing at the Behest of Lin Biao, which took place in Shanghai from 2 to 20 February 1966. This forum was significant in three ways. First, it symbolized the formation of a political alliance between Mao and Lin Biao. Second, the forum outline entitled "Summary of the Army Forum on Literature and Art Work called by Jiang Qing at the Behest of Lin Biao" (also referred to as the "February Summary") intended to oppose the February Outline (full title "Outline of a Report of the Group Five to the Center about the Current Literary Criticism", which promoted that criticism of Wu Han's new historical play Hai Rui Dismissed from Office should be confined to an academic context.) submitted on February 3, 1966. The February Summary pointed out that "since the founding of our People's Republic, we have been under the dictatorship of a sinister anti-party and anti-socialist line which is opposed to Mao Zedong Thought."[12] The Summary refocused its attack from Wu Han and other literary circles on Peng Zhen, Lu Dingyi, and their leadership of the Beijing Municipal Committee, the Propaganda Department of the CCP Central Committee, and the "Group of Five" led by Peng Zhen, which was formed in response to Chairman Mao's call for a revolution in the cultural field in 1964. Third, as Mao Zedong and Lin Biao formed a political alliance, the center of radical propaganda shifted from the local media tool, Shanghai's *Wen Hui Bao*, to a tool of national military propaganda, the *Liberation Army Daily*. From that time until 31 May 1966, when Chen Boda and his Work Team took over the *People's Daily*, the *Liberation Army Daily* served as an important tool of radical propaganda in mobilizing society and controlling ideology, conveying the voice of another command post from the Party Central Committee.

Looking at it from the angle of Mao Zedong's desire to control the leadership of the ideological sector, there may be significant political considerations behind the criticism of Wu Han's *Hai Rui Dismissed from Office*. In other words, when thinking about why *Hai Rui Dismissed from Office* was criticized, the important thing is not the content of the play but the political position of Wu Han, its author. Wu Han and his play *Hai Rui Dismissed from Office* were suitable breakthrough points. Specifically, in the Communist Party's administrative system, Wu Han, the deputy mayor of Beijing, was supervised by the First Secretary and Mayor of the Beijing Municipal Committee, the Secretary of the Central Secretariat, and the Group of Five leaders of the Cultural Revolution, Peng Zhen. In addition, almost all members of the Group of Five were from the Beijing Municipal Committee and the Communist Party's Central Propaganda Department, which manages the Communist Party's official newspaper, the *People's Daily*, the Communist Party's theoretical magazine, *Red Flag*, and Xinhua News Agency. At that time, Lu Dingyi was the

Minister of the Central Propaganda Department, and both he and Peng Zhen were supporters of the Liu–Deng line. The superior organization of the Central Propaganda Department was the Central Politburo, located at the center of state power. In addition, *Frontline*, which provided a platform for Wu Han's speech, was the political and theoretical magazine of the Beijing Municipal Committee. It was under the dual leadership of the Beijing Municipal Committee and its counterpart department, the Central Propaganda Department. Therefore, it can be considered that using Wu Han as a breakthrough point was a clever move to follow and investigate the responsibilities of the Beijing Municipal Committee, the Central Propaganda Department, and the Group of Five. Although no historical evidence supports this speculation, it is reasonable to ponder given Mao Zedong's strategic behavior.

From the end of March 1966, after forming a political alliance with Lin Biao, Mao launched a fierce attack on Peng Zhen, Lu Dingyi and the Beijing Municipal Committee, the Propaganda Department of the CCP Central Committee, and the members of the Cultural Revolution Group of Five, and the agencies and organizations that were under Peng Zhen and Lu Dingyi's leadership. Mao launched his first attack on Peng Zhen, severely criticizing and accusing him of not supporting the leftists. In early April, the attack extended to Wu Han and another two members of the "Three-Family Village" Deng Tuo (the Beijing municipal party secretary in charge of cultural and education) and Liao Mosha (minister of the United Front Work Department). On April 19, the Central Secretariat informed all units in Beijing to stop carrying out the orders of the Beijing Municipal Committee, essentially paralyzing them. Then, at the enlarged meeting of the Political Bureau of the CCP Central Committee held on May 4, 1966, Peng Zhen, Lu Dingyi and Luo Ruiqing, and Yang Shangkun, who had lost the political struggle within the military, were branded as anti-party groups and dismissed from their posts. In the meeting, it was also decided to abolish the Cultural Revolution Group of Five, cancel the February Outline, and set up a new Central Cultural Revolution Group, which comprised Chen Boda, Kang Sheng, Jiang Qing, and Zhang Chunqiao. This new Central Cultural Revolution Group was set up under the Standing Committee of the Central Political Bureau, a higher authority than the Cultural Revolution Group of Five under the Central Secretariat. By allocating his supporters to core positions in the power structure, Mao established a strong power base to implement his will within the power structure. The Propaganda

Department and the Beijing Municipal Committee were subsequently reorganized.[13] As a result, Mao took complete control of the management of information, propaganda, and ideology, and the leadership of the political center of Beijing also fell into his hands.

For Peng Zhen, the initial intention of avoiding the politicization of academic debates was to protect Wu Han and himself. From the perspective of the Communist Party's reward and punishment mechanism, if his subordinate Wu Han made a political mistake, he would also be held accountable. Therefore, from the end of 1965 to the beginning of 1966, he organized a writing team, using *Beijing Daily*, *Beijing Evening News*, and *Frontline* as media fronts. He tried to confine the debate to the academic sphere and avoid politicization.[14] But at the meeting of the Secretariat of the CCP Central Committee held from April 9 to 12, 1966, Peng Zhen was not only criticized by Mao Zedong and other radicals but even Zhou Enlai and Deng Xiaoping thought that Peng Zhen's line crossed against Chairman Mao. The meeting also decided that the Secretariat of the CCP Central Committee should draft a circular to correct and criticize the errors of the February Outline.[15] The content of the February Outline was initially approved by Liu Shaoqi, Zhou Enlai, and Deng Xiaoping at the Standing Committee of the Political Bureau of the CCP Central Committee held on February 5, 1966.[16]

March and April 1966 were essential periods in the political struggle of the central leadership. During this crucial period, Liu Shaoqi went abroad for a visit, while Lu Dingyi inspected local areas to avoid the investigation of an anonymous letter concerning his wife, Yan Weibing. In such circumstances, the criticism from top leaders like Deng Xiaoping and Zhou Enlai dealt a fatal blow to Peng Zhen. As a member of the Central Political Bureau and its alternate members, Zhou Enlai enjoyed high political prestige. His decisions greatly influenced the centrist faction, which was neither Maoist nor Liu Shaoqi supporters.

To defend himself, Peng Zhen, who was then isolated in the political struggle of the central leadership, launched a criticism in the *Beijing Daily* on April 16, 1966, of Wu Han, Deng Tuo, Liao Mosha, and their writings, *the Three-Family Villages,* and *Yanshan Night Talk*. However, his turn to abandon his subordinates for self-preservation was ineffective and criticized by the opposition as a "pseudo-criticism and real cover-up."On May 8, the *Liberation Army Daily* published an article by Gao Ju (Jiang Qing's pseudonym), "Firing at the Anti-Party and Anti-Socialist Black Line." On May 10, Yao Wenyuan also published a review of *Three-Family Villages* in the *Liberation Army Daily* and *Wen Hui bao* in Shanghai. During the nationwide political campaign to criticize *Three-Family Villages* in May, key members of the Beijing Municipal Committee were accused of being "anti-party and anti-socialist black lines." They became the first targets of the Cultural Revolution.

1.2 The May 16 Notification and the Red Guard Organization

One of the key documents of the Cultural Revolution, the "Notification of the Central Committee of the Communist Party of China," or the "May 16 Notification," adopted at the enlarged meeting of the Central Committee of the CCP held in May 1966, stipulated the tasks facing the Cultural Revolution, stating: "The entire Party must follow the instructions of Comrade Mao Zedong, hold high the banner of the proletarian cultural revolution, thoroughly expose the reactionary bourgeois stand of those so-called 'academic authorities' who appose the party and socialism, thoroughly criticize and repudiate the reactionary bourgeois ideas in the academic, educational, journalistic, literary and publishing circles, and seize the leadership in these cultural fields. To achieve this, it is necessary to simultaneously criticize those representatives of the bourgeoisie who have sneaked into the party, the government, the army, and various spheres of culture, to purge them and, in some cases, to transfer them from their posts."[17]

The May 16 Notification clearly shows that two categories of people, bourgeois academic authorities and those in power who support them, were the targets of criticism during the Cultural Revolution. This may be considered similar to the "anti-rightist" movement. Still, the authorities in power were criticized more than the academic authorities, completely different from the anti-rightist movement. The designation of bourgeois academic authorities and the ruling factions that shelter them as targets of criticism implies two points: first, a segment of intellectuals and party cadres will become targets of the movement because they are designated as such; second, for various social groups, making trouble for party organizations at all levels is usually considered a counter-revolutionary act, but in the Cultural Revolution, they are called upon to make trouble, and making trouble becomes a legitimate revolutionary act. In other words, some social groups were empowered as the main body of the movement, and they had the right to criticize those party cadres and organizations considered revisionist. This was precisely the objective of the Cultural Revolution.

Although the definition of the movement's subject in the May 16 Notification was somewhat ambiguous, its scope was limited to some extent. There was no explicit statement in the May 16 Notification about which social groups could be the subject of the movement except for the abstract concept of "the proletarian left." Perhaps Mao did not have a clear idea before the emergence of the Red Guards. However, shortly after the publication of the May 16 Notification, the cultural and educational fields became at the forefront of criticizing party organizations, and young people who were still in high school at the time quickly stepped onto the political stage as the main force of the movement, as exemplified by the big-character posters at Beijing University.

Why did the Cultural Revolution begin in the field of education in the first place? In other words, why was it not the peasants and workers but a section of intellectuals and secondary school students who first rose up in the pre-Cultural Revolution hierarchy? There are two main reasons for this.

First, most people understand the Cultural Revolution as a movement to criticize the bourgeois academic authorities in the field of culture and education.[18] The Cultural Revolution, which had been underway since November 1965, when Yao Wenyuan published his "Comment on the new historical play Hai Rui Dismissed from Office" in Shanghai's *Wen Hui Bao*, was marked by a sharp confrontation at the top of the central government over whether to confine the movement to an academic debate or to elevate it to a political struggle. However, since the criticism of Wu Han and the *Three-Family Villages* was, after all, a matter for cultural and educational circles and the bourgeois academic authorities, one of the targets of the criticism stipulated in the May 16 Notification was likewise concentrated in cultural and educational circles, it is easy to understand that people in those circles were more concerned about the Cultural Revolution than workers and peasants. Moreover, at the policy level, the unfolding of the Cultural Revolution was also primarily confined to the literary and educational circles up to the summer of 1966.[19]

Second, there is a relationship with the political mobilization carried out by literary and educational circles. The publication of the May 16 Notification brought to the surface the political struggle brewing at the top of the central government. The mass mobilization mode of political participation provided opportunities for gathering political forces. Specifically, due to the large-scale personnel changes in the central leadership in May 1966, one faction of political forces sought support from society to defeat other political forces. In particular, the fledgling Central Cultural Revolution Group had to obtain outside assistance as it had no base in the central leadership other than Mao's support and Lin Biao's cooperation. Under these circumstances, Peking University was the first chosen by the Central Cultural Revolution Group.

After the publication of the May 16 Notification, Kang Sheng sent his wife, Cao Yiou, to Peking University to guide the direction of the movement. Before Cao Yiou's departure, Kang instructed her to "light the fire at Peking University and work upwards."[20] Lu Ping, the Party Secretary and President of Peking University, whom radical university members had criticized during the "Four Clean-ups" campaign in January 1965, had been supported by Peng Zhen. Thus, for those who had opposed Lu Ping, criticism of Peng Zhen meant there was a possibility of re-evaluating Lu Ping and the social education movement at Peking University.[21] In this context, on May 25, 1966, seven people, including Nie Yuanzi, a Peking University lecturer indirectly hinted at by Kang Sheng, posted a big-character poster at Peking University criticizing President Lu Ping and their leadership.[22] Afterward, Mao Zedong highly praised the big-character poster, elevated it to "the manifesto of the Paris Commune of China in the 1960s." Mao instructed that it be immediately broadcast throughout the country. The big-character poster was conveyed to the entire country on the night of June 1 through the Central People's Radio, and the next day the *People's Daily* also gave a glowing report.

The most striking aspect of Nie Yuanzi's big-character posters was that they directed criticism of the political campaign at the party organizations around them. In particular, Mao Zedong's high praise for this action, which was dramatically reported in the *People's Daily*, the Communist Party's organ, was tantamount to a political mobilization order from the highest leadership level in China. It also was equivalent to announcing that attacks on party branches or high-level Party officials at the grassroots level encouraged Revolutionary action. The fact that the Beijing Municipal Committee was reorganized on June 4 further reinforced this trend. In this wider context, schools across the country were plunged into chaos from the moment the big-character poster was broadcast on the radio. The criticizing party cadres and teachers in schools quickly became a frequent occurrence on several occurred frequently campuses.

On June 2, the day after Nie Yuanzi's big-character poster was broadcast nationwide, a big-character poster with the headline "Carrying the Great Proletarian Cultural Revolution to the End" was posted at the High School Affiliated with Tsinghua University (Qinghua Fuzhong). The big-character poster was signed by the "Red Guards." Since then, the Red Guards have become widely known in Beijing's high schools. In less than a week, schools such as the Middle School Affiliated with the Beijing Geosciences Institute (Beijing Dizhi Xueyuan Fuzhong), the Middle School Affiliated with the Beijing Petroleum Institute (Beijing Shiyou Xueyuan Fuzhong), the Middle School Affiliated with the Beijing Mining and Technology Institute (Beijing Kuangye Xueyuan Fuzhong), the Middle School Affiliated with Peking University (Beida Fuzhong), and the Beijing No. 25 Middle School (Beijing 25 zhong) also set up student organizations with the names like "Red Guards," "Red Flag," and "Dongfeng."[23] According to the recollection of Bu Dahua, one of the founders of the Red Guards, on May 29, 1966, the cadre children of the secondary school affiliated with Tsinghua University agreed to write big-character posters in the future, all signed as the "Red Guards." This date has since been recognized as the founding date of the Red Guards.[24]

As mentioned earlier, the inducing factors contained in the May 16 Notification, the social effect of the big-character posters at Peking University, and, more importantly, Mao's support for the big-character posters at Peking University all played significant roles as external factors in the formation of the Red Guards by young people in opposition to the school leadership. In this sense, the emergence of the Red Guards may result from top-down mobilization. However, it is essential to emphasize that young people were already divided into groups before the Red Guards were formed. With this division, the embryo of political factions had given birth. In other words, the Red Guards were formed by politicizing factional interests among young people.

2 The Politicization of Competing Interests Among Young People

2.1 Formation of Factions

As will be discussed in detail later, the formation of the Red Guards is closely related to the formation of factions among young people, but it should be noted that the two did not arise simultaneously. The Red

Guards were formed after the publication of the May 16 Notification. They were a product of the Cultural Revolution, while factions existed before the Cultural Revolution and resulted from the state's hierarchical distribution of citizenship.

In the case of Tsinghua University High School, the birthplace of the Red Guard movement, for example, factional rivalry among young people had already developed before the Red Guard was formed. The initial factional rivalry occurred between the cadre children, led by Bu Dahua, Wang Ming, Luo Xiaohai, and Kuang Taosheng, and the middle-class children (in between Red and Black family backgrounds, such as Capitalist, Petty bourgeois), led by Zheng Guangzhao (now known as Zheng Yi, from a capitalist family background) and Zhong Weiguang (who claimed to be from a bad family background with details unknown).[25] The two disagreed on the opportunities for advancement in education and membership in the league, and there was constant conflict from 1964 onwards. According to the memoirs of Zhong Weiguang, Zheng Yi, and Zhang Chengzhi, such conflicts occurred five times before the Red Guards emerged.[26] Zheng Yi also recalled this period in history, saying that the factional structure formed before the Cultural Revolution was carried over into the Cultural Revolution.[27]

From the author's research on the Red Guards organization at Tsinghua University Affiliated Middle School and the recollections of eyewitnesses mentioned earlier, it can be seen that the direct cause of the factional differentiation and opposition between the children of cadres and the middle class before the Cultural Revolution was a competition over opportunities for upward mobility. However, it is necessary to explain why competition for upward mobility would lead to factional opposition. This question needs to be understood in the context of the emphasis on class struggle that began in 1962. As explained in earlier chapters, in the 1960s, opportunities for upward mobility were very scarce for young people. This was true not only in rural areas but also in cities. The state completely controlled scarce upward mobility resources, and the standards for their distribution were highly politicized. In this situation, the competition for benefits that came with upward mobility was easily combined with ideological competition, making the competition highly politicized. Competitors were also very likely to elevate their interests to the level of pursuing the lofty goals of the state, making their pursuit of personal interests noble and legitimate due to overlapping with national goals. In a specific political context, even the competitors might confuse their pursuit of personal interests with the state's goals, leading to a sense of sublime nobility.

In this context, the emphasis on class origin among the three criteria for promotion to higher education, namely grades, class origin, and political performance, has intensified competition among young people, especially between the cadre children and those from middle-class family backgrounds with intellectual orientation, resulting in an escalation of class tensions. The middle-class children resent being labeled as "white-collar students"[28] and that the cadre children are given priority in the entrance examinations because of their class origin. On the other hand, cadre children found it difficult to compete with middle-class children, who often had parents who were elites before the founding of the PRC, had good educational backgrounds, and better conditions for the transmission of knowledge than the new elites of cadres' families. Under such circumstances, the cadre children felt that using academic performance as a criterion for admission was preferential to the middle-class children. They considered it a severe violation of class policy established after the country's founding. Based on differences in perceptions of equality, this opposing structure among young people was carried over into the Cultural Revolution and directly influenced the contentious and divisive structure within the Red Guard organization and its behavioral characteristics.[29]

The above situation is not unique to the high schools affiliated with Tsinghua University but also exists in other Beijing high schools where both the cadre and middle-class children were concentrated. One example is the "4, 6, 8" student incident in June 1965 in Beijing's no. 4, no. 6, and no. 8 middle schools.[30] In addition, a joint study by Anita Chan, Stanley Rosen, and Jonathan Anger on the Guangzhou Red Guards and a questionnaire survey by Rosen on the Guangzhou Red Guards show nearly the same results.[31] As mentioned earlier, the initial factional structure in the Tsinghua University-affiliated high schools arose before the Cultural Revolution and existed between some middle-class and cadre children. In March 1966, as the political environment changed, the competition for interests among young people gradually became politicized. This politicization of interest competition resulted in the involvement of previously uninvolved parties and the formation of new factions, further complicating the factional structure.

Specifically, beginning in mid-October 1965, a critical discussion took place in the secondary school affiliated with Tsinghua University due to the controversy over Yao Wenyuan's article, "Comment on the new historical play Hai Rui Dismissed from Office." Then the confrontation between the central leadership over whether it was an academic debate or a political struggle was also reflected in the school, bringing about a conflict between the cadre children and the party branch of the school. The cadre children advocated the elimination of the "black market" behind the "three-family villages" through criticism of the "three-family villages." According to the cadre children, the so-called "black market" refers to the children of the bourgeoisie and intellectuals with bourgeois habits who had a strong resonance with the three-family villages and were depicted as its social base. Therefore, they objected to the party branch limiting the discussion to academic arguments, accusing it of not highlighting politics and of favoring the children of the bourgeoisie.[32]

In late March 1966, the February Outline, which advocated academic arguments, was criticized by Mao as being utterly wrong in content. At the same time, the position of the February Summary was adopted by the Central Military Commission. In such a political environment, the arguments of the cadre children became even sharper. The *Liberation Army Daily*[33] and the Little Red Book, *Quotations from Chairman Mao*, which was difficult for the average student to get hold of, became the weapons with which the cadre children won the argument. They were either influenced by the editorials in the *Liberation Army Daily* or learned about the situation within the central leadership through their parents. Thus, they aimed their criticism at the school's party branch. The criticism of the Party branch by the cadre children led the progressive students, centered on student cadres and members of the Communist Youth League, to rise up against the cadre children to protect the party branch. As a result, the progressive student faction, centered on student cadres/Communist Youth League members, became entangled with the factional structure centered on the cadre and middle-class children.

From late April to late May 1966, the cadre children directed their criticism at the party branch of the school among the three factions of the Tsinghua University Affiliated Middle School. While the middle-class children and progressive students did not ally with them at that time, they nevertheless stood on the same side as the party branch. The main reason for their protection of the party branch was a judgment about the situation rather than political identity. They lacked a reliable channel to obtain internal information. Therefore, they likely acted based on the traditional experience that is in line with the party branch was safe. This explains why, after June 1966, the strongest anger toward the school's party cadres came not from the Red Guards composed of party cadre children but rather from the young people who had just been supporting the party branch.[34] The only way to remove the political disadvantages of the previous miscarriage of justice was to show the party's "revolutionary aspirations" through intense anger. This is a concrete expression of the psychological structure that transforms loyalty into interest, as discussed in Chapter 2. Hu Ping points out that people advocated a particular political view during the Cultural Revolution, not for their purposes in most cases but to claim that this view could prove their revolutionary nature and show that they were more revolutionary than others.[35] This judgment is not only applicable to the Cultural Revolution. Still, it is also crucial for understanding the behavior of Chinese people in the highly politicized society of the Mao Zedong era.

One could say that the pre-existing factional structure before the outbreak of the Cultural Revolution was carried over directly into the movement.[36] As the Cultural Revolution progressed, it developed into three main factions: the "Old Red Guards," composed mostly of cadre children, the "Conservative Red Guards," composed mainly of progressive students; and the "Rebel Red Guards," composed primarily of middle-class students. This factional structure became the core of the division and differentiation within the Red Guards during the Cultural Revolution.

2.2 *Types of Politicized Factional Behavior*

Although the Red Guards at the beginning of the Cultural Revolution considered themselves red five most of their members were children of the red three (revolutionary cadres, revolutionary soldiers, and revolutionary martyrs).[37] Why, then, were the cadre children, rather than the middle-class children and the progressive students, who were mainly members of the Communist Youth League, the first to rise up in rebellion against the rival factions that had been formed before the outbreak of the Cultural Revolution? In fact, in the rapidly changing political situation at the time, all students were striving to participate in political movements to become advanced elements. However, their pre-Cultural Revolution social status influenced and constrained their actions during the early stage of the Cultural Revolution, resulting in different behaviors when participating in politics. The actions of the cadre children were most in line with Mao's philosophy of the Cultural Revolution and therefore received Mao's support. Considering the political environment at the time, the only people who could act according to Mao's expectations were the children of cadres. Several main reasons for this are explained below.

First, in a socialist country, the proletariat is the master of the state, and, except for the special period of the Hundred Flowers Campaign, it has been the movement's main body since the country's founding. Therefore, in an environment where class struggle had been emphasized since 1963, only those who could confront the school leadership without any class concern were those of red five class origin. Moreover, from the ideological point of view of the Cultural Revolution, it is unlikely that those who first rose up in rebellion were from outside the red five class. This is because the uprising was positioned to prevent revisionism during the launch and initial stage of the Cultural Revolution. This did not mean rebelling against authority and institutions but against the revisionists who had broken away from the orthodox Marxist line. As such, the only group qualified to accuse others of unorthodoxy were those from the red five class because orthodoxy is a prerequisite to raising an accusation of revisionism. From another angle, the educational reforms advocated in secondary schools before the outbreak of the Cultural Revolution, which emphasized class struggle and education along class lines, placed middle-class and black five children at an even more significant disadvantage in furthering their education. It would not make sense for such students to cooperate actively.

After the Peking University posters were broadcast nationwide, young people all over the country began to realize that attacking the party organization at school with posters was the right thing to do in this political action. As a result, young people of middle-class origin also began to actively criticize the party organization. For example, the middle-class students of Tsinghua University High School posted big-character posters criticizing the party organization at their school on the night of the national broadcast, one day before the cadre children. But there was a big difference between the two in terms of content. The cadre children criticized the party branch of the school for failing to carry out the class line and for implementing the bourgeois educational line. The middle-class children, however, were addressing non-political issues such as hygiene, waste of water and electricity, and relations with the surrounding peasants.[38] The critique of middle-class children was not in line with the political direction of the time and, therefore, could hardly have been chosen as the main body of the movement, no matter how aggressive the action was. In his memoirs, Zhong Weiguang, then a student at the Tsinghua University High School, explained the criticism launched by the middle-class children against the Party on June 1, saying, "in fact, no matter what view we held, June 1966 was bound to fail, it was doomed. Because the political situation at that time was to emphasize class struggle and the class line and to promote the dictatorship of the proletariat."[39] The Great Proletarian Cultural Revolution was a movement with the proletariat as the main body. Those of non-red five class origins were excluded from the beginning.

Second, it can be said that the rebellion of the cadre children resulted from Mao's call. In fact, since 1964, Mao had frequently been calling on the cadre children to revolt, and in 1964 there was a talk between Mao, Mao Yuanxin, and Wang Hairong, in which Mao said, "Class struggle is one of your main objectives" and "you must revolt against the existing educational structure...The cadre children should be the successors of the revolution and should have a high sense of responsibility and mission."[40] In a conversation with Wang Hairong, he instructed that the school system should be overturned, saying, "Schools must be allowed to revolt. You should go back and lead the rebellion."[41] These talks greatly encouraged the cadre children and made them realize two things. First, revolt against the current educational system was revolutionary; second, cadre children's special status entailed a "sense of mission and responsibility" as "successors of the revolution."[42] Against this background, there were already rebellions among the cadre children from 1965 onwards. In Beijing, the children of the red five, mainly the cadre children, in the no. 4, no. 6, and no. 8 middle schools staged a "strike" to protest against the school's guidelines. They demanded to leave school and join the Four Clean-ups campaign in the countryside. This 4, 6, 8, Student Incident soon spread to the Girls' High School of Beijing Normal University, the First Affiliated High School of Beijing Normal University, and the Affiliated High School of Peking University.[43] In March 1965, a group of students from the red five class, mainly cadre children, demanded the abolition of university entrance examinations in Shanghai.[44] Moreover, in an article entitled *Long Live the Spirit of Rebellion*, Luo Xiaohai of the Tsinghua University Affiliated Middle School in Beijing on October 28, 1965, clearly expressed his opposition to education. "The existing educational regulations are a great constraint on learning, and a rebellion on this point alone is extremely necessary."[45] In the words of Zhang

Chengzhi, one of the founders of the Red Guards, "From Luo Xiaohai's Rebellion to the Red Guards' *Three Essays on the Rebellion Spirit*, the direct source comes from Mao Zedong's conversation with his niece Wang Hairong in the 1960s.".[46]

Unlike the cadre children, the progressive students, composed of student cadres and members of the Communist Youth League, although also of red five class origin, opposed the cadre children. When the cadre children criticized the party branch, they chose to take the side of the party branch against the cadre children, and in the process of defending the authority of the party branch, they began to form factions. As mentioned earlier, the behavior of the student cadres was not entirely based on political identification but more likely on previous political habits and experience and perhaps also on a cautious attitude of not daring to take any risks. Experience since the country's founding shows that when a political action occurs, choosing to be on the same political side as the party cadres and actively cooperating with their efforts is more likely to be rewarded. As successive political campaigns have demonstrated, for each participant, political campaigns are an important opportunity to gain political resources and create an arena for activists. For such young people, who have already been elected as student cadres and members of the Communist Youth League, and who have an advantage in the upward path, the choice to align themselves with the party organization is undoubtedly the safest, as seen by the experience of previous campaigns.

On the surface, the behavior of the progressive students and the cadre children before and at the beginning of the Cultural Revolution appears opposed. Still, the two can be identical regarding the political resources they appropriated through active participation in the movements. It is only that the pre-Cultural Revolution movements were conducted under the party's leadership. In contrast, the Cultural Revolution was an attempt by the supreme leader Mao Zedong to reshape the bureaucracy through the "Great Democracy" method by directly delegating to the masses the right to supervise party cadres and party organs. The consistency between the party's and Mao Zedong's will was lost. Thus the actions of the progressive students who were used to relying on the party organization could not keep up with the needs of the situation and were considered conservative. On the contrary, the actions of the cadre children, who were supposed to be punished for rebelling against the party organization, became the protagonists of the first movement of the Cultural Revolution because they were in line with Mao's ideology and provided him with a method of political mobilization from the bottom up, which he then enthusiastically supported.

2.3 *The Struggle of the Two Commands and the Three Essays on the Rebellion Spirit*

As the above analysis shows, the rise of the Red Guard organization, centered on the cadre children, was no accident when considered in light of the political situation of the time. Likewise, the confrontation between the Red Guards and the Work Team was arguably unavoidable in the political circumstance of the time. This was mainly because the two wanted to achieve opposing goals. The Red Guards wanted the Work Team to affirm the correctness of their actions and recognize the legitimacy of the Red Guard organization. On the contrary, Liu Shaoqi's instructions to the Work Team were to "Digest the Red Guards and soften them."[47] That is to say: while recognizing the leftist status of the Red Guards, they were to be gradually absorbed into the Communist Youth League organization. With conflicting goals, it is only natural that antagonism between the two should arise. In addition to this, and more importantly, the two acted on the instructions of two opposing command centers.

Liu Shaoqi decided to dispatch Work Teams. After the big-character posters at Peking University were reported in the media, secondary schools and universities everywhere were thrown into pandemonium, with frequent criticism of school party cadres and teachers. Soon, some school party organizations were paralyzed and unable to conduct regular classes. In addition, with the reorganization of the Beijing Municipal Committee, there were incidents in all schools in which there was distrust in carrying out instructions of the old Beijing Municipal Committee. Schools demanded that the new Beijing Municipal Committee send a Work team.[48] In such a political environment, Liu Shaoqi dispatched a Work Team after asking Mao Zedong for instructions.[49]

The Work Teams had two tasks. The first was implementing the May 16th Notification to expose and criticize the revisionists and bourgeois academic authorities within the government agencies and schools. The second was to restore the functions of the party organizations and place the student movement under the party's leadership.[50] Controlling the direction of the Cultural Revolution by harnessing the student movement was the main purpose of sending the Work Team. The Work Team entered the school in the High School affiliated with Tsinghua University. It immediately announced that the party branch had made a significant mistake and organized the students to criticize it.[51] The stubborn Red Guards strongly opposed their organization's placement under the party's leadership. From June to early July, there were frequent incidents of the expulsion of Work Teams from universities and high schools across the country.[52]

Following the confrontation with school leadership, the "anti-Work Team" activities became the second wave of conflict among the Red Guards. It is worth noting that the Red Guards did not initially oppose the Work Team. When the Work Team was first stationed at the Tsinghua University Affiliated Middle School, it maintained good relations with the Red Guards. The Work Team recognized the members of the Red Guards as activists who criticized the "revisionist line" and spoke highly of their "enthusiasm for the revolution." However, this good relationship did not last long. Soon, differences of opinion between the Red Guards and the Work Team over the "Cultural Revolution guiding principles" became public.

The Red Guards felt that the Work Team's approach placed too much emphasis on policy and too much on unity at the expense of struggle.[53] On the other hand, as the Red Guards acted without authority in defiance of the Work Team's command, the Work Team concluded: "In the movement, the left-wing organizations established by students who left the party and league organizations to work independently played a positive role in the early stages of the movement, but in the long run, the consequences were not good and were not conducive to the proletarian dictatorship and the party's absolute leadership." Following Liu Shaoqi's instructions to "digest the Red Guards and soften them" and "gradually dissolve them into the league organization," the Work Team asked the Red Guards to disband their organization, give up their radical slogans and actions and participate in the Cultural Revolution with other students in their classes and league organizations.[54] But some Red Guards claimed, "our leader is the Party Central Committee and Chairman Mao,"[55] obstinately insisting on their organization. As a result, there was constant friction between the Work Teams and the Red Guards in many schools, deepening the confrontation between the two. However, the situation in the Tsinghua University Affiliated Middle

School was slightly different. The relationship between the Red Guards and the Work Team did not develop into extreme antagonism. Of course, examples like Tsinghua University Affiliated Middle School were very few, and most Red Guard-like student organizations in other schools were declared "counter-revolutionary organizations" by the Work Teams and were banned.[56]

The main reason for this situation can be the result of the confusion caused by the simultaneous transmission of instructions from the two Party Central Committees. Specifically, the Work Team emphasized party leadership and stability of order. However, after Chen Boda led the Work Team into the *People's Daily* on May 31, 1966, the newspaper published a series of inflammatory editorials calling for the "complete destruction of 'black gangs,' 'black organizations,' and 'black discipline.'"[57] After the Work Team entered the schools, some young people within the Red Guards believed that the Work Team represented the will of the Party Central Committee and began to act by its instructions.[58] But a few quoted the *People's Daily* article to debate with the Work Team on the legitimacy and survival of the Red Guard organization. Later, the nationally renowned *Three Essays on the Rebellion Spirits* by the Red Guards at the Tsinghua University Affiliated Middle School resulted from the confrontation between the Work Team and the Red Guards in the process of "control and counter-control."

The Red Guard's actions were criticized by the Work Team for breaking away from the party leadership but received "ardent support" from Mao Zedong.[59] In late July 1966, Mao and the Central Cultural Revolution Group accused the Work Team of suppressing the masses while expressing their support for the Red Guards' actions. At the 11th Plenum of the 8th Central Committee of the National Congress of the Chinese Communist Party (hereinafter, the 11th Plenum of the 8th Central Committee of the CCP), the Red Guards and their actions were affirmed as "revolutionary youth," "breakthroughs," and "the general direction of the revolution is correct." The Red Guards thus replaced the Communist Youth League as the most important student organization in schools and became the leading force in the political action of the Cultural Revolution. Mao Zedong's meeting with the Red Guards in Tiananmen Square on August 18 further raised the status of the Red Guards in the Cultural Revolution. It gave the impression that the Red Guards had been able to influence the political center of China.[60]

So, why did Mao Zedong support the Red Guard's rebellion? The main reasons for this can be summarized in the following two points. First, the Red Guards' actions aligned with Mao's idea of launching the Cultural Revolution from the bottom up to bring governance to the world through "chaos under heaven." As mentioned, Mao Zedong had frequently called on the cadre children to revolt since 1964. Before the outbreak of the Cultural Revolution, Mao's mobilization orders for rebellion became even more explicit. On March 18, 1966, on the eve of the Cultural Revolution, Mao Zedong openly encouraged an attack on the central government: "Whenever the central government does something bad, I will call for local rebellions and attacks on the central government. More Sun Wukongs (Monkey King)should come out everywhere to create chaos."[61] The Red Guards, who declared that "the rebellion is justified" and "the proletariat will make havoc in the Palace of Heaven and create a new world for the proletariat,"[62] were exactly the "Sun Wukongs" that Mao was expecting.

Second, Mao's support of the rebellion was motivated by the internal political struggle of the Party Central Committee. As shown in the May 16 Notification, Mao Zedong believed that revisionists had appeared in the Party Central Committee before the Cultural Revolution. In the Circular, Mao criticized the representatives of the bourgeoisie and asked party organizations at all levels to pay full attention to "the bourgeois characters sleeping around us." At the enlarged meeting of the Standing Committee of the Political Bureau on May 23, it was decided to stop Peng Zhen, Luo Ruiqing, and Lu Dingyi as Secretaries of the Central Secretariat and Yang Shangkun as Alternate Secretary of the Central Secretariat. Removed from their posts were Peng Zhen as First Secretary and Mayor of the Beijing Municipal Party Committee and Yang Shangkun as the Minister of the Propaganda Department of the Central Committee of the Communist Party of China.[63] This chain of events shows that exposing the revisionists within the leadership of the CCP Central Committee was the focus of this political campaign. Yet Liu Shaoqi changed the spearhead of the political movement through the Work Team he dispatched. According to Harding's research, Liu Shaoqi explicitly instructed the Work Team that many bourgeois spokesmen had clandestinely entered the government and party organizations of universities and high schools. They should be criticized and, if necessary, removed from their posts. Under such instructions, 90 percent of the cadres of the Ministry of Finance were attacked, and most of the cadres of the Ministry of Culture were removed from their posts.[64] As a result, the campaign's target to expose the revisionists shifted from the central leadership to the lower party and government agencies. In the month following the dispatch of the Work Team, many party cadres and intellectuals in the party and government organs at the central ministerial level, local party and government organs at all levels, and educational organs such as universities and high schools were branded as "bourgeois agents" and "reactionary academic authorities." People were also politically induced by the Work Team to launch a fierce criticism of the "reactionary figures" within their units. On the surface, the May 16 Notification had been thoroughly implemented under Liu Shaoqi's leadership. Still, there was a discrepancy between this and Mao's campaign to expose the revisionists within the central leadership.

In addition, another task of the Work Team, to lead an "orderly" Cultural Revolution by harnessing the student movement, also deviated from Mao's "chaos to achieve governance." Specifically, the Work Team classified students into "revolutionary leftists who could be trusted," "centrists with low political consciousness," and "counter-revolutionary rightists," according to their attitude toward the group, and the "left-wing" students were mobilized to struggle against rightist students. This further intensified the contradictions within the student body.[65] The result was not the intended goal of mobilizing the masses to expose the party's revisionists but rather a movement of denunciation among the people. Against this background, in late July 1966, the Central Cultural Revolution Group criticized the Work Team for inciting student-on-student violence and committing a major error.[66] To affirm the Red Guards' actions meant negating the Work Team. Since the Work Team was dispatched on Liu Shaoqi's orders, its removal eliminated obstacles to the bottom-up launch of the Cultural Revolution and prepared the theoretical groundwork for future criticism of Liu Shaoqi, making it a dual-effective move.

Against the above background, the 11th Plenum of the 8th Central Committee of the CCP was held in Beijing on August 1. At this meeting, Mao Zedong explicitly expressed his support for the actions of the Red Guards. He distributed the big-character poster *Bombard the Headquarters* as the reading material for the meeting. Although not named, the big-character poster criticized the Central Committee's first line, headed by Liu Shaoqi. It characterized the confrontation of views within the Party Central Committee as a struggle between two lines.[67] Compared to the

May 16 Notice, Mao's big-character poster targeted the campaign directly at the "command" of the Party Central Committee.

3 Acts of Rebellion by the "Old Red Guards"

3.1 The Object of Criticism

After August 1966, the Red Guards gained legitimacy with Mao's support. They overnight replaced the Communist Youth League as the most highly regarded student organization in schools, significantly increasing their political status in the Cultural Revolution. After Mao received the Red Guards in Tiananmen Square on August 8, the Red Guard organization expanded nationwide. Most conservative Red Guard organizations centered on Progressive students were founded during this period. Conservative Red Guard organizations in secondary schools are not the focus of this book and are omitted here.

August and September 1966 were the heydays of the Old Red Guards. Their rebellious activities during this period were mainly to "Smash the Four Olds" (old ideas, culture, customs, and habits),"[68] conduct "the Great Link-Up" (exchange of revolutionary experiences, *da chuanlian*), criticize "black gangs," and sweep away "monsters and freaks." They also advocated educational reform and the "birth bloodline theory." These actions were instrumental in mobilizing the masses to expand the rebellion to the whole country. In particular, the direct involvement of the Beijing students in the "Southward exchange of revolutionary movement" in local rebellions contributed to local factionalization and led to the paralysis of local power authorities. Nonetheless, the Red Guards centered on the cadre children, failed to play the role of "Bombard the Headquarters" that Mao had hoped for because the goals of their rebellion differed markedly from Mao's objectives, i.e., the direction of their struggle was not within the party, but mostly outside it. Specifically, the Old Red Guards' actions of "breaking the Four Olds," "sweeping away all monsters and freaks," and criticizing the revisionist education line aimed at people outside the red five class and those related to them. Although they also took part in activities to criticize the "black gangs" of the old Beijing Municipal Committee, the Propaganda Department of the CCP Central Committee, and the cultural and educational circles, generally speaking, that was not the focus of their rebellious activities. The downfall of the Beijing Municipal Committee's main leadership shows that almost all the "black gangs" were not exposed by the Red Guards but were criticized from within and ejected by the party.[69]

Undeniably, the behavior of the "Old Red Guards," criticized by those of non-red five class origins, was related to the ideology of class struggle and the political mobilization at the beginning of the Cultural Revolution. But more importantly, their position in the hierarchy before the Cultural Revolution and interest in seeking promotion determined their behavioral characteristics. This was particularly evident in the birth bloodline theory.

3.2 Behavioral Theory—Birth and Bloodline Descent Theory

In the rebellion of the cadre children, the most notable thing was the promotion of the "birth and bloodline theory." In June 1966, the couplet "If the father is a revolutionary hero, the son also is a hero; if the father is a reactionary, the son is a bastard" appeared in the high school affiliated with Peking University and soon became widely popular in society. Red Guards from cadre families used the couplets to emphasize "lineageism" for the following four purposes.

First, political privilege is sought based on bloodline. As the saying goes, "If the fathers have made the country, the children must pass it on from generation to generation,"[70] the bloodline theory centers on power transmission. In other words, the children of revolutionary cadres, especially senior cadres, do not need to study hard, accumulate knowledge and experience, and compete fairly to inherit social power and wealth. For the reason that "who fought to win the country? It was our 'red five' fathers and brothers!"[71] Considering the power structure in China, the demand for hereditary power is of great importance. In Mao-era China, gaining political power was the only way to advance. Land and factories were state-owned, and knowledge, while necessary, was not an important way to advance in power. Moreover, in the ideology of socialist China, knowledge is also class-based. In this discourse, "lineageism" in the couplet meant that the cadre children could achieve promotion by their birthright. They sought not the heredity of personal and family property but the heredity of power, social status, and rank based on "class" as a group.

Second, middle-class children were excluded from the path of promotion. Although the most important indicator of class policy in Mao-era China was class division, it was not the only criterion, and political performance was also highly valued. The emphasis on political performance meant that for middle-class children, even if they were not well-born, there was still hope for promotion if they were willing to work hard. However, the rise of middle-class children, whose intellectual qualifications were superior, inevitably threatened the cadre children.[72] In this context, adding a bloodline element to class policy meant that the cadre children monopolized promotion.

Third, they monopolized the right to participate in the Cultural Revolution. The Red Guards, given legitimacy by Mao, became the most revolutionary youth organization of the early Cultural Revolution. For the youth, joining the Red Guards meant gaining the most important political resource. Access to this political resource determined political status and position. Therefore, the cadre children tried desperately to monopolize the right to participate in the Cultural Revolution. For example, to prevent their actions from being imitated by other organizations, in early August 1966, they concealed Mao's Letter to the Red Guards of Tsinghua Middle School from the outside world, which they had obtained from unofficial sources and which had not yet been made public. This meant that they concealed the news that Mao supported the Red Guards. In addition, the Regulations on the Organization of the Red Guards state: "Members of the Red Guards must be progressive members of the red five, and those whose parents, relatives, or friends are backward elements of the red five may not join the Red Guards."[73]

Fourth, between late July and August 1966, the popularity of the birth bloodline theory was closely linked to the competition within the school for leadership during the Cultural Revolution. On 25 July, Mao ordered the withdrawal of the Work Team. The retreat of the Work Team created a vacuum in the leadership of the Cultural Revolution in schools. In August of the same year, the 11th Plenum of the 8th Central Committee of the CCP adopted the Decision on the Great Proletarian Cultural Revolution (hereinafter referred to as the Sixteen Points), which defined the "Cultural Revolution Groups," "Cultural Revolution Committees," and "Cultural Revolution Congresses," as the power bodies of the Cultural Revolution. It stipulated that the representatives of these powerful bodies would be elected directly, as in the case of the Paris Commune. On this basis, it was instructed that the "Cultural Revolution Group," the "Cultural Revolution Committee," and the "Cultural Revolution Congress" in the schools would be composed mainly of revolutionary students, plus some revolutionary teachers.[74] However, the criteria for determining "revolutionary" were not clearly defined in the Sixteen Points. A series of events after

May 1966 showed that the original criteria of party members, league members, activists, and the general public were no longer applicable. The Red Guard organization, which was opposed to the school's party organization and Work Team and centered on the children of cadres, was supported by Mao Zedong. Although the Sixteen Points did not name the Red Guards, they were highly praised for their actions: "Despite all their shortcomings, their revolutionary orientation is correct."[75] At the same time, the Sixteen Points also state that the "minority" whom the Work Team suppressed "must also be protected, for the truth is sometimes in the hands of the minority."[76] Evaluating both simultaneously may have included Mao's intention to reconcile the contradictions between the young people of the Work Team period. Still, it confused the criteria for determining "revolution." In this context, based on the affirmation of class origin, the birth bloodline theory proposed a new criterion of revolution based on the parents' merits. In this respect, the doctrine of the birth bloodline was of great importance at the time. It touched on the core of citizenship distribution and caused widespread societal controversy. From another perspective, the debate between the birth bloodline theory and the family origin theory represented by Yu Luoke is not just a dispute over political claims. Behind these claims, it involves the distribution of citizenship closely related to every Chinese citizen. Therefore, the debate between the two has considerable depth and breadth, to the extent that their dispute overshadowed the primary goals of the Cultural Revolution.

In fact, between the retreat of the Work Team and September 1966, the leadership of the Cultural Revolution in almost all secondary schools and universities fell into the hands of the cadre children. This was mainly because from June to July, there were only Red Guard groups comprised of cadre children in most high schools. These groups had the support of Mao and, together with the Work Team's retreat, became the de facto leader of the schools. In this context, the birth bloodline theory provided cadre children a basis for legitimacy in taking the leadership of the Cultural Revolution.

The birth bloodline theory was arguably an early Red Guard proposal for dissatisfaction with and improving class policy. Still, it did not differ fundamentally from the class policy of the Communist Party. The class policy was difficult to balance between class divisions and political attitudes. As Yu Luoke, the author of *Class Origin Theory*, said, "'Looking at family background and performance' will inevitably slip into the quagmire of 'only looking at the family background, not performance.'"[77]

However, the initial Red Guards' claim of birth bloodline theory did not gain the support of the Central Committee of the CCP. There were two main reasons for this. First, as Mitsuyuki Kagami pointed out, the inner principles of the ideology of the Cultural Revolution contradicted the birth bloodline theory.[78] In other words, "the top cadres of the party, the 'biggest powers that be,' i.e., Liu Shaoqi, Deng Xiaoping, and Peng Zhen, were all revolutionary cadres and representatives of the red five. There was no leader of a superior pedigree than theirs. Despite this, they were accused by Mao Zedong, the highest authority in the party, of being capitalist roaders within the party and of being criminals in restoring capitalism in Red China. More importantly, unlike previous campaigns, the criticism of them was not only carried out within the party but also called on people outside the party, including the youth, to participate in the criticism. In other words, in terms of class origin, it meant that people outside the party in low positions could criticize the top cadres in high positions."[79] Second, the debate on birth bloodline theory intensified the factionalism among young people. Like the Work Team, the Red Guards' actions and claims targeted the movement outside the party rather than the central leadership.

Mao Zedong and the Central Cultural Revolution Group tried to steer the Red Guards, mainly the cadre children, toward criticizing the "capitalist roaders within the party."[80] When Jiang Qing, head of the Central Cultural Revolution Group, met with the Red Guards on August 6, 1966, she clarified: "What is the task of our Great Proletarian Cultural Revolution? First, to uncover capitalist roaders within the party, and second, to break old ideas, culture, customs, and habits."[81] However, for the Red Guards of the cadre children, the priority was to exclude the middle-class children, such as intellectuals and capitalists, and to establish a pure revolutionary class composed of the red five.[82] The Red Guards, the cadre children, used this as the aim of the Cultural Revolution. Their criticism was aimed at those outside of the party, which went against Mao Zedong's intention to overthrow the "capitalist roaders within the party" and hindered the progress of the Cultural Revolution. Therefore, starting from October 1966, they gradually lost their position as the main force on the political stage.

In fact, given their family origins, it was only natural that the Red Guards, who were the children of cadres, would not be able to undertake

Mao's task of "cannonballing the commanders." Since the core of the Old Red Guards was made up of the children of senior cadres, it was undoubtedly difficult for them to accept the idea that targeting the party for rebellion meant bringing down their parents.

3.3 Fragmentation and Disintegration

Both Lin Biao's National Day speech on October 1, 1966, and the editorial in *Red Flag* No. 13 on October 3 called on the people to "thoroughly criticize the reactionary bourgeois line." The spearhead of the Cultural Revolution was finally oriented toward the party and was eventually directed at the "capitalist roaders within the party." The criticism of the "bourgeois reactionary line" affected the parents of the Old Red Guards, and many became children of "black gangs" overnight. As a result, the Old Red Guards began to resist the criticism of the bourgeois reactionary line and gradually became suspicious of the goals of the Cultural Revolution. On the other hand, the Rebel Red Guards, who began to form after August 18, 1966, and who had gradually grown in size during the debate on the birth bloodline theory, actively participated in the criticism of the bourgeois reactionary line and received the support of the Central Cultural Revolution Group. Because of the support from the Central Cultural Revolution Group, the Rebel Red Guards expanded its size and power, replacing the Old Red Guards as the most revolutionary force.

After October 1966, the Old Red Guards were divided into three broad categories.[83]

First, following the incidents in October 1966, in addition to the "Old Red Guards" group, which was in the process of splitting up, there were also "conservative" groups centered on progressive students and "rebel" groups centered on middle-class children. Some of the Old Red Guards who left joined these two organizations.[84]

Second, some joined the breakaway movement of the so-called "dropouts" (*xiao yao pai,* namely, those who lost interest in collective action, distancing themselves from politics). These "dropouts" will be discussed in detail in Chapter 9.

Third, the Capital Red Guards Joint Action Committee (hereinafter referred to as Liandong) was formed to confront the Central Cultural Revolution Group. Liandong was the first organization in the Cultural Revolution to openly question Mao's philosophy and the actions of the

Central Cultural Revolution Group. Criticism of Mao and the Central Cultural Revolution Group appeared explicitly in Liandong's statements. For example, at a rally on December 20, 1966, Liandong chanted slogans such as "Down with the Central Cultural Revolution Group" and "Long live Liu Shaoqi." In Circular 003, issued on 1 January 1967, they also proposed "abolishing the dictatorship," "loyalty to Marxism and pre-1960 Mao Zedong Thought," and "protection of outstanding and loyal revolutionary cadres." Liandong was significant in openly questioning the Cultural Revolution and criticizing Mao's philosophy of the Cultural Revolution.

Song Yongyi summarizes Liandong's ideas in three points. First, it demanded a return to the Chinese political system before the Cultural Revolution; second, it demanded that the bureaucratic system established 17 years before the Cultural Revolution be maintained; and third, it opposed the critique of the birth bloodline theory.[85] In addition, Song Yongyi, by comparing it with such "ultra-leftist" heretical trends as the Shengwulian (the Great United Committee of Proletarian Revolutionaries of the Hunan Provincial Assembly), highly valued the Liandong trend as a heretical trend from the "right" that criticizes the Cultural Revolution.[86]

The ideology of Liandong had little impact on society at the time. There were two reasons for this. First, the Liandong critique of the Cultural Revolution was as strongly characterized by the pursuit of group interests as their involvement in the Cultural Revolution. Second, Liandong's demand for a return to the pre-Cultural Revolution Chinese political system was unattractive to most Chinese in early 1967, between late 1966 and 1967 when the urban underclasses expressed their views on the pre-Cultural Revolution policies. The period between late 1966 and 1967 was the height of a rebellion among the urban underclass to express their dissatisfaction with pre-Cultural Revolution policies and to pursue economic interests. Moreover, for young people who did not come from the red five, it was only during the trend of the movement to criticize the bourgeois reactionary line that they could obtain their "revolutionary rights." Under such circumstances, the demand of Liandong to stop the movement and return to the pre-Cultural Revolution system is a voice out of step with the trend of the times.

Like the Old Red Guards, Liandong was organized on the principle of birth bloodline, even more so than the Old Red Guards. This is fully reflected in its full name "The Joint Action Committee of the Revolutionary Children of the Communist Party Central Committee, the Beijing

Municipal Party Committee, the State Council, the Standing Committee of the National People's Congress, the Revolutionary Children of the Chinese People's Liberation Army's Generals, Commanders, and Cadres from Sixteen Provinces and Cities."[87] The emphasis on the birth bloodline and the sense of privilege isolated them from others. Therefore, due to the influence of the birth bloodline theory and their sense of privilege, their criticism of the Cultural Revolution was seen as an expression of frustration. The opposition rebel groups and the Central Cultural Revolution Group fiercely criticized Liandong. Its core members were eventually arrested in January 1967, leading to its decline.

Notes

1. Mi Hedu, *Hongweibing zhe yidai* (*The Generation of the Red Guards*) (Sanlian Bookstore, 1993), p. 180.
2. For the political conflict between Mao Zedong and Liu Shaoqi in the 1960s, consult the following studies: Roberick MacFarquhar, translated by Wang Xiaoge, *wenhuadageming de qiyuan, 1958–1960, di 2 juan* (Hebei People's Publishing House, 1991); the English version: Roderick MacFarquhar, *The Origins of the Cultural Revolution, Part 2: The Great Leap Forward, 1958–1960* (Cambridge University Press, 1983); Harry Harding, *Organizing China: The Problem of Bureaucracy, 1949–1976* (Stanford, CA: Stanford University Press, 1981); James. R. Townsend, translated by Kojima Tomoyuki, *Modern China: A Comparative Analysis of Political Systems* (Keio Correspondence, 1980); the English version: James. R. Townsend, *Politics in China* (Boston: Little Brown, 1974).
3. Stuart R. Schram, the Chinese version: "1949nian dao 1976nian de Mao Zedong sixiang", in Roberick MacFarquhar et al., eds., *The Cambridge History of the People's Republic of China*, volume 2, pp. 82–83. The English version: Roderick MacFarquhar, and John K. Fairbank eds., *Cambridge History of China*, volume 15: *The People's Republic, Part 2: Revolution Within the Chinese Revolution, 1966–1982* (Cambridge University Press, 1991).
4. James. R. Townsend, op. cit., p. 301.
5. Wang Nianyi, "Ping Wenhua dageming shinianshi" ("Review of the Ten-Year History of the 'Cultural Revolution'"), Zhang Hua and

Su Caiqing, eds., *Looking Back at the "Cultural Revolution,"* Part One (Chinese Communist Party History Press, 2000), p. 572.

6. Satoshi Amako, "The Chinese Proletarian Cultural Revolution", *Iwami Lectures on World History II.6* (Iwami Shoten, 1999), p. 189.
7. Id.
8. According to Ding Wang's classification, before August 1966, the 16 members of the Central Political Bureau, Mao Zedong, Liu Shaoqi, Zhou Enlai, Zhu De, Chen Yun, Lin Biao, Deng Xiaoping, Dong Biwu, Peng Zhen, Chen Yi, Li Fuchun, Liu Bocheng, He Long, Li Xiannian, Li Jingquan and Tan Zhenlin, and the six alternate members, Ulanwu, Zhang Wentian, Lu Dingyi, Chen Boda, Kang Sheng and Bo Yibo, can be roughly divided into four groups. The first group was the Radical faction led by Mao Zedong, with Lin Biao, Chen Boda and Kang Sheng as members. The second group was the Moderate faction led by Liu Shaoqi, with Deng Xiaoping, Peng Zhen, He Long, Li Jingquan, Tan Zhenlin, Ulanfu, Lu Dingyi, Bo Yibo, and Zhang Wentian as members. In addition to these two factions, there was also the State Council faction with Zhou Enlai, Chen Yi, Li Fuchun, Li Xiannian as members, and the faction of the old military cadres with Zhu De, Chen Yun, Dong Biwu and Liu Bocheng as members. Since there was no clear opposition between the State Council faction and the old military cadre faction, as well as between these two factions and the other factions, the State Council faction and the old military cadres faction could be called centrist factions. Ding Wang, *Zhonggong "wenge"yungongzhong de zuzhi yu renshi wenti: 1965–1970* (*Organizational and Personnel Issues in the CCP's "Cultural Revolution" Movement: 1965–1970)* (The Institute of Contemporary China Studies, 1970), pp. 2–4.
9. Satoshi Amako, "The Chinese Proletarian Cultural Revolution", op. cit., p. 189.
10. Harry Harding, "The Cultural Revolution: China in Chaos", Roberick MacFarqugar and John K. Fairbank eds., *Cambridge History of the People's Republic of China*, Part Two, pp. 121–129.
11. Satoshi Amako, "The Chinese Proletarian Cultural Revolution", pp. 189–190.
12. Zhongyang wenxian yanjiushi (The Documentary Research Office of the Central Committee of the Communist Party of China),

Guanyu jianguo yilai dangde ruogan lishiwenti de jueyi zhushiben (*Annotated Text of Resolutions on Several Historical Issues of the Party since the Founding of the Nation*) (People's Publishing House, 1985), p. 375.

13. *1965nian 9yue dao 1966nian 5yue wenhua zhanxianshang liangtiao luxian douzheng dashiji* (*A Chronology of the Struggle on the Two Roads on the Cultural Front from September 1965 to May 1966*), in The Party History, Party Construction and Political Work Teaching and Research Department of the National Defence University of the People's Liberation Army ed.,'Wenhuandageming' yanjiuzhiliao (Shang) *Research Materials on the Cultural Revolution (Part One)* (National Defence University of the People's Liberation Army Press, 1988), pp. 4–12.
14. For the specific actions of Peng Zhen between late 1965 and early 66 in organizing a writing group to resist in order to avoid politicizing the criticism of Wu Han, consult the following source: Li Yun, "Wo yu sanjiachun"("I and Three Family Village"), in *Zhongguogongchangdang dangshiziliao,zongdi6ji* (*General Sixth Series of Materials on Communist Party History)* (The Communist Party History Press, June 1988), pp. 58–81.
15. "1965nian 9yue dao 1966nian5yue wenhua zhangxinashang liangtiao daolu douzheng danshiji", ("A Chronology of the Struggle on the Two Roads on the Cultural Front from September 1965 to May 1966"), op. cit., *Research Materials on the Cultural Revolution (Part One)*, p. 11.
16. Chen Qingquan and others, *Ludingyi zhuan* (A Biography of *Lu Dingyi*) (The Chinese Communist Party History Press, 1999), p. 481.
17. "Zhonggong Zhongyang tongzhi, beijinghuagongxueyuan bian, wuchanjieji wenhuadageming huibina"("Notice of the Central Committee of the Communist Party of China, Compilation of the Great Proletarian Cultural Revolution"), (The Beijing Institute of Chemical Technology, ed., 1967), pp. 1–7.
18. Wang Shaoguang, *Failure of Charisma: The Cultural Revolution in Wuhan*, pp. 28–31.
19. "Zhonggongzhongyang guowuyuan guanyugongye jiaotong qiye he jibenjianshe danwei ruhe kaizhan wenhua dageming de tongzhi (1966nian7yue2ri)" ("Notice of the Central Committee of the Communist Party of China—State Council on How to Carry Out

the Cultural Revolution in Industrial and Transportation Enterprises and Capital Construction Units (2 July 1966)", *Research Materials on the "Cultural Revolution"* (Part One), p. 54.

20. Lin Haoji, "Beijing diyizhang dazibao shi zenyang chulongde" ("How the First Big-Character Poster Came Out of the Cage at Peking University (January 1981)"), *Research Materials on the "Cultural Revolution" (Part One)*, p. 33.
21. Yin Hongbiao, "Wengezhongde 'diyizhang maliezhuyi dazibao'" ("The First Big Character Poster of Marxist-Leninist Fact Posters During the Cultural Revolution"), in Lin Qingfeng et al., eds., *The Cultural Revolution: Facts and Analysis*, pp. 3–14.
22. Masashi Ando and others, *The Cultural Revolution and China* (Iwami New Books, 1986), p. 44.
23. Jiang Pei, *Hongweibing kuangbiao* (*Red Guard Rampage*) (Henan People's Publishing House, 1994), p. 11.
24. Liang Liang, "Yige hongweibing faqizhe de zishu" ("Autobiography of a Red Guard Initiator"), *Cultural Revolution Research Materials (Part One)*, p. 66.
25. Zhong Weiguang, "Qinghua fuzhong hongweibing dansheng shishi" ("The History of the Birth of the Red Guard Group in Tsinghua High School"), *Beijing Spring*, October 1996, p. 13.
26. Yang Lijun, "The Social Factors of the Red Guard Movement: Class and Stratum Perspectives" (Japan: *Modern China* No. 74, September 2000), pp. 219–22.
27. Zheng Yi, "Qinghua fuzhong, hongweibing he wo"("Tsinghua High School, the Red Guards, and Me"), *Beijing Spring*, November 1996, p. 2.
28. Refers to students who are not concerned with politics and national affairs and bury their heads in their studies.
29. Yang Lijun, op. cit., 2000, pp. 219–22.
30. Mi Hedu, op. cit., pp. 118–119.
31. Anita Chan, Stanley Rosen and Jonathan Unger, op. cit., 1980, pp. 397–446; Rosen, op. cit., 1982.
32. Zhong Weiguang, op. cit., pp. 9–10.
33. The *Liberation Army Daily* played a great role in the launching of the Cultural Revolution. In particular, from January to June 1966, the *Liberation Army Daily* surpassed the *People's Daily*, the organ of the Party Central Committee, and played a role in inducing public opinion on the Cultural Revolution. After June 1966, as the

Central Cultural Revolution Group took over the leadership of the *People's Daily*, the style of reporting in the *People's Daily* began to take on a gunpowder flavor. During the long period between then and the fall of the Gang of Four, the *People's Daily* replaced the People's *Liberation Army Daily* and played a role in inducing public opinion on the Cultural Revolution.

Main editorials in the *Liberation Army Daily* from January to June 1966

Date	*Headline*
1 January	Raising the Great Red Banner of Mao Zedong Thought Higher and Fighting for the Continued Prominence of Politics and the Firm Implementation of the Five Principles
3 February	Always Highlight Politics
9 February	Stressing the Word "Public"—Again on Highlighting Politics
14 February	The Most Important and Fundamental Strategy—The Third Essay on Highlighting Politics
18 February	Politics Commands the Military and Politics Commands Everything—The Fourth Essay on Highlighting Politics
2 March	Take Mao's Books as the Supreme Instruction for All the Work of the Army—The Fifth Essay on Highlighting Politics
23 March	Promoting Materialism and Grasping Living Ideas—The Sixth Essay on Highlighting Politics
5 April	The Key Lies in the Leadership of the Party Committee—The Seventh Essay on Highlighting Politics
18 April	Hold High the Great Red Flag of Mao Zedong Thought and Actively Participate in the Great Socialist Cultural Revolution
4 May	Don't Ever Forget the Class Struggle
21 May	Mao Zedong Thought Feeds the Strongest Soldiers
7 June	Mao Zedong Thought Is the Telescope and Microscope of Our Revolutionary Cause

34. Zhang Chengzhi, op. cit., p. 66.
35. Hu Ping, "bisai geming de geming" ("The Revolution of the Race Revolution"), *Beijing Spring, June* 1996, pp. 18–23.
36. The conservative Red Guards were formed after Mao Zedong supported the Red Guard organization between 1 and 18 August 1966. Its core members were student cadres, members of the Communist Youth League, and activists among students. Unlike the cadre children, the advanced student groups did not have access to accurate information. They, therefore, had the characteristic of acting on party's instructions in response to the chaotic political

situation. When the cadre children criticized the Party organization on the eve of the Cultural Revolution, they took a stand to protect the Party organization. Subsequently, when the Work Team took the leadership of the Cultural Revolution in the school, they seriously criticized the Party organization while cooperating with the work of the Work Team. After that, when Mao Zedong supported the Red Guards, they formed the Red Guard organization. Since they were always consistent with the Party in their political views, after October 1966, the Red Guards were called "conservatives" by the rival "rebel" groups.

There is no fundamental difference between the "conservative" Red Guards and the Red Guards of the cadre children. If they differ, the biggest difference between the two is that the Red Guards of the cadre children were the founders of the Red Guard organization and the main body of the political movement at the beginning of the Cultural Revolution. At the same time, the Conservative Red Guards were the followers of the Red Guard organization and participants in the political movement at the beginning of the Cultural Revolution. After the "conservative" Red Guards were formed, they participated with the Red Guards of the cadre children in revolutionary actions, including the "Smash of the Four Olds," the criticism of the "black gangs," the "Great Link-up," the critics of the educational sector. Although the "conservative" Red Guards could not fully agree with the "birth bloodline theory" put forward by the cadre children, they did not criticize it as fiercely as the "rebel" groups did. After October 1966, the two groups began to converge with the rise of the rebel groups. It can be argued that the perception of belonging to the same "red five" class made the Red Guards of the cadre children and the conservative Red Guards merge in the campaign of "sweeping away all monsters and freaks".

As the "Conservative" Red Guards and the Red Guards of the cadre children took joint action, the decline of the Red Guards of the cadre children also affected the "Conservative" organization. "In late 1967, as the rebel groups strengthened, many Conservatives members left the organization and joined the Rebels. However, while members within the "conservative" group left the organization, new members continued to join. The new members during this period were basically members of the Red Guards of the children of former cadres.

Among the three organizations, the Red Guards of the cadre children, the conservative Red Guards, and the Red Guards of the "rebels," both the Red Guards of the cadre children and the Red Guards of the "rebels" became the main body of the movement at different times of the Cultural Revolution. However, the "conservative" Red Guards, composed of progressive students, the main body of the political movement before the Cultural Revolution, were only followers of the movement from the beginning to the end of the Cultural Revolution. "The most important reason for this is that the party organizations that had previously provided shelter and political opportunities to the "conservative" Red Guards became the target of the movement during the Cultural Revolution".

37. Ying Hongbiao, op. cit., "Hongweibing yundong de liangda chaoliu" ("Two Currents of the Red Guard Movement"), p. 243.
38. Zhong Weiguang, op. cit., p. 11.
39. Ibid.
40. The Nankai University Weidong Resource Office, ed., *Zuigao zhishi* (Supreme Instructions, 1967), pp. 152–160.
41. The Beijing Film Studio Maoist Commune, ed. and printed, *Zhanwubusheng de maozedong sixiang wansui;* (*Long Live the Invincible Maoist Thought*), p. 121.
42. Ren Zhichu, *Hongweibing yu xipishi* (*The Red Guards and the Hippies*) (Mirror Media Group Press, 1996), p. 266.
43. Mi Hedu, op. cit., pp. 118–119.
44. Xu Youyu, "Wenhuadageming Zhong hongweibing xingwei dongyin de diaocha yu fenxi" ("An Investigation and Analysis of the Motivations of Red Guard Behavior in the Cultural Revolution"), *The Chinese Social Science Quarterly* (City University of Hong Kong Press, 1996), Summer volume, p. 109.
45. *Shoudu hongweibing diyiqi, Jinian 8 18 zhuankan* (Capital Red Guards, Issue 1, Special Issue in Commemoration of August 18), 1967, p. 27.
46. Zhang Chengzhi, op. cit., p. 38.
47. Yin Hongbiao, "Hongweibing yundong de zhuyao liupai"("The Main Genres of the Red Guard Movement") *China Youth*, vol. 4, 1997, p. 36.

48. "Zhou Enlai, Dong Biwu, Chen Boda tongzhi 8yue5ri zai qinghua daxue de jianghua", Hebei Beijing shifan xueyuan "douzheng shenghuo" bianjibu "*wuchanjieji wenhuadageming ziliao huibiao*" ("Speeches by Comrades Zhou Enlai, Dong Biwu and Chen Boda at Tsinghua University on August 5"),The Editorial Department of the Life of Struggle, Beijing Normal College, Hebei, ed., *Compilation of Materials on the Great Proletarian Cultural Revolution*, December 1966, p. 111.
49. For the process of deciding on the dispatch of the Work Team, please refer to the following studies: Wang Shaoguang, *Failure of Charisma: The Cultural Revolution in Wuhan,* op. cit., pp. 32–33; Wang Nianyi, *The Age of Great Turmoil*, op. cit., pp. 29–35.
50. Harry Harding, "The Cultural Revolution: China in Chaos", op. cit., p. 141.
51. Zhang Chengzhi, op. cit., p. 66.
52. For examples of student resistance to the Work Team, please refer to Wang Nianyi, *The Age of Great Turmoil*, op. cit., pp. 33–46.
53. Shi Yun, "Hongweibing yundong shuping" ("A Review of the Red Guard Movement"), *China Youth Movement*, no. 1 (1989), p. 22.
54. Ibid.
55. Qinghuadaxue fushu zhongxue (Tsinghua University High School), "Zailun wuchanjieji de zaofan jingshen wanshui (1966nian7yue4ri)" ("Further Discussion on Long Live the Proletariat's Spirit of Rebellion (July 4, 1966)", *Red Flag*, No. 11, 1966.
56. Shi Yun, op. cit., p. 22.
57. "Huanhu beida de yizhang dazibao" ("Hailing a Big Character Poster from Beijing University"), *People's Daily*, 2 June 1966.
58. Harry Harding, "The Cultural Revolution: China in Chaos", op. cit., p. 142.
59. "Mao Zedong gei Qinghuadaxue fushu zhongxue hongweibing de xing (1966nian8yue 1ri)" ("Letter from Mao Zedong to the Red Guards of the Tsinghua University Hign School (1 August 1966)"), op. cit., "*Cultural Revolution*" *Research Materials (Part One)*, p. 63.
60. Ren Zhichu, op. cit., pp. 40–41.
61. "Zhonggong Zhongyang zhengzhiju kuoda huiyi jueding (1966nian 5yue 23ri), (*Decision of the Enlarged Meeting of the Political Bureau of the Central Committee of the Communist*

Party of China (May 23, 1966), op. cit., *"Cultural Revolution" Research Materials (Part One)*, p. 10.

62. Qinghuadaxue fushu zhongxue hongweibing (Red Guards of the Tsinghua University Affiliated Middle School), "Wuchanjieji de geming zaofan jingshen wanshui" ("Long Live the Revolutionary Rebellion Spirit of the Proletariat"), *"Cultural Revolution" Research Materials (Part One)*, p. 63.
63. "Zhonggongzhongyang zhengzhiju kuoda huiyi jueding" (Decision of the Enlarged Meeting of the Political Bureau of the Central Committee of the Communist Party of China (May 23, 1966), op. cit., p. 24.
64. Harry Harding, op. cit., p. 141.
65. For the Work Team's method of classifying students, please refer to the following studies: Mithuyuki Kagami, *gyaku sethu toshite no tyuugoku kakumei (The paradoxical Cultural Revolution)* (Tabata Bookstore, 1986), p. 41; William Hinton, translated by Haruna Akira, *Hundred Day War: The Cultural Revolution in Tsinghua University* (Heibonsha, 1976), p. 69 (The English version: William Hinton, *Hundred Day War: The Cultural Revolution in Tsinghua University* (New York: Review Press, 1972).
66. "Guan Feng, Yao Wenyuan, Zhang Chunqiao 8yue 2ri zai guowuyuan jiedaishi de jianghua" ("Speeches by Guan Feng, Yao Wenyuan and Zhang Chunqiao at the State Council Reception Room on August 2", op. cit., Collection of Materials on the Great Proletarian Cultural Revolution, p. 103.
67. Wang Nianyi, op. cit., 1988, pp. 54–55.
68. For a study of the Red Guards' "Smash the Four Olds" rebellion, please refer to: Wang Nianyi, op. cit., pp. 69–71; Jiang Pei, op. cit., pp. 49–81, 132–153.
69. Ying Hongbiao, op. cit., "The Main Genres of the Red Guard Movement," p. 33.
70. Beida fuzhong hongqi zhandoudui (The Red Flag Combat Team of Peking University High School), "Zilaihongmen zhanqilai le" (1966nian 7yue), "The Born-Reds Have Stood Up (July 1966)," in Beijing shoudu maozedong sixiang bingtuan (Beijing Capital Maoist Thought Corps), *Bingtuan zhanbao* (*Corps War Journal*), November 1966.
71. Beijing hangkong xueyuan hongweibing, (Red Guards of the Beijing Aviation College), "Hongweibing yongyuan genzhe Mao

Zedong geming" ("Red Guards will always follow Mao Zedong in his revolution!"), *Guangming Daily*, August 26, 1966.

72. Beida fuzhong hongqi zhandou xiaozhu (Red Flag Combat Team), "Zilaihongmen zhanqilai le" ("The Born-Reds Have Stood Up (July 1966)"), op. cit.
73. Mitsuyuki Kagami, "History of the Chinese Cultural Revolution", Iwanami Shoten, 2001, p. 63.
74. Zhongguo gongchandang Zhongyang weiyuanhui (The Central Committee of the Communist Party of China), "Guanyu wuchanjieji wenhuadageming de jueding" ("Decision on the Great Proletarian Cultural Revolution"), op. cit., p. 75.
75. Ibid.
76. Ibid.
77. Yu Luoke,"Chushen lun" ("On Class Origins (November 1966)", *Inaugural Issue of the Capital High School Cultural Revolution Newspaper*, 18 January 1967.
78. Kagami Mitsuyuki, *gyaku sethu toshite no tyuugoku kakumei (The paradoxical Cultural Revolution*), op. cit., p. 23.
79. Ibid., p. 23.
80. In a letter to the Red Guards of the Tsinghua University Middle School which had argued that "We only allow the leftists to rebel, not the rightists to betray. If you dare to rebel, we will immediately suppress it. This is our theory. The state organs are already in our hands," Mao Zedong pointed out: "Not only in Beijing, but also in the Cultural Revolution movement throughout the country, we send our warm support to all those who share your attitude. We support you and also ask you to unite all those who can be united" (Red Guards of the Tsinghua University Affiliated Middle School, "Long Live the Proletariat's Spirit of Revolutionary Rebellion (July 4, 1966)", op. cit., "Cultural Revolution", p. 64).
81. "Jiang Qing tongzhi zai tianqiao juchang de jianghua" ("Comrade Jiang Qing's speech at the Tianqiao Theatre" (a leaflet from that time).
82. Ying Hongbiao, "Hongweibing yundong de liangda chaoliu"; op. cit., "Two Currents of the Red Guard Movement," p. 235.
83. Liang Liang, "Yige hongweibing faqizhe de zishu" ("Confessions of a Red Guard Initiator"), op. cit., p. 69.
84. The conversion of the old Red Guards into a conservative Red Guard organization was not the first time, but had already taken

place in August 1966 during the debate on the "Birth Bloodline theory."

85. Song Yongyi, "Wenhuadageming zhong de yiduan sichao" ("Heretical Thinking in the Cultural Revolution"), *The Cultural Revolution: Facts and Analysis*, op. cit., p. 260.
86. Ibid.
87. "Zhongyang Beijing dangzhengjun ganbu zidi (nv) lianhe weiyuanhui tonggao zhongfa mizi 003" (Circular of the Joint Committee of the Central Committee, Beijing Party, Government and Military Cadres childen Zhong Fa Secret Word 003, op. cit., *Hongweibing kuangbiao* (*Red Guard Rampage*), pp. 103–106.

CHAPTER 6

The Rise and Fall of the Tsinghua University *Jinggangshan* Corps

From October 1966 onwards, the importance of the Old Red Guards in secondary schools gave way to the Rebel Red Guards in universities, and the leadership of the Cultural Revolution in universities was transferred from the conservative students and students with party membership to the general youth. From this period forward, the Cultural Revolution entered the second sub-movement phase. This chapter focuses on the rise and fall of the rebel groups in universities that were the main force behind the second sub-movement.

Nationwide, the rebel groups developed as the Old Red Guard groups declined. Most of them began to form after August 18, 1966. In October, in response to the call to criticize the bourgeois reactionary line, after receiving the support of the Central Cultural Revolution Group and Mao Zedong, they replaced the Old Red Guards as the main force of the movement.[1] According to research, members of the rebel groups were mainly in the lower levels of the party's power apparatus. The Rebel Red Guards in schools usually had three characteristics. First, there were only a few cadre children, almost no senior cadre children, and very few party or youth league members. Second, most of the faction leaders were labeled "counter-revolutionaries," "anti-party elements," "rightists," "fake leftists, real rightists," "white students," and so on before the Cultural Revolution or during the Work Team period at the beginning of the

Y. Lijun, *Contest for Citizenship and Collective Violence during China's Cultural Revolution*, IPP Studies in the Frontiers of China's Public Policy, https://doi.org/10.1007/978-981-97-0906-9_6

Cultural Revolution. Third, most members were middle-class children (employees, teachers, freelancers, small business owners, shopkeepers, mid-level peasants, etc.). The Rebel Red Guards were almost exclusively of middle-class origin, particularly in major high schools and universities, where few students were children of workers or peasants.[2] Conversely, rebel groups in the factories were mainly made up of "backward elements" and those criticized in various movements and other social discontents, with very few party members and activists participating.[3] In short, the rebel groups were a collection of non-vested interest groups in society at the' time. Because of their members different class origins and social statuses, their advocations varied. The behavior of secondary school, university, and factory rebel groups also varied considerably. As analyzing all the rebel groups in the Cultural Revolution is impossible, this discussion will concentrate on the Tsinghua University *Jinggangshan* Corps (hereinafter referred to as "the Corps"). University rebel groups, like the Corps, dominated the second sub-movement of the Cultural Revolution. This chapter focuses on exploring the driving force of the Cultural Revolution by analyzing the process of the rise, development, and decline of the Corps.

This chapter consists of four sections. The first section examines the historical background of the development of the Corps into a national rebel organization. It analyzes the role of the state and society in the process of the Corps' emergence. The second section examines the division, alliance, and re-division within the Corps after the Cultural Revolution entered its third phase in 1967, with the workers' rebel groups in Shanghai as the movement's mainstay. It will analyze the reasons for these changes. The third section will look at the rivalry between the two factions of the Corps after the summer of 1967 in their efforts to obtain more citizenship from the state and the resulting disintegration of the organization. Finally, in the fourth section, the author will summarize the main arguments of the chapter.

Like the other chapters, this chapter focuses on the interaction between factions, and between factions and the state as the main points of analysis. Specifically, the key questions this chapter seeks to answer and explain include: How do state policies and the speech and actions of state agents affect the collective behavior of factions? How do different factions utilize the political opportunities provided by the state to compete for citizenship? To what extent does the competition between factions influence state policies and the speech of state agents?

1 The Rise of the Corps

The Corps was established at Tsinghua University and was based on the Tsinghua University *Jinggangshan* Corps. To examine its rise, we must first discuss the historical background, formation, and development of the Tsinghua University *Jinggangshan* Red Guard organization. The Tsinghua University *Jinggangshan* Red Guards (hereinafter referred to as the "*Jinggangshan* Red Guards") were founded on September 24, 1966. Their leader, Kuai Dafu, was a third-year university student of poor peasant origin. Labeled a counter-revolutionary during the Work Team period, he did not give in and protested with hunger strikes and other actions. He was highly regarded as a hero of the anti-Work Team after the Work Team retreated.[4] "When the *Jinggangshan* Red Guards were formed, only about a dozen members, all of whom, like Kuai Dafu, had been severely suppressed by the Work Team.[5] Because they were labeled counter-revolutionaries, they were shut out of all the Red Guard groups on campus while the Work Team was there. However, this small organization destroyed opposing conservative organizations within three months, formed a union with other rebel factions that shared their political stance, and developed into a large organization with over 5000 members.[6] The Corps played a central role in the national rebel movement, not only in Beijing but growing to more than 8000 members nationally in the six months after its founding.[7] This section focuses on two questions: how did the *Jinggangshan* Red Guards develop their organization, and why were they able to transform from a minority faction to a majority faction in a short period of time?

To put it briefly, the most important reason for the rapid expansion of the *Jinggangshan* Red Guards was the change in state policy, which brought about a change in the distribution of citizenship and a reorganization of vested and non-vested interest groups. Suppose we make changes in state policy as the axis. In that case, we can divide the period from the formation of factions within Tsinghua University in June 1966 to the establishment of the Corps in December 1966 into three stages.

The first stage was the formation of factions from early June to the end of July. During this period, there were differences of opinion regarding whether to support or oppose the Work Team, and Tsinghua University students were divided into two major camps: the conservative and rebel factions. The Work Team judged the most militant students in the rebel faction as "counter-revolutionaries" and became the main target of attack.

The second period was the confrontation between the conservatives and the rebels that lasted from the end of July to the end of September. State policy was extremely ambiguous during this period. After the retreat of the Work Team, the Central Cultural Revolution Group began to appear on the front line, openly leading the Cultural Revolution. Despite the Central Cultural Revolution Group's demand for unity, the ambiguous national policy expanded the range of interpretations of the Cultural Revolution among the masses. As a result, conflicts between the two factions occurred frequently.

The third period, from late September to the end of December, was when the rebel factions achieved a comprehensive victory. During this period, the focus of the Cultural Revolution shifted from attacking the "lower-level" targets such as sweep away "monsters and freaks" and "smash the Four Olds" to targeting the "capitalist roaders within the Party." The Central Cultural Revolution Group clearly expressed its support for the rebel factions. Changes in state policy and the Central Cultural Revolution Group's support for the rebel faction organizations led to a shift in the balance of power between the conservative and rebel factions. The former's organizations ultimately collapsed and rapidly disintegrated, while the latter quickly grew.

The main focus here is to analyze how the changes in national policies and the words and actions of state agents during these three periods impacted the collective behavior of different factions.

1.1 Formation of the Faction

Before the *Jinggangshan* Red Guards were formed, there were already two rival groups at Tsinghua University, the 8–8 Faction, which advocated exposing the bourgeois reactionary line of the Work Team, and the 8–9 Faction, which supported exposing the "black gangs" in the university Party Committee. The 8–8 Faction was founded on August 8, 1966, and was named the "8–8 Revolutionary Link-up Association" (*8–8 geming chuanlian hui*). Four days after Mao Zedong received the Red Guards in Tiananmen Square on August 18, they changed their name to the "Mao Zedong Thought Red Guards." Although the core members of the 8–8 Faction also opposed the Work Team, they were not classified as "counter-revolutionaries." Some had joined Kuai Dafu in opposing the Work Team in late June 1966 but backed off after being pressured.[8] In addition, judging from the class origin of its members, most were students of non-red five classes.[9] On August 9, 1966, the 8–9 Faction was formed, calling itself the "8–9 Pulling the Black-Gangs Link-up Association" (*8–9 jiu heibang chuanlian hui*), and its name changed later to the "Red Guards of Tsinghua University" on August 19. The 8–9 Faction consisted of students who supported the Work Team, including many children of the top party leaders and students with party and league membership. The faction was led by Liu Tao (the daughter of Liu Shaoqi), He Pengfei (the son of Marshal He Long, the vice chairmen of the Central Military Commission), Liu Jufen (the daughter of Liu Ningyi, director of the International Liaison Department of the CCP), Li Lifeng (the son of Li Jingquan, the first secretary of the Southwest Bureau), Qiao Zongguan (the son of Vice Foreign Minister Qiao Guanhua), and many other senior cadre's children.[10]

Why did the *Jinggangshan* Red Guards, the 8–8 Faction, and the 8–9 Faction form such a divisive structure? It is necessary to illustrate this issue in June 1966, when Nie Yuanzi's big-character poster was reported in the press on June 1. A wave of criticism of the Party Committee emerged on the campus of Tsinghua University. Once the Work Team was on campus, the issue of how to get reappointed to the Work Team became critical to the students. This is because the closer one is to the center of power, the more political-economic benefits one can obtain. Changes in the center of power often trigger new competition for political resources within the sphere of influence. As many political movements since the founding of the PRC have shown, those who are the first to act on the instructions of the Party Central Committee in a political movement are often selected as activists. Of course, there are also cases like the anti-rightist movement, where those who took the lead were instead branded as anti-establishment as the movement changed dramatically during the process, so there is a particular risk. But as with business competition, the higher the risk, the greater the reward. Moreover, as noted earlier, two elements are essential in the distribution of citizenship: state policy and the personal will of the state agents who hold the power to distribute citizenship. Changes in state policy on the distribution of citizenship bring about changes in how citizenship is distributed nationwide. Changes in the state agents with distributive power restructure the "patron–client" relationship within that person's sphere of authority. In other words, to enter new patron–client relationships, the masses compete. In this sense, after the school's leadership was transferred from the Party Committee to the Work Team, students competed with each other to get closer to the center of power.

Upon entering Tsinghua University, the Work Team found fault with the Party Committee. It led students and faculty to launch a criticism of the Party Committee, from which students turned to criticize the Party Committee cadres. To show their revolutionary zeal to the Work Team, some students took the radical action of putting three-cornered hats on Party Committee cadres and sending them to the criticism platform. But instead of being well received by the Work Team stopped, this action, with the radical students being severely criticized. A few days later, in the wave of the anti-Work Team fervor that spread throughout Beijing, the target of the radical students' criticism began to turn on the Work Team. As a result, the Work Team branded the most radical students, such as Kuai Dafu, as counter-revolutionaries.[11]

Between late June and late July 1966, the central task of the school movement shifted from criticizing the school Party Committee to criticizing Kuai Dafu. Thus, for the students, criticizing Kuai Dafu became one of the ways to show their loyalty to the Work Team. Especially for those who had supported Kuai Dafu, turning the tables was the best way to save themselves, and many of Kuai Dafu's supporters turned to criticize him. Many threw themselves more fervently into criticism than anyone else, even spreading rumors to mitigate or escape responsibility for their radical actions.[12] Nevertheless, the Work Team treated the senior cadre children and students of the party and the youth league who had been heavily involved when the Party Committee was in power as leftist students who could be trusted even after the Party Committee was criticized. This was especially true for Liu Tao and He Pengfei, who acted directly as spokespersons for the Work Team's advisor Wang Guangmei in carrying out political instructions among the masses.[13]

According to William Hinton, cadre children such as Liu Tao and He Pengfei were no match for the rising rebel leader Kuai Dafu regarding leadership or argumentative ability. However, at the beginning of the movement, the general public could hardly have expected Liu Shaoqi to be the target of attacks, so such status as the daughter of the president and the son of the vice chairman of the Central Military Commission was seen as having some degree of leadership authority and received support from the top leaders and many students.[14] Liu Tao and He Pengfei's support from the upper echelon included Liu Shaoqi and Wang Guangmei and had certain connections with Bo Yibo and Wang Renzhong.[15] Also, many students who acted on the Work Team's instructions supported the two from a lower level.

As mentioned earlier, the Work Team adopted the pre-Cultural Revolution criteria to distribute citizenship, relying on the students with party and league membership, especially the children of senior cadres. In other words, the factionalized structure of Tsinghua University is the result of the citizenship distribution and competition that took place when the Work Team was in control of the university. But this result lasted only for a brief period.

1.2 *The Confrontation Between the 8–8 Faction and the 8–9 Faction*

August to September 1966 was a period of great confusion in state policy. This was mainly reflected in the Sixteen Points, the landmark document of the Cultural Revolution adopted at the 11th Plenum of the 8th Central Committee of the CCP.

First, the provisions and objectives of the Cultural Revolution were very ambiguous. The Sixteen Points stipulates that the tasks of the Cultural Revolution are mainly threefold: first, to struggle against the "capitalist roaders within the party," second, to criticize the bourgeois reactionary academic authorities and ideological tendencies of the bourgeoisie, and third, to reform education, literature, and art, and all other things that are not in line with socialism. Three tasks were "struggle, criticize, and transform (*dou*, *pi*, *gai*)." In the case of the Work Team, the Sixteen Points pointed out that the group had mistakenly treated some of the genuine revolutionary activists and had made mistakes in direction and line.[16] But in the specific implementation stage, that is, after the retreat of the Work Team, it is not explicitly stated whether the mass movement should be launched first to expose the mistakes of the Work Team and investigate its causes or to initiate the struggle against the "capitalist faction in power within the Party." Perhaps for those plotting the Cultural Revolution, there was no difference between the two, as they both pointed to the Liu Shaoqi–Deng Xiaoping line. Still, for the ordinary participants, the difference was enormous.

Second, the provisions concerning the leadership of the movement by the party and the relationship between the party and mass organizations were ambiguous. The Sixteen Points stipulate that the Cultural Revolution must, in principle, be led by party organizations at all levels, thus requiring them to provide correct leadership for the Cultural Revolution.

However, at the same time, it advocated the liberation of the masses, that is, recognizing the autonomy of the mass movement. Moreover, the Sixteen Points needed to address the relationship between party organizations and mass organizations such as the Red Guards. This brought about confusion in the leadership of the mass movement by the party organizations, with some manipulating the mass organizations to maintain their power base and others automatically giving up their leadership in the movement.[17] Furthermore, the Sixteen Points gave the masses a reasonable basis to act according to their own will. The result of unrestricted political mobilization and political participation subsequently caused the movement to spiral out of control and contributed to its radicalization.

The third point is that the criteria for identifying the revolutionary leftists of the movement were very ambiguous. The Sixteen Points stipulated that the main body of the movement was the revolutionary leftists but did not elaborate on who exactly could be considered revolutionary leftists. Additionally, for the students who had formed a rivalry, the Sixteen Points affirmed that the general direction of the revolution was consistently correct for the majority of the youth and, on this basis, called for the protection of the dissenting minority since "the truth is sometimes in the hands of the minority." Such a provision could be seen as an attempt to reconcile the contradictions among the students. Still, the result was that each faction considered itself to be the revolutionary left.[18] As the movement developed, contradictions widened, and rivalry between factions deepened.

Why was the document conveyed by the CCP so ambiguous? According to Harding's research, the Sixteen Points, while legitimizing Mao's attack on Liu Shaoqi and other bureaucrats through the Cultural Revolution, also reflected the divergent views within the Central Committee.[19] In other words, the Sixteen Points resulted from a compromise between the various factions in the leadership of the Central Committee. This can also be seen from the composition of the Central Political Bureau after the 11th plenum of the 8th Central Committee of the CCP.

As shown in Table 1, the positions of Lin Biao, Tao Zhu, Chen Boda, Kang Sheng, and other members of the Cultural Revolution faction were significantly elevated in the new Central Politburo. In contrast, Liu Shaoqi dropped from second place to eighth. Additionally, Peng Zhen, Lu Dingyi, also Peng Dehuai and Zhang Wentian (Peng and Zhang were accused of being part of the "Peng, Huang, Zhang, Zhou Anti-Party

Group" at the Eighth Central Committee of the Chinese Communist Party in 1959), were officially dismissed from their posts. They were replaced by seven new members, who can be classified into two types. The first type consists of individuals who served as the first secretaries of the Central Bureaus, such as Tao Zhu (Central South Bureau), Li Xuefeng (North China Bureau), Song Renqiong (Northeast Bureau), and Liu Lantao (Northwest Bureau). The second type consists of military leaders, such as Xu Xiangqian, Nie Rongzhen, and Ye Jianying. While the former's promotions appeared to be a rise in official ranks, they were likely preparatory moves for criticism under the guise of executing the "capitalist reactionary line." That is, to criticize the top leaders of each Central Bureau, they must first be separated from local political power to prevent them from joining regional political forces against the center. However, this hypothesis requires further verification once the relevant historical records are released in the future. As for the latter group of new members, they were likely added to the Politburo for two reasons. First, they could be deployed into the PLA when the movement spun out of control. Second, they could be used to constrain Lin Biao's expanding power. As discussed in earlier chapters, Lin Biao's relationship with Mao Zedong was a political alliance that differed entirely from Mao's relationship with the Central Cultural Revolution Group. Therefore, it was critical for Mao to utilize Lin Biao's military power while limiting his power expansion. In this sense, it was perhaps most beneficial for Mao to rely on forces within the PLA other than Lin Biao.

In addition, among the 27 members of the new Central Political Bureau, nine of them—Liu Shaoqi, Deng Xiaoping, Tao Zhu, He Long, Li Jingquan, Ulanhu, Bo Yibo, Song Renqiong, and Liu Lantao—stepped down during the Cultural Revolution. At the same time, Tan Zhenlin, Chen Yi, and Ye Jianying were criticized during the "February Backlash." From this fact, it can be inferred that the understanding of the Cultural Revolution among the members of the new Central Politburo during the preparation of the Sixteen Points may not have been consistent.

In summary, the reshuffling of the Central Politburo formed a command center led by Mao Zedong, consisting of the People's Liberation Army, the Central Cultural Revolution Group, and the State Council bureaucracy to promote the Cultural Revolution. However, In this power structure, the People's Liberation Army, the Central Cultural Revolution Group, and the State Council bureaucracy were not only internally

Table 1 Changes in the membership of the Central Political Bureau (August 1966)

	The 5th Plenum of the 8th Central Committee (May 1958)	*The 11th Plenum of the 8th Central Committee (August 1966)*
1	▽ Mao Zedong	▽ Mao Zedong
2	▽ Liu Shaoqi	▽ Lin Biao
3	▽ Zhou Enlai	▽ Zhou Enlai
4	▽ Zhu De	▽ Tao Zhu
5	▽ Chen Yun	▽ Chen Boda
6	▽ Lin Biao	▽ Deng Xiaoping
7	▽ Deng Xiaoping	▽ Kang Sheng
8	Dong Biwu	▽ Liu Shaoqi
9	Peng Zhen	▽ Zhu De
10	Chen Yi	▽ Li Fuchun
11	Li Fuchun	▽ Chen Yun
12	Peng Dehuai	Dong Biwu
13	Liu Bocheng	Chen Yi
14	He Long	Liu Bocheng
15	Li Xiannian	He Long
16	Li Jingquan	Li Xiannian
17	Tan Zhenlin	Tan Zhenlin
18	◇ Ulanhu	Li Jingquan
19	◇ Zhang Wentian	Xu Xiangqian
20	◇ Lu Dingyi	Nie Rongzhen
21	◇ Chen Boda	Ye Jianying
22	◇ Kang Sheng	◇ Ulanhu
23	◇ Bo Yibo	◇ Bo Yibo
24		◇ Li Xuefeng
25		◇ Song Renqiong
26		◇ Liu Lantao
27		◇ Xie Fuzhi

Note ▽ full membership of the Political Bureau; ◇ alternate membership of the Political Bureau

divided, but also in conflict and antagonism with one another. Moreover, there was no complete agreement between them and Mao Zedong, and complex relationships of mutual benefit, restriction, dependence, and opposition existed. This significantly impacted the subsequent progress of the Cultural Revolution, and as internal conflicts emerged, its influence extended to the mass organizations' rebellion movement.

Furthermore, in addition to the issues with the composition of the Central Politburo, as mentioned earlier, the guidelines for the Cultural

Revolution in the Sixteen Points were not clear, providing both opposing factions with a basis to legitimize their interests, which made it difficult for the central leadership to lead the movement. After the withdrawal of the Work Teams at the end of July 1966, the Central Cultural Revolution Group became the official leadership body of the movement. In late July, the Central Cultural Revolution Group and Zhou Enlai, as directed by Mao Zedong, gave speeches at various universities, which can be summarized into three main points. First, admitting that the Work Teams had made mistakes and should make self-criticism in front of the masses. Second, restoring the reputation of students who had been suppressed during the Work Team period. Third, advocating for the liberation of the masses and not relying on outside leadership, such as the Work Teams.[20] However, the answers they gave in those speeches were as confusing and ambiguous as those given in the Sixteen Points as to whether they should first criticize the Work Team or the "capitalist faction in power within the Party" and the relationship between the party–league organization and the mass organizations, which were then the subject of much national debate. From the end of July to the beginning of August, to avoid being accused of suppressing the mass movement like the Work Team, the Central Cultural Revolution Group, and the central leadership, including Zhou Enlai, did not so much solve the problem as they did let it go as "not being able to take responsibility for everything" (*buneng baoban daiti)*.[21] For example, when Tao Zhu met with local students on August 21, 1966, the Xi'an students demanded that "we do not trust the provincial committee and that the provincial committee should be reorganized" Tao Zhu replied, "This is a matter for you to decide."[22] Similar statements were often made by Central Cultural Revolution Group members, Zhou Enlai, and other central leaders in August. This could be seen as the beginning of the "anarchist" trend that spread among students in late 1966.

In response to the confrontation within the student body, the central leadership called for unity between the two factions.[23] Uwe Richter notes that at this point in August 1966, the support of the Central Cultural Revolution Group for the rebels "had not yet been publicly demonstrated to be more ideological than organizational."[24] This means that, as discussed in the previous chapter, the Old Red Guard organizations in middle schools were still highly valued by Mao Zedong, Zhou Enlai, and the Central Cultural Revolution Group, and direct cooperation between the Central Cultural Revolution Group and the rebel organizations had not yet been established.

In such a chaotic political situation, the 8–9 Faction and the 8–8 Faction were pitted against each other. The 8–8 Faction advocated that the future campaign should focus on exposing the bourgeois reactionary line of the Work Team and restoring the reputation of the students persecuted by the Work Team. On the contrary, although the 8–9 Faction admitted that the Work Team had made a mistake in its line, they advocated reversing the movement back to its original direction and renewing the criticism of the Party Committee. In August, the 8–8 Faction refused to allow Kuai Dafu and his core supporters to join them. The reason was mainly that Kuai Dafu had been criticized and labeled as a "counter-revolutionary." Most of the 8–8 Faction, though sharing his ideology, did not want to risk uniting with him.[25] This later influenced the division within the rebel faction.

On the surface, the confrontation between the 8–8 Faction and the 8–9 Faction was a confrontation over differences of opinion on how to deal with the Work Team. However, there was still competition for citizenship in the new political environment of the period centering on becoming a student of the revolutionary left. As mentioned earlier, the Sixteen Points did not specify the criteria for judging revolutionary leftists. When facing students' questions, Tao Zhu of the Central Cultural Revolution Group summarized the standards for revolutionary leftists as follows: first, obey Mao Zedong and the party's central leadership; second, participate in the revolution most resolutely and courageously; third, unite 95 percent of the people; and fourth, always serve as a model.[26] When evaluating themselves, according to Tao Zhu's explanation, either side of the opposition considered themselves to be the revolutionary leftists.

Liu Tao and other cadre children deliberately avoided criticism of the Work Team. Because they had embraced the Work Team, protecting themselves was a natural aim. But more important than that was the judgment of Liu Tao and the other cadre children about the political situation at the time. Liu Tao explained the rationale for her actions in her self-criticism essay written in late December 1966, stating that "the central government has not yet reached a conclusion on the Work Teams, and it is still unknown whether criticizing them is right or wrong." She had failed to recognize that criticizing the Work Teams was a matter of ideological struggle and a central issue of the Cultural Revolution. She believed that criticizing the "black elements" was the correct and revolutionary direction of the movement, rather than criticizing the Work Teams, which she considered only an incidental aspect. "Therefore, in August, the organization headed by Liu Tao and other cadre children adopted the "preferring the left to the right" course of action,[27] believing that they were the most revolutionary and that everyone else was not because criticizing the 'black gangs' was the only way to demonstrate revolutionary spirit."[28]

While Liu Tao's self-criticism tended to shift responsibility away from herself, it also reflected the awareness of the children of cadres of the political situation at the time. In mid-August, He Pengfei, Liu Tao, and other cadre children posted several big-character posters on the campus criticizing Wang Guangmei,[29] in which the "revolutionary will" displayed was no less than that of the opposing 8–8 Faction. On August 18, Liu Tao even intended to post big-character posters criticizing Liu Shaoqi but was prevented by Wang Renzhong.[30]

Why did the 8–8 Faction persist in criticizing the Work Team? It can be considered that they mainly had two purposes. First, they attempted to restore their reputation. This was not only a matter of the right to participate in the Cultural Revolution but also related to the citizenship they could enjoy in the future. The second purpose was to become the revolutionary left. In the new round of competition, having experienced suppression could be used as an important political resource to defeat competitors. It can be considered that they recognized that criticizing the Work Team was more advantageous to achieving their goals than criticizing the bourgeois reactionary line, and thus they acted.

In short, the confrontation between the two factions was a disagreement over the Work Team and a competition for benefits. So, no matter how much the Central Cultural Revolution Group, Zhou Enlai, and others appealed to the students for unity, the confrontation between the two factions continued to deepen. On August 24th, the opposition between the two factions escalated from a war of words to a physical confrontation.[31] The immediate cause of the "August 24" incident was the posting of big-character posters criticizing Mao Zedong in the school. He Pengfei and other cadre children believed that was the work of the 8–8 Faction rebels. He and his 8–9 Faction thus united with the cadre children of Beijing Forestry College and Beijing Normal University High School to launch an attack on the 8–8 Faction.[32] In the ensuing struggle, the conservative group put forward a slogan: "Only the leftists are allowed to rebel."[33]

As mentioned earlier, the political leadership authority of the cadre children, such as Liu Tao and He Pengfei, the leaders of the conservative organization, came from the top. As this power collapsed, they lost the legitimacy of their leadership and could no longer obtain support from their subordinates. At this point, the 8–9 faction was frequent for members to quit, and the organization abandoned Liu Tao and He Pengfei. When the central government judged the Work Team problematic at Tsinghua University, Liu Tao's supporters began to withdraw. Although Liu Shaoqi was severely criticized by Mao Zedong in August 1966 for sending the Work Team and was relegated from the second to the eighth position in the reorganized Political Bureau of the 11th Plenum of the 8th CCP Central Committee, he remained in the Political Bureau, so Liu Tao still had the support of many people on campus, and by September the 8–9 Faction was still the majority group on campus.[34] Furthermore, the direct support Liu Tao received in August from Wang Renzhong, deputy head of the Central Cultural Revolution Group, was one of the reasons why she maintained her majority position.[35] The reversal of power between the conservatives and the rebels began in late September 1966.

Nationally, as Hinton notes, "neither the rebel nor the conservative organizations were fixed, and each individual shifted from one camp to the other in response to changes in the political situation before the two organizations were formed, and some continued to cross the line even after the organizations were formed. Generally speaking, the period from summer to fall was characterized by the rebellion's expansion and gradual loss of power of the conservatives. By December, the conservatives had finally disintegrated. However, during the struggle, the conservative organizations also saw an increase in membership and remained a force to be reckoned with until the end of the year. In June, the conservatives were the unquestioned majority. In August, their strength was evenly matched by the rebels. On October 6, when rebel organizations from across the country organized a march in Tiananmen Square to expose the bourgeois reactionary line, the rebels began to dominate, and conservatives began to split sharply."[36]

1.3 Composition of the Corps

As mentioned earlier, the mass movement after August failed to respond to Mao's expectations, and the confrontation within the student body overwhelmed the original purpose of the Cultural Revolution, which was to criticize "capitalist roaders within the party." The Red Guards in secondary schools were entangled in the theory of birth bloodline, while the conservative and rebel groups in universities fought desperately for the position of revolutionary leftists. No matter how much the central leadership called for unity among the "ninety-five percent leftists," the division within the student body continued to grow deeper and deeper. More importantly, some party cadres used the birth bloodline theory to divert the movement's spearhead away from themselves.[37] Against this backdrop, the central leadership adjusted its policy to favor the then minority, the rebel groups. A series of events from late September to early October reflected this change in state policy.

On September 26, Zhou Enlai received the Third Headquarters of the Red Guards of Beijing Universities and Colleges (hereinafter referred to as the "Capital Third Headquarters"). This organization, formed on September 6, was a coalition of students from twenty-four universities in Beijing who had been excluded from the movement because of the severe crackdown by the Work Team, with Kuai Dafu of Tsinghua University as its chief. This was the first time the central leadership had received the organization since its founding. Compared to the First–Second Headquarters[38] and the Red Guards in secondary schools, which frequently received direct instructions from the central leadership, this organization had previously received a lukewarm reception. As such, this reception was of great significance. During the meeting, Zhou Enlai clearly expressed his intention to support them and, in the name of the Party Central Committee, announced that the students whom the Work Team had suppressed would be rehabilitated, that bad records would be expunged, and that the Capital Third Headquarters would enjoy the same political and economic treatment as other student organizations.[39]

Before this, the Capital Third Headquarters students were excluded from the Red Guards and disqualified from participating in the movement. Since the leadership of the mass movement in the schools was in the hands of the red five, a group centered on the cadre children, even if they had organized, it would have been almost impossible for them to obtain the funds needed for communication and printing equipment to propagate their views. This was a problem for the Capital Third Headquarters and the 8–8 Faction of Tsinghua University, and other rebel groups, which suffered unequal treatment. Therefore, Zhou Enlai's instructions were essential for improving their political status. After October, Zhou's instructions were transmitted as an official central document.[40]

On October 1, 1966, in his National Day speech, Lin Biao stated that the struggle between the revolutionary proletarian line represented by Mao Zedong and the "reactionary bourgeois line" still existed. Two days later, the 13th issue of the *Red Flag* magazine published an editorial calling on the people to "thoroughly criticize the reactionary bourgeois line." At the subsequent Central Working Conference, Liu Shaoqi and Deng Xiaoping were accused of being reactionary bourgeois agents and were forced to criticize themselves.[41]

On October 6, rebel organizations from all over the country gathered in Beijing under the leadership of the Central Committee. They held a large parade in Tiananmen Square with about 12,000 participants to expose the bourgeois reactionary line. Zhou Enlai and key members of Central Cultural Revolution Group attended the event, with Zhou Enlai, Chen Boda, and Jiang Qing expressing their support for the rebels. Zhang Chunqiao read the central document on restoring the reputation of those persecuted by the Work Team.[42] Kuai Dafu was the leader and organizer of this event. His "*Jinggangshan* Red Guards" organization at the university played a central role in the rally, despite only being established two weeks prior.

This chain of events, especially the fact that Zhou Enlai and the Central Cultural Revolution Group attended the activities of the rebel groups, clearly conveyed a message: that the rebel groups, led by the *Jinggangshan* Red Guards, were the main force of the movement to criticize the bourgeois reactionary line. The widespread dissemination of this message led to a change in the balance of power between conservative and rebel groups nationwide. At the end of September 1966, when the central government made it clear that it supported Kuai Dafu and that criticism of Liu Shaoqi was the movement's goal, members of the 8–9 Faction began to withdraw in large numbers. Liu Tao and He Pengfei lost their leadership authority and were forced to resign on September 29. Most of those who left the 8–9 Faction left Beijing and went to localities for a "revolutionary experience exchange" (the Great Link-up).[43] Some others joined Kuai Dafu's *Jinggangshan* Red Guards.[44] The *Jinggangshan* Red Guards saw rapid development in October. On December 19, 1966, under the leadership of the Central Cultural Revolution Group, they united with all the left-wing organizations of Tsinghua University, including the 8–8 Faction, and the old name was replaced by a new, more ambitious moniker, "Tsinghua University *Jinggangshan* Corps" (hereinafter referred to as the "Corps").[45] At the same time, the seal of its rival faction, the 8–9 Faction, was frozen, and the Corps seized its offices.[46] As a result, some members of the 8–9 faction were incorporated into the Corps. Most students previously passive toward the movement joined the Corps and started participating in its activities.[47]

The process of disintegration of conservative organizations was not peaceful. Some core members fiercely resisted retaining their lost political positions. However, this time the conservatives did not target the opposing rebel groups but instead directed their anger toward the central Cultural Revolution Group that supported the rebels while severely criticizing the Old Red Guards in middle schools and conservative groups in universities. At the Central Work Conference on October 16, Chen Boda pointed out that the birth bloodline theory and the Theory of the Rule of Senior Cadre Children had deviated from the track of proletarian dictatorship and that Tan Lifu's speech in mid-August in support of the birth bloodline theory was in opposition to the central leadership.[48]

Against this background, in October, conservative groups began to post big-character posters fiercely attacking the Central Cultural Revolution Group, pointing out that it was a new bourgeois reactionary line and a promoter of the "leftist but rightist" line.[49] Ironically, the theory used by the conservative group to criticize the Central Cultural Revolution Group was precisely the same as the theory used by the Central Cultural Revolution Group when it mobilized the masses to attack the Work Team three months earlier. The Work Team was criticized for imposing its views on the masses ("*Baoban daiti*") and provoking the masses to fight each other. The conservatives had criticized the Central Cultural Revolution Group for the same reasons.[50] Moreover, the Central Cultural Revolution Group mobilized the masses by saying: "You don't need a nanny."[51] The conservatives now call for "independence from the Central Cultural Revolution Group; we don't need a nanny." Hinton commented on the behavior of the conservative groups: "Once it became impossible for them to stay on the right, they suddenly turned to attack from the left."[52] It was in this political environment that the Joint Action Committee (*liandong*), mentioned in the previous chapter, was formed. At Tsinghua University, such rebellions were the core members of the 8–9 Faction, mainly dominated by He Pengfei and Li Lifeng.[53]

The conservatives' criticism of the Central Cultural Revolution Group further deepened the confrontation between them and the rebels. In retaliation for previous ostracism by the conservative organizations and to show their loyalty to the Central Cultural Revolution Group, rebel groups launched a full-scale attack on conservative organizations. On the other hand, the attacks on the Central Cultural Revolution Group by the core members of the conservatives accelerated the organization's collapse. Under the leadership of the Central Cultural Revolution Group, the Capital Third Headquarters used the opportunity to have the public security departments arrest the most radical members of the conservatives.[54] This way, the confrontation between the two ended in a complete victory for the rebel groups. It should be noted that although the Conservatives as an organization no longer existed, the social base behind its formation remained. Therefore, as discussed later, the antagonism between the two factions continued in the form of internal factional struggles within the rebel factions during the later stages of the Cultural Revolution.

2 Fragmentation and Competition Around "Revolution"

December 1966 was the peak period for the Corps. On the 25th, Kuai Dafu, under the direction of Zhang Chunqiao,[55] documented over 5000 members of the Corps and radical faculty and staff at Tsinghua University to hold a large-scale demonstration in Tiananmen Square to criticize the bourgeois reactionary line represented by Liu Shaoqi and Deng Xiaoping. This began the criticism of Liu and Deng, which spread throughout the country. Kuai Dafu and the Capital Third Headquarters, under his leadership, then became the leaders of the national rebel groups. It should be noted that even during this peak period, the Corps had already harbored factors of division and decline, which were mainly manifested in the following three aspects.

First, as Hinton pointed out, "What the rebels gained in December 1966 was not the leadership of the university, but only the power to lead the mass movement." That "The January storm in Shanghai raised the level of struggle throughout the country, and the major question of who should lead the party and the state was brought to the forefront as a practical issue of power." For a short time, the *Jinggangshan* Corps held real power but failed to establish a legitimate administrative structure for the university. In January, members of the *Jinggangshan* Corps confiscated and burned the official seal of the conservative Provisional Revolutionary Committee and established a new committee to run the administration.

But this was only an informal body. Lacking formal authority, it was not recognized as a legitimate revolutionary committee by the party's Beijing Committee and the Party Central Committee.[56]

Second, as the rebellion grew in strength, the power of the organization expanded in terms of the "quasi-governmentalization" of the organization. According to Tang Shaojie's research, from the end of 1966 to the beginning of 1967, the Corps developed a government-like command structure, organization, propaganda, and communication system.[57] These systems were internally divided in great detail. Through its operation, the Corps controlled the political, financial, and propaganda powers of Tsinghua University and influenced rebel groups throughout the country. The Corp's influence over rebel groups nationwide was exercised through establishing local liaison stations, the dispatch of reporters,[58] and the publication of an internal journal.[59] With liaison stations and the propaganda-intelligence system, local events reached the Corps headquarters almost in real time. Information about the instructions of the central leadership and the power struggle within the central government was also quickly and widely disseminated to rebel groups nationwide. It became a unique phenomenon during the Cultural Revolution that mass organizations could learn about the content of central government documents even before the local party and government organs.

During the campaign to criticize the bourgeois reactionary line, the local party and government power system were on the verge of collapse due to attacks by mass organizations. The quasi-governmentalization of the Corps posed a significant threat to the communist regime. When Mao Zedong launched the Cultural Revolution, he said: "Great chaos will lead to great governance." However, considering the unfolding of the Cultural Revolution in 1967, the "chaos" Mao referred to as the "chaos" within his control to remove Liu Shaoqi and Deng Xiaoping and to rebuild the regime after the removal of the bourgeois reactionary line. In other words, Mao did not wish to see the mass organizations building a power structure outside the party and government power system. Therefore, in January 1967, Mao ordered military training in secondary schools and universities and instructed that student organizations be restricted from entering society.

The expansion of the Corps raised not only external questions about the legitimacy of quasi-governmental organizations but also brought internal issues regarding the management of lower-level organizations by the headquarters and factionalization within the organization itself. The Corps was formed by absorbing various organizations and their members with different opinions, so factionalization within the organization became a direct cause leading to its division.

Third, as with the Old Red Guards and the 8–9 Faction, the actions of the Corps were radicalized during the Cultural Revolution. This further affected the already chaotic social order and contributed to the spread of the national anarchist trend in late 1966. For example, after October 1966, the Corps dispatched liaison stations nationwide to lead local rebel groups to storm local party and government institutions. In a desperate attempt to defend their power and political status, local party and government cadres resorted to mobilizing collaborating mass organizations to confront the Corps.[60] As a result, there were conflicts between different mass organizations in various regions, and in some areas, there were even fatalities and injuries.[61] On November 8, an incident in Beijing occurred where Kuai Dafu organized rebel groups to storm *Zhongnanhai* (the CCP leadership's headquarters) and the Ministry of National Defense.[62]

On a nationwide scale, the 1966 rebellion went beyond the scope of the cultural and educational sector, as defined at the beginning of the Cultural Revolution, and extended to sectors affecting economic and productive activities. The students' "great Link-up" had already placed a significant burden on finance and transportation, and industrial worker participation in the Cultural Revolution exacerbated negative impacts. Under such circumstances, it was necessary to adjust policy again to avoid repeating the Great Leap Forward's mistakes and ensure normal production activities. The change in the affiliation of student organizations in 1966 shows that the change in state policy brought about a change in the distribution of citizenship and affected the reorganization of factions. Similarly, the split within the Corps in 1967 was undoubtedly linked to changes in state policy, with the establishment of the Three-In-One Regime being the direct cause of the factional split.

2.1 *Fragmentation of the Corps*

In early 1967, to control the factional struggles of the Red Guards and limit their authority, the CCP Central Committee took a series of measures to drive the Red Guards from society back to the schools to stabilize the social order. For example, on December 31, 1966, the CCP Central Committee and the State Council issued the "Circular on

Short-Term Military Training in Universities and Secondary Schools."[63] In February and March 1967, the Central Cultural Revolution Group instructed the leaders of Red Guard groups in universities, such as Kuai Dafu, on the "Grand Coalition" of Red Guard groups. On March 7, the CCP Central Committee issued the "Regulations on the Proletarian Cultural Revolution in Universities and Colleges." On the same day, the CCP Central Committee conveyed through the *People's Daily* the instructions on "resuming classes in primary and secondary schools to make a revolution, etc."[64] These measures temporarily restricted the authority and actions of the Red Guards but failed to fundamentally resolve the contradictions within the Red Guard organization.

In January 1967, Mao Zedong praised the January Revolution in Shanghai. With this as an opportunity, the nationwide rebel movement entered a phase of regime rebuilding in which representatives of the revolutionary masses, military cadres, and revolutionary cadres of the organs jointly formed the Three-in-One Provisional Leadership (herein referred to as the "Three-in-One"). Around the establishment of the Three-in-One, differences of opinion arose within the Corps, which finally split into the 4–14 Faction and the Regiment Faction after April. Contrasting between the two organizations lies in the treatment of old cadres. In other words, which of the senior cadres could be assessed as revolutionary when establishing the Three-in-One? The 4–14 Faction advocated that most cadres who held power in the university before the Cultural Revolution were revolutionary and that the cadres of the Three-in-One should be selected from among those suspended by the Work Team. Their basis was the instruction issued by the CCP Central Committee on March 1, 1967, stating that "most cadres are good." In contrast, the Regiment Faction (*tuanpai*) argued that the leadership of school education during the seventeen years between the founding of the PRC and the Cultural Revolution was in the hands of the "bourgeois reactionary line." Therefore, the cadres of the Three-in-One should be selected from those who had been oppressed for 17 years.[65] The former were separated from the Corps on April 14 and called themselves the 4–14 Faction, while the latter used the old name of the Corps and were called the Regiment Faction.

The differences between the 4–14 and the regiment factions appeared from the new political situation. However, in reality, the differences were simply a new manifestation of the opposition between conservatives and rebels that had emerged in the second half of 1966. This can be understood from the following two perspectives.

First, the faction's composition is very much related to the structure of the division between the conservatives and the rebels at the beginning of the movement. Hinton pointed out that most of the members of the 4–14 Faction were former members of the 8–9 Faction, who defected to the Corps after the breakup of the 8–9 Faction and joined the 4–14 Faction when the Corps split, while most of the Regiment Faction were former members of the *Jinggangshan* Red Guards, such as Kuai Dafu. In addition, the former rebel faction, the 8–8 Faction, split into two parts, one joining Kuai Dafu's Regiment Faction and the other transferring to the 4–14 Faction.[66] Why did some of the 8–8 Faction join the 4–14 Faction, which had a majority of former conservative members? According to Hinton's observation, the main reason is that most of these people were followers of Kuai Dafu's rebellion during the Work Team in 1966. Still, when they came under pressure from the Work Team, they broke away from the Kuai Dafu camp and changed their stance. As a result, they were criticized by Kuai Dafu as rightist waverers within the Corps Faction and were mistreated compared to Kuai Dafu's loyal followers.[67]

Second, the difference in advocation between the two is similar to the opposition between the conservatives and the rebels on the bloodline theory and the class origin theory. The central argument of the 4–14 Faction is the "unchanging class relations theory" (*jieji guanxi bubian lun*). In other words, they believed that the class relations and social structure of the seventeen years before the Cultural Revolution remained unchanged, that the progressive people were still the leading class, and that the landlords, rich peasants, counter-revolutionaries, bad elements, and rightists were still the classes being suppressed. Whether to rebel or not is not a criterion for judging the revolution.[68] In short, the 4–14 Faction advocated a return to the pre-Cultural Revolution class division and social structure. Therefore, they demanded that the cadres be reinstated. On the contrary, the Regiment Faction held the view of "changing class relations theory" (*jieji guanxi biandong lun*). In other words, they believed that China's class and social structure had significantly changed with the country's development over the past seventeen years. The contradiction between the bureaucratic privileged class and the workers had become the principal contradiction in the socialist country. The purpose of the Cultural Revolution was to bring down the bureaucratic privileged class, redistribute property and power, and bring about revolutionary changes in society.[69] The aim of both can be said to be the same, i.e., to put oneself in a higher position in the power structure by reinstating cadres friendly to one's side. The old vested interest groups supported restoring the bureaucratic system, and the old non-vested interest groups advocated rebellion. The genealogy of this structure is the same as the genealogy of the opposition between the conservatives and rebels.

Just as the confrontation between the rebels and conservatives at the beginning of the Cultural Revolution extended to the whole country, the structure of the conflict between the 4–14 Faction and the Regiment Faction did not exist only at Tsinghua University but everywhere in the country. Similar situations existed in various organizations during this period. For example, there was opposition between the 4–3 Faction and the 4–4 Faction in Beijing's high schools, the opposition between the Mountain Faction (*shanshang pai*) and the Under-Mountain Faction(*shanxia pai*) in Heilongjiang, the Three New Faction (*sanxin pai*) and the Three Steel Faction (*sangang pai*) in Hubei Province, the confrontation between the Followers(*jingen pai*) and Resisters(*kangju pai*) in Guangzhou, and the confrontation between the General Headquarters and the General Station in Yuci, Shanxi Province, and so on. As mentioned, this was an internal confrontation within the rebel faction, but it could be described as a confrontation between the conservatives and the rebels. This was because, with the disintegration of the conservative organizations in late 1966, a large number of conservative members joined the rebel groups, thus preserving the power of the conservatives within the rebel faction. So why did the weakened old conservative faction re-emerge and form the opposition with the Red Guards across the country during the spring and summer of 1967? The reasons for this mainly lie in the following two points.

First, establishing the Three-In-One Regime intensified interfactional rivalries over citizenship, as the confrontation between the 4–14 Faction and the Regiment Faction at Tsinghua University illustrates. The two pre-Cultural Revolution camps of vested and non-vested interests reappeared.

Second, the revival of old conservative forces was linked to the support of the PLA. As mentioned, in early 1967, Mao took a series of measures to stabilize the chaotic social order, one of which was to order the PLA to intervene in the Cultural Revolution. The establishment of the Mao–Lin alliance in 1966 played an essential role in Mao's launching of the Cultural Revolution. When the radicals were denied publication in Beijing, the PLA newspaper provided them with a stage for propaganda. Nevertheless, although the military academies participated in the

Cultural Revolution before 1967, the PLA avoided direct involvement as much as possible. However, on January 23, 1967, the CCP Central Committee, the State Council, the Central Military Commission, and the Central Cultural Revolution Group issued the Decision of the PLA to Resolutely Support the Revolutionary Leftist Masses, in which Mao Zedong criticized the PLA's posture of non-intervention and ordered "active support for the revolutionary leftist masses in their struggle for power. Wherever genuine proletarian leftists ask the army to assist them, the army should send troops to support them actively" and "resolutely suppress counter-revolutionary elements and counter-revolutionary organizations that oppose the revolutionary proletarian left. If they resort to arms, the army should resolutely return fire."[70] After the publication of this document, the army participated in the local "three supports and two militaries" (support for the leftists, workers, and peasants, and military training and control) as an essential part of the Three-in-One formation.

As most party and government institutions were paralyzed in early 1967, the PLA was given enormous power to restore social order and reestablish authority. The military not only had the authority to judge whether mass organizations were revolutionary or counter-revolutionary, but Mao also instructed them to have the right to shoot if warnings against attacking military facilities were ignored.[71] However, due to the lack of clear standards for judging the "leftists," the PLA (many of whom were of red five class origin) tended to treat them as "leftists" and support organizations with a large number of party–league members, progressive students, and red five class students.[72] This is mainly because the PLA comprised the children of the five red classes. Moreover, military cadres belong to the same bureaucratic privileged class as party and government cadres. In particular, the cadres of the security, defense, and personnel departments of local parties and government agencies and enterprises were close to the military cadres because of their working relations, so few among the military cadres approved of the rebel faction's words and actions that dismissed the cadres as a bureaucratically privileged class. In this way, conservative forces expanded in power with the support of the army.

Once the army supported the conservative forces (hereinafter referred to as "conservatives"), the radical rebel forces (hereinafter referred to as "rebels") began to attack the army. The army could not openly suppress the rebel groups, so it manipulated conservative groups to confront rebel groups. As a result, intense conflicts erupted between rebel and conservative factions within mass organizations and between the military and rebel organizations. Incidents of rebel organizations attacking the People's Liberation Army command structures frequently occurred throughout the country.[73] In January 1967, the Central Committee of the CCP and the Central Military Commission issued a directive forbidding all attacks on military zones. Between February and early March, the PLA had many armed counter-attacks.[74] The rebels referred to these incidents as the "February Crackdown."

In February 1967, almost simultaneously with the frequently armed crackdowns on rebel groups by the PLA, the Central Committee of the CCP experienced the "February Countercurrent," in which Tan Zhenlin and other army generals criticized the radical practices of the Central Cultural Revolution Group. These events led Mao to adjust his policy to favor the rebel groups. As a result of this policy change, the student organizations that had returned to school rushed back into society. This was played out at Tsinghua University, where Mao's policy adjustment directly impacted the student rebel movement. The confrontational setting between the 4–14 Faction and the Regiment Faction shifted from the school into society.

2.2 *Centered Around Revolutionary Competition*

In late March 1967, the Central Committee of the CCP issued a series of documents to point out that the PLA had made a directional mistake in inciting the masses to fight each other in the "support for the left" movement and ordered that the PLA be strictly forbidden from judging mass organizations as counter-revolutionary.[75] At the same time, it was immediately instructed to restore the reputation of those mass organizations they had suppressed. Also, on April 6, the Central Military Commission issued the "Ten Articles of the Military Commission," which restricted the army's authority. According to these articles, the power granted to the military in January 1967 to decide whether a mass organization was revolutionary or counter-revolutionary was henceforth withdrawn. In addition, the Central Committee ordered the military to prohibit shooting at mass organizations.[76] This chain of events freed the rebel organizations from the army's control and immediately directed their struggle against cadres in the military.

Perhaps Mao anticipated that adjusting the central government's policy would bring about an attack by the rebels on the army cadres. Mao did two symbolic things in April to ease the attacks and appease the anger of senior military leaders like Tan Zhenlin, who was criticized during the February Countercurrent. First, he released the arrested members of the Joint Action Committee (*Liandong*). Most of the JAC members were children of senior military cadres, and their release could be seen as a way to please military cadres. In addition, the JAC was a nationally known conservative organization, and the release of its members has led some conservative groups to believe that future policy will change in their favor. The second was to invite Tan Zhenlin and other senior military officials to a meeting at his residence the day before National Day and to allow Tan Zhenlin to attend the National Day parade the day following. These two actions greatly encouraged the conservative factions and the military leaders.[77] As a result, both the radical and conservative factions believed that Mao would support them. However, Mao's actions intended to ease the conflict and maintain balance exacerbated the confrontation between the two factions.

The Central Cultural Revolution Group again used student organizations to counter the February Countercurrent. Between May and August 1967, there was an intense confrontation between the rebel and conservative organizations around the "drag out the armed capitalist roaders" among military cadres. There were frequent incidents of rebel organizations attacking the military across the country. In such cases, conservative groups usually took a stand to protect the military. The army, forbidden to shoot at mass organizations, supported the conservatives against the rebels in several ways. This included informing them of the location of weapons and creating the illusion that mass organizations were snatching weapons. If the rebel organizations came to seize weapons, they were not allowed to succeed.[78] Under these circumstances, the conflict between the rebel and conservative organizations escalated into violent clashes, which began as early as May and June, and widespread violent clashes occurred throughout the country after July and August.

The main reasons for the massive armed struggle throughout the country were the mass organizations' competing interests. Still, the incendiary editorials of the *Red Flag* and the *Liberation Army Daily* also played a key role. The editorial in the July 26, 1967 issue of the *Liberation Army Daily* elaborated that "we must drag out a small handful of power-holders within the party and the military who are taking the capitalist road." The article in the *Red Flag* of 1 August even called on revolutionaries throughout the country to make "rooting out the small group within the military" the current central task.

The Red Guards of Kuai Dafu's Regiment Faction at Tsinghua University, which received direct instructions from the Central Cultural Revolution Group, were active throughout the country and became the main force behind the campaign to "drag out a handful in the military." Against this background, the confrontation between the 4–14 Faction and the Regiment Faction over who was more revolutionary took on a new political significance, focusing on who was more revolutionary. This was evident in events such as the Zhongnanhai Sit-In with the slogan "Expose Liu Shaoqi," the attack on the British Embassy, and the criticism of Xu Shiyou incident in Nanjing.

The Zhongnanhai Sit-In

From July 10 to August 6, 1967, the Beijing Red Guards, mainly from Tsinghua University, staged a month-long sit-in rally in front of Zhongnanhai to "drag out Liu Shaoqi." According to Hinton's research, the beginning of the event was straightforward. In protest against Liu Shaoqi's persecution of all students during the Work Team, a group from the Beijing College of Engineering and Architecture held a demonstration outside Zhongnanhai, shouting: "Drag out Liu Shaoqi," with only four organizations initially participating. Since Zhou Enlai had made it clear that Liu Shaoqi should not be publicly attacked in front of the masses, it was sufficient to use rallies, print, radio, and big-character posters. However, the Central Cultural Revolution Group, especially Guan Feng and Qi Benyu, saw an opportunity to arrest and detain Zhou Enlai from this small protest movement. Under Guan Feng's instigation, the Regiment Faction of Kuai Dafu and the rival 4–14 Faction staged a competition in front of Zhongnanhai to "Drag out Liu Shaoqi." They competed over who could mobilize more members to join the campaign to drag out Liu Shaoqi, who could first occupy the most strategic location closest to Zhongnanhai's west gate and see who had the most powerful loudspeaker to make their accusations. Ultimately, the two factions fought over the site, causing hundreds of injuries, and the protest movement was disbanded in failure.[79] The competition between the Regiment Faction and the 4–14 Faction was inconclusive.

The Storming of the British Consulate

The main actor in the storming of the British Consulate on August 23, 1967 was not the Regiment Faction but the 4–14 Faction. The incident began when the British authorities arrested a Xinhua News Agency reporter after a Cultural Revolution-incited riot broke out in Hong Kong. As a result, the Anti-Imperialist and Anti-Revisionist Liaison Station, a rebel group in the Ministry of Foreign Affairs, delivered an ultimatum to the British authorities in Hong Kong, demanding the release of the arrested Chinese journalists within 48 hours. On the day the ultimatum expired, the Foreign Ministry's rebel group invited the 4–14 Faction to participate. Unknown to most, the purpose behind this incident, like the Zhongnanhai-Sit-In, was to bring down Zhou Enlai.[80] However, most of the participants were unaware of this. As the rebel group's act of giving an ultimatum to the British Hong Kong authorities was widely reported by Xinhua as a revolutionary act,[81] it is possible that the 4–14 Faction, less well-known as the Regiment Faction, participated in the campaign to raise their profile and make a name for themselves. However, the action was later qualified as counter-revolutionary, and the 4–14 Faction not only failed to achieve their goal but had to self-criticize for their involvement.[82]

The "Criticism of Xu Shiyou" Incident

The criticism of the Xu Shiyou incident was part of a series of operations to "drag out a small handful in the military," handled by the Regiment Faction under the direct instigation of the Central Cultural Revolution Group, namely Wang Li, Guan Feng, and Qi Benyu, and directed by Kuai Dafu. Military officials identified by Wang Li included Xu Shiyou of the Nanjing Military Region, Han Xianchu of the Fuzhou Military Region, and Chen Xilian of the Shenyang Military Region as the figures that must be defeated. Between July and August 1967, Kuai Dafu dispatched the Regiment Faction personnel to rally local rebels to overthrow Xu Shiyou, Han Xianchu, and Chen Xilian. To compete with the Regiment Faction in the drag out of a small handful in the military, the 4–14 Faction also placed hundreds of personnel in over ten liaison stations throughout the country. Compared to the Regiment Faction, which had over 2,000 personnel and 47 liaison stations, the 4–14 Faction was less influential and operated at a disadvantage. Although the Nanjing 4–14 Faction also wanted to join the ranks of those criticizing Xu Shiyou, they were accused of "pseudo-criticism" by the Regiment Faction. The attacks on the 4–14 Faction came from the Regiment Faction and the local Excellent Faction (*youxiu pai*), which also opposed Xu Shiyou. As such, the 4–14 Faction decided to join the Fengbo Faction, rivaling the Excellent Faction, and join the side of protecting Xu Shiyou. Engaged in fierce armed fighting, the two sides stopped only when Zhou Enlai and Jiang Qing intervened. As Zhou Enlai and Jiang Qing opposed the campaign to expel Xu Shiyou, and Mao Zedong criticized the "Drag out a small handful in the military" as an act to "destroy the Great Wall of China," this time the 4–14 Faction got the upper hand because it happened to be on the right side.[83]

Nationally, July and August 1967 were the most chaotic months of the Cultural Revolution. Even Mao Zedong said in his meeting with Edgar Snow in 1970: "July and August 1967 were extraordinary; the world was in chaos."[84] Such pandemonium resulted from the simultaneous outbreak of antagonisms that existed on many fronts and manifested in two main ways.

First, there was a confrontation between the bureaucratic cadres headed by Zhou Enlai and the Central Cultural Revolution Group, for the Central Cultural Revolution Group, which lacked support from the central leadership, one of its goals was to use the Cultural Revolution to develop its power base. They used mass organizations to bring down as many old bureaucratic cadres as possible and inserted their forces into the power hierarchy. This naturally created a confrontation with Zhou Enlai, who desperately tried to maintain the bureaucracy and protect the cadres. Thus, in the summer of 1967, the Zhongnanhai sit-in, the storming of the British Consulate, and the "exposure of Wu Hao's betrayal" all involved attempts by the Cultural Revolution Group to bring down Zhou Enlai.[85]

Second, there was a confrontation between the Central Cultural Revolution Group and the PLA. The army's suppression of the rebels during the "Support the Left" movement not only brought about the confrontation between the PLA and the rebels but also went against the wishes of the Central Cultural Revolution Group, which supported the rebels, thus causing dissatisfaction among the group. Additionally, Mao's policy of limiting the authority of the PLA weakened the army's ability to quell the rebel faction. Hence, a military section manipulated its supporting mass organizations to confront the Red Guards. An excellent example of this played out in Wuhan, where the military region openly supported and rewarded the actions of the conservative mass organization Million Heroes (*baiwan xiongshi*). This led to violent clashes between Million Heroes and the Workers' Headquarters (*gongren zongbu*).[86]

In addition to the above two causes, power struggles within the army, among the bureaucratic cadres, and between military and local cadres at local levels also contributed to large-scale factional conflicts. Harding points out that Mao's power during the Cultural Revolution was composed of the military forces under Lin Biao, the Central Cultural Revolution Group and radical mass organizations, and the bureaucratic forces of the State Council headed by Zhou Enlai. In the summer of 1967, splits emerged within and among these three forces and between them, also between each of these three forces and Mao.[87] The complication of the power struggle within the upper hierarchy greatly affected the factional struggle within the mass organizations. Forces at the top of the hierarchy manipulated the mass organizations to overcome rival groups. The mass organizations were actively involved in the movement to gain the central government's support in the factional struggle. The result was frequent armed conflict and, in the words of Mao Zedong, "a full-scale civil war unfolded throughout the country."[88] The CCP Central Committee issued several statements resolutely prohibiting armed struggle.[89] Still, since the bureaucracy overseeing the implementation of the Central Committee's orders was paralyzed and the army was involved in the chaos, the Central Committee's directives were utterly ineffective. Answering a call for "civil attack and military defense," Jiang Qing met with rebel groups.[90] Yao Wenyuan published an article in the *Red Flag* magazine entitled "The Proletariat Must Firmly Grasp the Barrel of the Gun,"[91] further fanning the flames of conflict.

As mentioned earlier, Mao's key policy from 1967 onwards was to reestablish power, and to this end, he involved the PLA in the Cultural Revolution. In April 1967, Mao weakened the authority of the PLA to gain more control over it, but this did not mean he favored attacks on the PLA by mass organizations. With the system of maintaining social order destroyed, its restoration could only depend on the PLA. Therefore, in August 1967, Mao Zedong, Zhou Enlai, and Lin Biao, to reestablish the authority of the PLA, renewed the establishment of the Three-In-One Regime and adjusted their policies in the following four areas.

First, as the instigator of "dragging out a small handful in the military," the Central Cultural Revolution Group was reorganized, and Wang Li, Mu Xin, Lin Jie, and Guan Feng, who directly prompted the rebellion, were removed from their posts. At the same time, the party's theoretical journal, the *Red Flag*, which had long published the radical statements of the Central Cultural Revolution Group, was temporarily suspended in

August 1967 after publishing an editorial entitled "Expose a Handful of People in the Military."[92]

Second, the "Support the Army and Cherish the People" campaign was launched nationwide.[93] This was the second time during the Cultural Revolution that the slogan "Support the PLA" was put forward. The first time was in April 1967, when Mao Zedong prevented the rebels from attacking the army.[94] However, the "Support the Army and Cherish the People" campaign could not be carried out due to the Wuhan incident, which intensified fighting.

Third, the mass organizations were reorganized, and their activities were restricted. From September to October 1967, the CCP Central Committee issued a series of documents to limit the activities of mass organizations.[95] The main provisions were as follows: (1) prohibition of revolutionary experience exchange activities; (2) demand for "resumption of classes and revolution"; (3) prohibition of mass organizations owning broadcasting and communication equipment; (4) promotion of a grand coalition of organizations; (5) prohibition of the seizure of weapons from the PLA; and (6) accountability for members of the May 16 Corps, which aimed to overthrow Zhou Enlai."

Fourth, the resumption of party organization activities. On October 27, 1967, in the name of the Central Committee of the CCP and the Central Cultural Revolution Group, a document entitled "Instruction on the Restoration of Party Organizational Life in Units with Established Revolutionary Committees" was issued.[96] Although its actual implementation took place after 1968, the very existence of this document was of great importance at the time.

After September 1967, following the instructions of the Central Government to "resume classes and make revolution," student organizations gradually returned to school. The factional struggle within student organizations continued to revolve around the issue of cadres in the Three-In-One Regime on campuses.

3 Radicalization and the Disintegration of Organizations

As mentioned before, during the "drag out a handful of people in the military" campaign, the 4–14 Faction and Regiment Faction of Tsinghua University competing in public to prove who was more revolutionary through summer social activities while deepened their split. After

September 1967, the 4–14 Faction and Regiment Faction returned to campus to rebuild the Three-In-One Regime. As before, the struggle focused on which cadres should be reinstated into the Three-In-One Regime. The criteria for both sides were simple: to oppose those cadres who supported the opposing faction. Both factions gathered secret materials on old cadres who supported the opposing side and immediately attacked the opposing side for harboring counter-revolutionaries once a problem was discovered.[97]

As mentioned earlier, constructing the Three-In-One Regime is also a process of redistribution of power. In China, where the private sphere does not exist, the seizure of power is the highest level of access to citizenship. The possession of power prioritizes all aspects of political, economic, and cultural rights. Therefore, the seizure of power was a matter of future political status for either side. In September, Tsinghua University held several criticism rallies against old cadres.[98] Still, these rallies had nothing to do with the ideological reform of the cadres and could be described as a competition within the student organizations to gain power.

Wang Shaoguang pointed out that although the Three-In-One Regime was designated as the form of government during the Cultural Revolution, there were no regulations regarding the selection criteria for candidates and the methods for selecting elected officials.[99] Such an unregulated reconstruction of government would naturally lead to struggles for power. In the case of Tsinghua University, since the central government stipulated that establishing the Three-In-One Regime must be based on the "Grand Alliance" of mass organizations, the 4–14 Faction and Regiment Faction temporarily achieved this goal by establishing the Joint Headquarters. Still, eventually, they split, and, with the escalation of their actions, their opposition ultimately turned into violent conflict.

3.1 Realization of the Grand Coalition

The period from late September to November 1967 was a great unity between the 4–14 Faction and the Regiment Faction. Based on Kuai Dafu's recognition of the 4–14 Faction as a legitimate organization and the principle that both sides had the same number of seats and equal status in the joint headquarters, the two sides entered into a coalition on September 22. The new organization was called the Joint Headquarters (Hereinafter referred to as Joint HQ). On 8 November, a Joint

Headquarters Committee was formed, consisting of 17 members of the Regiment Faction and 16 members of the 4–14 Faction.[100]

Why did the opposing sides unite? There are three main reasons for this.

First, as mentioned above, since realizing the Grand Coalition was a prerequisite for establishing the Three-in-One alliances, both for the 4–14 Faction and Regiment Faction, failure to join the coalition would have deprived them of the legitimacy of joining the Three-In-One Regime. Performative alliance for the Three-In-One Regime. This was true for Tsinghua University and many other regions, including Wuhan, where similar situations arose.[101]

Second, Mao's call for a Grand Coalition and the Central Cultural Revolution Group's warnings were crucial factors in the union. In September 1967, the *People's Daily* published Mao's latest instructions, which centered on the Grand Coalition and the end of factional struggle. Mao pointed out: "Within the working class, there is no fundamental conflict of interest. Within the workers under the dictatorship of the proletariat, there is even less reason for a split into two rival factions."[102] This was followed by a nationwide propaganda campaign to realize the Grand Coalition and end factional struggle. Almost at the same time, the Central Cultural Revolution Group put pressure on the leaders of the mass organizations at Tsinghua University: "The whole country is engaged in unity; how long are you going to struggle? If you don't establish some form of unity and cooperation system, we won't be able to invite you to the National Day celebrations."[103]

The third reason was related to Kuai Dafu's political interests. The Grand Coalition of Tsinghua University was made possible by the concessions made by Kuai Dafu in terms of the name of the coalition and the allocation of seats on the committee.[104] However, his compromise was not for the organization's benefit but for his political interests. In September 1967, Mao Zedong proposed to convene the Ninth National Congress of the Communist Party of China (hereinafter referred to as the "Ninth Congress"). As a result, the question of how party delegates were to be elected and what kind of people could become party delegates aroused concern not only in the central leadership but also among mass organizations.[105] The big-character posters in late December 1967 revealed that, Kuai Dafu, already a standing member of the "Beijing Revolutionary Committee," had set his sights on the position of central committee member of the Ninth Congress.[106] The 4–14 Faction ridiculed this, saying, "How could Kuai Dafu become a central committee member when he's not even a Communist Party member?" However, this was not necessarily impossible during the Cultural Revolution. To become a member of the Central Committee, Kuai Dafu had first to realize the Grand Coalition within Tsinghua University.

As mentioned above, the Grand Coalition of Tsinghua University was externally pressured to gain power. This kind of alliance failed to ease the antagonism and competition between the two factions, and the alliance itself soon fell apart.

The Joint HQ showed signs of division from its establishment, as there was a dispute between the two factions over the old cadre Tan Haozheng. Tan was a cadre of the university's Communist Youth League and was one of the first university officials to criticize Liu Shaoqi. Both the 4–14 Faction and the Regiment Faction wanted to recruit him, but Tan preferred the 4–14 Faction. When Tan revealed his preference for the faction, the Regiment Faction began criticizing him. As a result of the Regiment Faction's attack, Tan received full support from the 4–14 Faction. After establishing the Joint HQ in September, the 4–14 Faction hoped that he would become a cadre of the headquarters, but the Regiment Faction rejected this. Neither side was willing to compromise on this issue, and their struggle continued until the Joint HQ split.[107]

In addition to the disagreement about Tan specifically, there are other, deeper reasons for the split between the two factions. They are discussed here in terms of both internal and external causes.

Deeper reasons beyond the dispute over Tan Haozheng caused the split between the two factions. Firstly, the alliance between the two factions was not an organizational alliance but a strategic one among the top leadership. In other words, the union did not have the widespread approval and support of the members. Before the ceremony to celebrate the Grand Coalition, some members of the 4–14 Faction used the excuse that the earlier decided name General Headquarters of the *Jinggangshan* Corps (*jinggangshan bingtuan zongfuwuzhan*) did not show the existence of the 4–14 faction forced their representatives to request the Regiment Faction to change the name to the Joint HQ.[108] Furthermore, some members of the Regiment Faction were also dissatisfied with Kuai Dafu for agreeing to form the Joint HQ for personal interests. On October 5, they set up a Liaison Station for Fighting Privateers and Criticizing Revisionists (*dousi pixiu lianluozhan*) to counter Kuai Dafu's selfish behavior and the revisionism of the 4–14 Faction.[109] In November, opposition from the lower strata became more substantial around the election of the members of the Joint HQ Committee. An article titled "Who Will Serve as Members of the Headquarters Committee," in *Jinggangshan Newspaper*, a publication of the Regiment Faction, on November 17, highlighted this situation.

"Headquarters' members should be for the people, elected by the masses and subject to their supervision. It should not be left to a few people to hide in a small room and bargain for seats as if they were deal-making," and "what kind of people can be members and what kind of people cannot be members should be decided by the masses by setting criteria and by discussion among the masses. Neither should one identify oneself as a 'leftist' nor a member."[110] The organizational breakdown of a coalition lacking a mass base such as this is also evident from the split in several lower-level organizations, such as the Joint Office and the Joint Information Room.[111]

From an external perspective, the provocative speech by Qi Benyu from the Central Cultural Revolution Group destroyed the only basis of the Grand Coalition. As early as October, Qi criticized the coalition at Tsinghua University as fence-sitting and incited Kuai Dafu by saying, "Have you achieved unity so quickly? Didn't Marx call for workers to unite a hundred years ago, and hasn't it been achieved yet?"[112] In December, he said, "The Grand Coalition of Tsinghua University must be centered on the Regiment Faction, and that the 4–14 Faction must recognize the leadership of the Regiment Faction."[113] As mentioned before, the Grand Coalition was based on equal representation and political equality, and the ultimate goal of the two factions was to gain power. Therefore, Qi Benyu's speech intensified the conflict between the two factions and destroyed the Grand Coalition.

3.2 The Hundred Day War

In December 1967, Qi Benyu's speech catalyzed the 4–14 Faction and the Regiment Faction to split again. As early as October, Mao Zedong had already criticized the factional "egocentrism" seen in the Grand Coalition. Accordingly, the 4–14 Faction began to counter the Regiment Faction and regarded its "egocentric" arguments as anti-Maoistic. However, since Qi Benyu rather than the Regiment Faction first proposed that the Grand Coalition should be centered on the Regiment Faction, the Regiment Faction retaliated by attacking the 4–14 Faction's actions as anti-revolutionary, specifically targeting Qi Benyu and the Central Cultural Revolution Group. The Regiment Faction then took the lead and arrested Chen Chusan, the leader of the 4–14 Faction, turning him over to public security authorities as a counter-revolutionary. To secure Chen's release, the 4–14 Faction held a demonstration of more than 1,000 people on December 21.[114] The two factions began attacking each other, with small-scale armed conflict frequently breaking out.

Chen Chusan was released in late December, but this not only failed to end the struggle between the two factions but sharpened the confrontation between them. The political stance of the Regiment Faction became passive. In early 1968, two events directly related to Kuai Dafu occurred in the changing political situation: first, Qi Benyu, a member of the Central Cultural Revolution Group who had been directly commanding Kuai Dafu's actions since 1967, was removed from his position; second, the Party Central Committee declared Chen Lining, who had been dubbed the "modern madman" in 1967, a counter-revolutionary.[115] Due to the close relationship between Kuai Dafu and Qi Benyu and Kuai's full support for Chen Lining, this period presented a challenging situation for Kuai Dafu.[116]

In mid-April 1968, the Regiment Faction arrested old cadres Luo Zhongqi and Li Gang, who supported the 4–14 Faction, and the faction leaders, Wen Xuemin and Rao Weisi, in an attempt to restore its reputation. They were tortured and forced to make "confessions" attacking Mao Zedong and the proletarian headquarters based on the fabricated Luo, Wen, Li, Rao Counter-revolutionary Group. An attempt was made to link them to the Yang, Yu, Fu Counter-revolutionary Group (Yang Chengwu was the PLA's deputy chief of staff, Yu Lijin was the director of the Air Force Political Department, and Fu Chongbi was the commander of the Beijing Military District). On the other hand, in retaliation, the 4–14 Faction also arrested an old cadre who supported the Regiment Faction, Tao Sen. On April 23, just as the 4–14 Faction was preparing to hold a struggle session for Tao Sen, the Regiment Faction seized the venue and launched a violent counterattack. This action continued until July 27, expanding into a large-scale hundred-day-long-armed conflict.[117]

A detailed discussion of the "Hundred Days War" has been provided by Harding and Tang Shaojie, so it will not be repeated here.[118] During the conflict, 11 people were killed, and over 300 were injured.[119] In July, Mao sent the "Capital Workers' Propaganda Team for Mao Zedong

Thought" to stop the armed struggle in schools. In halting the clashes, five workers died and 731 were wounded. After August 1968, the "Capital Workers' Propaganda Team for Mao Zedong Thought" was officially stationed at the university. The Regiment Faction and the 4–14 Faction were automatically disbanded. The student movement, which had thus lasted for nearly two years, finally came to an end.

In September 1967, Mao Zedong called for the cessation of factionalism, and the CCP Central Committee published a series of documents to address factionalism. As it was becoming clear that the re-establishment of order and the restoration of production were central to state policy, and when most people were beginning to retreat from the rebel struggle, why did Kuai Dafu provoke armed struggle against Mao's wishes? There were five main reasons for this.

First, the party and government organs needed time to resume their functions to carry out the central government's instructions. As such, the central directives could not be fully implemented among the masses, and the functions of restraining the behavior of mass organizations could not be effectively carried out.

Second, while the central leadership called on the mass organizations to stop the factional struggle, the power struggle within the central leadership continued, which affected the factional strife at Tsinghua University. In March 1968, as part of the campaign to purge the 5–16 Corps, Yang Chengwu, Yu Lijin, and Fu Chongbi were accused of being a counter-revolutionary group and were criticized. In this context, the Regiment Faction alleged that the Yang, Yu, Fu Counterrevolutionary Clique was behind the 4–14 Faction. By linking the 4–14 Faction to the then-largest counter-revolutionary group publicly criticized, the Regiment Faction expected that the central leadership would intervene directly in the factional struggle at Tsinghua University.[120]

The third reason is that the social control system had not yet been restored, and the power struggle between factions was conducted in anarchy. Power struggles in an anarchic state would inevitably lead to the radicalization of actions.

Fourth, in terms of the faction leaders' strategy, intensifying confrontation by creating enemies was also a way to promote unity within the organization. In the words of Kuai Dafu, "Our ranks are facing division and disintegration," and "Without armed struggle, the old cadres will gradually leave us, and the members will lose their fighting spirit and break away from the ranks."[121]

Fifth, launching armed struggles was a way for the Regiment Faction's leaders to use the central leadership's power to defeat the 4–14 Faction. When Kuai Dafu persuaded members to join the struggle, he said, "We want to expand the situation by starting a struggle, and if the central government intervenes, it will benefit us."[122] Kuai Dafu did so because of the precedent of Peking University, where the New Peking University Commune, led by Nie Yuanzi, launched an armed struggle against its opposition, the *Jinggangshan* Faction of Peking University, on March 27, 1968. By obtaining the support of the Central Cultural Revolution Group, the New Peking University Commune finally defeated the opposition.[123] As a result, Kuai Dafu followed suit and planned the armed struggle with that intention.

3.3 Disintegration of the Organization

In the summer of 1968, the interfactional struggle ended with the intervention of the "Capital Workers' Propaganda Team for Mao Zedong Thought." However, it should be noted that the two factions' organizations had begun disintegrating before the armed struggle began. After October 1967, members of both factions gradually began to withdraw. Some started to work on educational reforms within the schools, while others left the political movement altogether and returned to their studies and private lives.[124] On the eve of the Hundred Days War, the Regiment Faction and the 4–14 Faction saw many departures due to infighting among their members. The result was that only the core members of the factions continued to struggle. Many left the university for safety, and those who remained in the university campus areas formed self-defense units that transcended the factions to safeguard local security.[125] In this way, the students who participated in the armed struggle were completely isolated. By the end of July 1968, the two factions involved in the armed struggle numbered only about 400, of whom about 300 were Tsinghua University students, including 200 from the Regiment Faction, and about 100 from the 4–14 Faction. The rest came from outside the university.[126]

The reasons for the withdrawal of participants can be considered in four main ways.

The first reason for the participants' withdrawal is that the national policy explicitly prohibited factional struggles. In 1968, a political campaign was launched nationwide to criticize factionalism and anarchism. Meanwhile, rebuilding party organizations, reorganizing class ranks, and restoring production activities became the main tasks of the newly established revolutionary committees. To some extent, the implementation of these measures played a role in preventing factional struggles.

Second, as the struggle for power within the state became more superficial, the manipulation of mass organizations by leaders for personal gain gradually became known to society as the movement progressed, resulting in the disengagement of many people from the political movement. For example, with the establishment of the Three-In-One Regime, many of the party and government cadres who were overthrown during the early years of the Cultural Revolution regained power. In addition, during the campaign to "criticize a handful in the military," conservatives and rebels engaged in large-scale armed struggles in various places, resulting in many deaths and injuries. Some cadres in the central leadership who instigated the campaign to "criticize a handful in the military" began to actively promote "support the army and cherish the people" only a month later. These behaviors, which fluctuated with the cycle of national policies and adjusted positions according to the political situation, brought questioning and weariness to the participants of the Cultural Revolution.

Third, the movement's objectives did not meet the interests of the participants. The construction of the Three-In-One Regime turned into a struggle for seats between a minority of faction leaders, which did not reflect the majority's interests. As a result, many participants chose to withdraw from the movement.

Fourth, the disengagement of participants from the movement can be attributed to the manipulation of the masses by faction leaders for personal gain. For example, as demonstrated by the Grand Coalition at Tsinghua University, the alliance formed by faction leaders for their power struggles ultimately led to a loss of cohesion among its members.

4 Concluding Remarks

This chapter analyzes the university rebel groups that were the main body of the second sub-movement of the Cultural Revolution, particularly the composition of the Tsinghua University *Jinggangshan* Corps, which was the main force influencing the movement. From the above analysis, it can be seen that the driving force for the sub-movement was both external and internal. The external thrust refers mainly to the influence of the state on society and can be considered in two ways.

The first is the direct influence of the central leadership. This was mainly through state policy changes and state agents' statements to consciously guide the direction of the Cultural Revolution or to induce the masses. For example, in October 1966, Mao Zedong consciously adjusted state policies in favor of the rebel groups to eradicate the Liu–Deng faction within the party. Additionally, Zhou Enlai and the Central Cultural Revolution Group frequently met with rebel groups. They provided direct support for developing the so-called "minority faction" of the rebel groups.

The second is the indirect influence of the central leadership. Changes in state policy, unclear instructions from the state and state agents, and power struggles within the central administration gave mass organizations opportunities to compete for factional interests. They had an indirect effect on the antagonistic structure of the mass organizations. For example, the ambiguous interpretation of the Sixteen Points broadened the scope of the masses' understanding of the Cultural Revolution. It brought about competition between conservatives and rebels for the position of being "leftist" students. In addition, the rebels used Mao's measures to restrict the army's authority in 1967 to attack the military. The power struggle within the central leadership in 1967 intensified the confrontation among mass organizations and led to nationwide armed struggles.

The two external driving forces from the state discussed above are crucial in understanding the emergence and development of this submovement of the Cultural Revolution led by university Red Guard organizations, especially the influential Tsinghua University Jinggangshan Corps and its formation, disintegration, and dissolution. However, it is not enough to fully explain the opposition between factions solely from the perspective of the central leadership's influence. As previously mentioned, Mao Zedong called on the masses to overthrow the capitalist roaders within the party but also actively promoted the reconstruction of social order and advocated an end to the violent struggle. Despite this, the 1967 Red Guard movement still spiraled out of the state's control and entered a phase of nationwide radicalization. To explain this, it is necessary to consider the internal driving forces of the social movement itself beyond the state elements. The internal driving forces that promoted the social movement can be classified into two categories.

First, intra-factional rivalry is one of the causes of radicalization. For example, the competition between the 4–14 Faction and the Regiment

Faction in 1967, which competing to showcase their "revolutionary" before the masses, is a typical example.

Second, as the example of Kuai Dafu's Hundred Day War shows, the leaders' interests within the faction, their understanding of the movement, and their strategic considerations also contributed to the radicalization of the movement.

In addition, the internal dynamics of the social movement not only affected sectarian actions but also had an impact on the adjustment of national policies and the behavior of national agents. As discussed in this chapter, the actions of conservative groups, such as the 8–9 Faction in 1966, impeded the movement. The government reoriented policy in a radical direction to recruit rebels. However, as the anarchism caused by the mass movement became more pronounced, state policy became increasingly focused on maintaining order. Thus, the sub-movement unfolded due to the state's and society's interaction.

In August 1968, with the introduction of the Capital Worker Mao Zedong Thought Propaganda Team into the university, the factional struggle at Tsinghua University declined. The organization subsequently disbanded on its own. On the surface, the state appeared to have played a leading role in ending the social movement. However, the fact that the government was able to send Mao Zedong Thought Propaganda Teams and the PLA Mao Zedong Thought Propaganda Teams to control the behavior of a small number of radicals was closely related to the decline of the social movement itself. In other words, in 1968, many participants in the mass organizations withdrew, leaving only a few radical members. The stability of social order became the desire of most Chinese people. Only in this case, where the social movement had declined independently, could the government control a few radicals' behavior by introducing worker Mao Zedong thought propaganda teams and the PLA Mao Zedong thought propaganda teams. It was not only the occurrence and development of the social movement but also its decline that could be regarded as a result of the interaction between the state and society.

After the Worker Mao Zedong Thought Propaganda Teams were deployed to universities and middle schools, social order gradually resumed, and the Cultural Revolution moved from disorder to order. As early as January 1967, Mao Zedong had envisioned conducting a "revolution" based on maintaining a certain degree of social order, as evidenced by his use of Shanghai as a model for the reconstruction of political power. This will be further elaborated on in the next chapter.

Notes

1. Shi Yun, op. cit., "Review of the Red Guard Movement" (*Hongweibing yundong shuping*), p. 27.
2. For the composition of school rebel organizations, please refer to the following studies: Stanley Rosen, op. cit. (1975), pp. 645–683; Yin Hongbiao, op. cit., "The Two Currents of the Red Guard Movement", pp. 243–244.
3. For the composition of factory rebel organizations refer to the following studies: Wang Shaoguang, op. cit., pp. 85–92; Li Xun, op. cit., pp. 82–128; Elizabeth J. Perry and Li Xun, *Proletarian Power; Shanghai in the Cultural Revolution* (Boulder, CO: Westview Press, 1997), p. 40.
4. William Hinton, op. cit., *Hundred Days of War: The Cultural Revolution at Tsinghua University*, pp. 76–79.
5. "Speech by Comrade Kuai Dafu at the Conference to Celebrate the Half Anniversary of the Founding of the *Jinggangshan* Red Guards" (Kuai Dafu tongzhi zai qingzhu Jinggangshan hongweibing chengli banzhounian dahuishang de jianghua), The Editorial Department of the Tsinghua University Jinggangshan Newspaper of the Red Generation, *Jinggangshan*, March 25, 1967.
6. "Firing Fiercely at Liu-Deng's Reactionary Line: Remembering the Great Action of the Jinggangshan Corps on 25 January and February" (Xiang Liu-Deng fandong luxian menglie kaihuo: ji jingguangshan bingtuan 12yue 25ri da xingdong), The Editorial Department of the Tsinghua University *Jinggangshan* Newspaper of the Red Generation, *Jinggangshan*, January 1, 1967.
7. Comrade Kuai Dafu's Speech at the Conference to Celebrate the Half Anniversary of the Founding of the Jinggangshan Red Guards, op. cit.
8. Tsinghua University *Jinggangshan* Red Guard Propaganda Team, ed, *Collection of Selected Large Character Posters of Tsinghua University 1966*, pp. 10–11; William Hinton, op. cit., p. 141.
9. Liu Tao, "My Preliminary Self-Criticism" (Wode chubu jiancha), The Editorial Department of the Tsinghua University *Jinggangshan* Newspaper of the Red Generation, *Jinggangshan*, December 31, 1966.
10. Ibid.

11. Kuai Dafu, “Where the Work Team Goes” (Gongzuozu wang nali qu), op. cit., *Collection of Selected Large Character Posters from Tsinghua University 1966*, pp. 1–4.
12. Kuai Dafu, “Statement on the June 27th Congress” (Guanyu 6yue 27ri dahui shengming), “To Comrades Liu Caitang, Wang Tiecheng, etc.” (Zhi Liu Caitang, Wang tiecheng tongzhi), op. cit. In. *Collection of Selected Big-Character Posters from Tsinghua University*, pp. 10–11.
13. *Collection of Selected Big-Character Posters from Tsinghua University*.
14. William Hinton, op. cit., pp. 80–82.
15. Kuai Dafu, “June 9–30 Activities”, (6yue 9ri -30ri huodong qingkuang), op. cit. “*Collection of Selected Big-Character Posters from Tsingghua University*, p. 44; Liu Tao, op. cit., 31 December.
16. “Decision of the Central Committee of the Communist Party of China on the Great Proletarian Cultural Revolution” (Zhongguo gongchandang zhongyang weiyuanhui guanyu wuchanjieji wenhua da geming de jueding), op. cit., *Research Materials on the Cultural Revolution (Part 1)*, p. 72.
17. Harry Harding op. cit., pp. 152–154.
18. Liu Tao, op. cit.
19. Harry Harding, op. cit., pp. 146–147.
20. Addresses by the Central Committee and Related Comrades (Zhongyang ji youguan tongzhi jianghua) op. cit., *Compilation of Materials on the Great Proletarian Cultural Revolution*, pp. 71–153.
21. “Speech of Comrades Jiang Qing and Kang Sheng at Peking University on the Evening of July 25” (Jiang Qing Kang Sheng tongzhi 7yue 25riwan zai beida de jianghua); “Speech of Comrades Guan Feng, Yao Wenyuan, and Ma Li at the Beijing Broadcasting Institute on July 27” (Guan Feng, Yao Wenyuan, Ma Li tongzhi 7yue 27ri zai Beijing guangbo xueyuan de jianghua), op. cit. in *Compilation of Materials on the Great Proletarian Cultural Revolution*, pp. 84, 90.
22. “Tao Zhu’s Speech to Students in the Field on August 21” (Tao Zhu 8yue 21ri dui waidi tongxue de jianghua), op. cit., in *Compilation of Materials on the Great Proletarian Cultural Revolution*, p. 136.

23. "Addresses by the Central Committee and Related Comrades" (Zhongyang ji youguan tongzhi jianghua), op. cit., *Compilation of Materials on the Great Proletarian Cultural Revolution*, pp. 71–153.
24. Uwe Richter, translated by Sadaaki Watanabe, *The Cutural Revolution of Peking University*, Iwanami shoten, 1993, p. 61.
25. William Hinton, op. cit., pp. 91–92.
26. "Speech of Tao Zhu at the Assembly of All Revolutionary Teachers and Students of Peking University to Celebrate the Party's Birthday on July 1" (Tao Zhu 7yue 1ri zai Beijing daxue quanti geming shisheng yuangong qingzhu dang de shengri de dahuishang de jianghua"), Tao Zhu's Speech" (Tan Zhu jianghua), cp. Cit., *Compilation of Materials on the Great Proletarian Cultural Revolution*, pp. 80, 157.
27. Liu Tao, op. cit., My Preliminary Self-Criticism.
28. Liu Tao, op. cit.
29. Ibid.
30. Ibid.
31. For details of the armed struggle that took place at Tsinghua University on August 24, refer to William Hinton, op. cit., pp. 94–104.
32. Liu Tao, op. cit.
33. William Hinton, op. cit., p. 99.
34. Liu Tao, op. cit.
35. Ibid.
36. William Hinton, op. cit., p. 91.
37. Song Yongyi and Sun Dajin, op. cit., p. 79.
38. The 1st HQ (the Red Guard Headquarters of the Capital University and College) was a joint organization formed on August 27, 1966 by conservative Red Guard organizations from various universities. The 2nd HQ (Headquarters of the Red Guards of the Capital Universities) was a joint organization formed on September 5, 1966 by some of the rebel organizations of the universities to confront the 1st HQ, with political views similar to those of the 3rd HQ. On November 7, 1966, a part of the 2nd HQ broke away from the parent body and established the "Revolutionary Rebel Liaison Station of the Capital University and College Red Guard Headquarters." Subsequently, the "Revolutionary Rebel Liaison Station of Capital University and College

Red Guard Headquarters" assumed the leadership of the 2nd HQ.

39. "Zhou Enlai's Speech to the Revolutionary Rebel Headquarters of the Capital's Colleges and Universities at the Workers' Gymnasium on September 26" (Zhou Enlai 9yue 26ri zai gongren-tiyuguan dui shoudu dazhuanyuanxiao geming zaofan silingbu de jianghua), *Compilation of Materials on the Great Proletarian Cultural Revolution*, pp. 215–220.
40. On 5 October 1966, the General Political Department of the Central Military Commission, in order to rehabilitate those who were suppressed by the Work Team, released a document entitled "Urgent Instruction on the Great Proletarian Cultural Revolution in Army Colleges and Universities." Although this instruction was issued directly to the military-affiliated schools within the PLA, the CCP added to it the "Official Instruction", extending this "Urgent Instruction" to universities, junior high schools, and high schools above the provincial level throughout the country, asking all students and teaching staff to resolutely implement it. The CCP central leadership made the following call: "All comrades who were branded as 'counterrevolutionaries', 'anti-party elements', 'rightists' and 'false leftists and true rightists' by the Party committees and Work Teams of the institutions at the beginning of the movement should be declared null and void and be rehabilitated in public. The materials which individuals are forced to write for self-criticism should all be returned to them for disposal, and the materials collated by the Party Committee and the Work Team and by others should be discussed with the masses for disposal, and with the consent of the masses, they may also be destroyed in public" ("Urgent Instructions on the Great Proletarian Cultural Revolution in Military Colleges and Universities") (Guanyu jundui yuanxiao wuchanjieji wenhua da geming de jingji zhishi), edited by the Headquarters for the Vindication of Revolutionary Mass Appropriation of Capital in Guangdong Province, *Compilation of Vindication Materials*, January 1968, pp. 1–2.
41. Chen Boda, "The Two Routes in the Great Proletarian Cultural Revolution" (Wuchanjieji wenhua da geming Zhong de liangtiao luxian), op. cit., "*Research Materials on the Cultural Revolution*" *(Part 1)*, pp. 133–141.

42. "Speech by Comrades Chen Boda, Jiang Qing and Zhou Enlai on 6 October at the 'National Pledging Conference for Revolutionary Teachers and Students in Beijing to Fire Fiercely at the Bourgeois Reactionary Line'" (Chen Boda, Jiang Qing,Zhou Enlai tongzhi 10yue 6ri zai "quanguo zaijing geming shisheng xiang zichanjieji fandong luxian menglie kaihuo shishi dahui" shang de jianghua), op. cit., *Compilation of Materials on the Great Proletarian Cultural Revolution*, pp. 247–248.
43. Liu Tao, op. cit.
44. "Speech of Comrade Guan Feng" (Guan Feng tongzhi jianghua), op. cit., *Compilation of Materials on the Great Proletarian Cultural Revolution*, p. 250.
45. William Hinton, op. cit., p. 100.
46. "Important Statement by the Tsinghua University Red Guard Rebel Corps on Freezing the Big Seal of the 'Tsinghua University Red Guards'" (Qinghua daxue hongweibing zaofan bingtuan guanyu dongjie "qinghua daxue hongweibing" dayin de zhongyao shengming), op. cit., *Jinggangshan*, 19 December 1966.
47. Willanm Hinton, op. cit., p. 103.
48. Chen Boda, "The Two Routes in the Great Proletarian Cultural Revolution", (Wuchanjieji wenhua da geming zhong de liangtiao luxian), op. cit., *Research Materials on the "Cultural Revolution" (Part 1)*, p. 135.
49. On October 12, 1966, the Maoist Red Guards of the Beijing Broadcasting Institute posted a big-character poster, "The Central Cultural Revolution Group should carry out self-criticism in our institute" (Zhongyang wenge xiaozu yinggai dui woyuan jinxing ziwo piping), criticizing the Central Cultural Revolution Group for distrusting the masses, suppressing them, and imposing its own opinions on them. In late November the 8–1 Column of the Beijing Aviation Institute posted several big-character posters criticizing the Central Cultural Revolution Group. On December 1, Li Hongshan of the Beijing Forestry Institute made a speech fiercely criticizing the Central Cultural Revolution Group.
50. "Four Refutations of Fallacies" (Bo miulun size), op. cit., *Jinggangshan*, December 19, 1966.
51. "Speech by Comrades Jiang Qing and Chen Boda at the Beijing Normal University on July 27" (Jiang Qing, Chen Boda tongzhi

7yue 27ri zai beishida de jianghua), op. cit., *Compilation of Materials on the Great Proletarian Cultural Revolution*, p. 89.
52. William Hinton, op. cit., p. 126.
53. William Hinton, op. cit., p. 127.
54. Minutes of the Discussion between Comrades Jiang Qing, Zhang Chunqiao, Guan Feng, Qi Benyu, Song Renqiong, Yao Wenyuan, and Other Comrades Receiving Representatives of the 1st HQ, 2nd HQ Rebel Liaison Station, 3rd HQ, Capital Corps, and the Revolutionary Faction of Some Colleges and Universities at the Great Hall of the People on December 18 (Jiang Qing, Zhang Chunqiao, Guan Feng, QI Benyu, Song Renqiong, Yao Wenyuan deng tongzhi 12yue 18ri zai renmin dahuitang jiejian yisi, ersi zaofan lianluozhan, sansi, shoudubingtuan yiji bufen yuanxiao gemingpai daibiao de zuotan jiyao), op. cit., *Compilation of Materials on the Great Proletarian Cultural Revolution,* p. 580; "Resolute Support for the Revolutionary Action of the Red Guard Rebel Corps of Tsinghua University" (Jiejue zhichi Qinghua daxue hongweibing zaofan bingtuan de geming xindong), op. cit., *Jinggangshan*, December 19, 1966.
55. Tang Shaojie, "The Rise and Fall of the Tsingghua *Jinggangshan* Corps" (Qinghua jingguangshan bingtuan de xingwang), op. cit., *The Cultural Revolution: Facts and Analysis*, p. 51.
56. William Hinton, op. cit., pp. 137–138.
57. For details on the power organization, organizational system, and propaganda system of the Corps, see Tang Shaojie, "The Rise and Fall of the Tsinghua *Jinggangshan* Corps", pp. 59–61.
58. "General Order No. 6 of the General Headquaters of the Revolutionary Rebellion of the Red Guards of the Capital University and Colleges – 'No Cheating and Deceiving'" (Shoudu dazhuan yuanxiao hongweibing geming zaofan zongsilingbu di 6hao tongling: buxu zhaoyao zhuangpian) was issued by the General Handquaters of the Revolutionary Rebellion of the Red Guards of the Capital University and Colleges (3rd HQ), *Capital Red Guards*, January 31, 1967. According to the "Capital Red Guards", on January 31, 1967, Kuai Dafu, in the name of the "3rd HQ", set up liaison stations in Shenyang, Chongqing, Changsha, Nanjing, Shanghai, Hangzhou, Xi'an, Nanchang, Anshan, Fuzhou, Hefei, Nanchong, Yinchuan, Changzhou, Yangzhou, Guizhou, Xuzhou, Tianjin, Nanning,

Qiqihar, Qingdao, Lushun, Guangzhou, Taiyuan, Zhangzhou, Jilin, Urumqi, Chengdu, Zhengzhou, Luoyang, Jinan, Tongxian, Tangshan, Wuhan, Kunming, Yantai, Shantou, Wuxi, Zhuzhou, Weifang, Handan, Qingdao, Harbin, Anxin, and other cities.

59. For details on the internal publications of the Corps, see Tang Shaojie, "The Rise and Fall of the Tsinghua *Jinggangshan* Corps", p. 51.
60. Harry Harding, op. cit., at 152–154.
61. Uwe Richter, op. cit., p. 61.
62. "Minutes of the Talk by Comrades Zhou Enlai and Jiang Qing on Receiving Representatives of the Red Guards in Some Schools on November 8" (Zhou Enlai, Jiang Qing tongzhi 11yue 8ri jiejian bufen xuexiao hongweibing daibiao tanhua jiyao), *Remarks by Comrades in Charge of the Central Committee on the Great Proletarian Cultural Revolution (sequel)* (Zhongyang fuze tongzhi guanyu wuchanjieji wenhua da geming de jianghua), 1967, pp. 15–17; "Instructions by Comrade Chen Boda on November 8" (Chen Boda tongzhi 11yue 8ri zhishi); "Remarks by the Head of the Central Committee on the Treatment of the Zhongnanhai Question on November 11" (Zhongyang shouzhang 11yue 11ri duidai Zhongnanhai wenti de jianghua), op. cit., *Compilation of Materials on the Great Proletarian Cultural Revolution*, pp. 334, 338–342.
63. "Notice of the Central Committee of the Communist Party of China on Short-term Military and Political Training for Revolutionary Teachers and Students in Universities and Secondary Schools" (Zhonggong Zhongyang guanyu dui dazhongxuexiao geming shisheng jinxing duanqi junzheng xunlian de tongzhi), op. cit., *Research Materials on the "Cultural Revolution" (Part 1)*, p. 195.
64. "Regulations of the Central Committee of the Communist Party of China on the Current Proletarian Cultural Revolution in Universities and Colleges" (Zhonggong Zhongyang guanyu dui dazhuanyuanxiao dangqian wuchanjieji wenhua da geming de guiding); "The Resumption of Classes and Revolution in Primary and Secondary Schools" (Zhongxiaoxie fuke nao geming), op. cit., *Research Materials on the "Cultural Revolution" (Part I)*, pp. 330, 331.
65. William Hinton, op. cit., p. 140.

66. Ibid.
67. Ibid.
68. Zhou Quanyang, "The 4–14 Trend Wave Must Win" (4–14 sichao bisheng), in Song Yongyi and others (eds.), op. cit., pp. 390–408.
69. Ibid., p. 393.
70. "Decision of the CCP Central Committee, the State Council, the Central Military Commission, and the Central Cultural Revolution Group on the People's Liberation Army's Resolute Support for the Revolutionary Leftist Masses" (Zhonggong Zhongyang, guowuyuan, zhongyangjunwei, zhongyangwenge xiaozu guanyu renminjiefangjun jiejue zhichi geming zuopai qunzhong de jueding, op. cit., *Research Materials on the "Cultural Revolution" (Part 1)*, pp. 258–259.
71. Wang Nianyi, op. cit., p. 201.
72. Ibid., p. 200.
73. "Notice of the Central Committee of the Communist Party of China on Not Directing the Struggle at the Military" (Zhonggong Zhongyang guanyu bude ba douzheng fengmang zhixiang jundui de tongzhi), op. cit., *Research Materials on the "Cultural Revolution" (Part 1)*", p. 250.
74. Wang Nianyi, op. cit., pp. 202–206.
75. "Decision of the CCP Central Committee, the State Council, the Central Military Commission, and the Central Cultural Revolutionary Group on the Qinghai Issue" (Zhonggong Zhongyang, guowuyuan, Zhongyangjunwei, Zhongyang wenhua da geming xiaozu guanyu Qinghai wenti de jueding) (7 May 1967); "Decision of the CCP Central Committee on the Anhui Issue" (Zhonggongzhongyang guanyu anhui wenti de jueding) (24 March 1967); "Decision of the CCP Central Committee on the Anhui Issue" (Zhonggong zhongyang guanyu anhui wenti de jueding) (1 April 1967); "Decision of the CCP Central Committee on the Fujian Issue" (Zhonggong Zhongyang guanyu Fujian wenti de jueding) (30 April 1967); and "Decision of the Central Committee of the CCP on the Sichuan Issue" (Zhonggong Zhongyang guanyu Sichuan wenti de jueding) (7 May 1967), op. cit., *Research Materials on the "Cultural Revolution" (Part 1)*, pp. 371, 372, 373, 389, 445, 446.

76. "Order of the Central Military Commission" (Zhongyang junwei mingling), op. cit., *Research Materials on the "Cultural Revolution" (Part 1)*, pp. 390–391.
77. Wang Shaoguang, op. cit., p. 115.
78. Content derived from the author's interviews with individuals who experienced the Cultural Revolution in Yuci, Shanxi, in 2000.
79. Willam Hinton, op. cit., pp. 154–157.
80. Ibid., pp. 169–172.
81. "The Red Guards in the Capital Acted Strongly against the British Charge" (Shoudu hongweibing dui yingdaiban caiqu qianglie xingwei) (August 22, 1967), op. cit., *Research Materials on the "Cultural Revolution" (Part 1)*, p. 555.
82. William Hinton, op. cit., p. 172.
83. Ibid., pp. 157–168.
84. "Summary of Chairman Mao's Meeting with American Friend Snow" (Mao zhuxi huijian meiguo youhao renshi sinuo tanhua jiyao), op. cit., *Research Materials on the "Cultural Revolution" (Part 2)*, p. 493.
85. For details of the "May 16 Corps", see Harry Harding op. cit., pp. 183–186.
86. Wang Shaoguang, op. cit., pp. 131–157.
87. Harry Harding, op. cit., pp. 166–173.
88. "Summary of Chairman Mao's Meeting with American Friend Snow", op. cit., *Research Materials on the "Cultural Revolution" (Part 2)*, p. 493.
89. "General Order of the CCP Central Committee, the State Council, the Central Military Commission and the Central Cultural Revolution Group" (Zhonggong Zhongyang, guowuyuan, zhongyangjunwei, zhongyangwengexiaozu tongling) (6 June 1967); "Notice of the CCP Central Committee" (Zhonggong Zhongyang tongzhi) (24 June 1967); "Notice of the CCP Central Committee on the Prohibition of Provoking Peasants to Enter Cities for Armed Fighting" (Zhonggong zhongyang guanyu jinzhi tiaodong nongmin Jincheng wudou de tongzhi) (13 July 1967); "Certain Decisions of the CCP Central Committee on the Hunan Issue" (Zhonggong zhongyang guanyu hunan wenti de ruogan jueding) (10 August 1967), etc.,

op. cit. *Research Materials on the "Cultural Revolution" (Part 1)*, pp. 492, 496, 500, 531–532.

90. "Jiang Qing's Speech on Cultural Attacks and Armed Conflicts at the Meeting with the Delegates from Henan" (Jiang Qing zai jiejian henan daibiao shi guanyu wengong wuwei de jianghua) (July 23, 1967), *Cultural Revolution Research Materials (Part 1)*, p. 501.
91. "The Proletariat Must Firmly Grasp the Barrel of the Gun" (Wuchan jieji bixu laolao zhangwo qiangganzi) (1 August 1967), op. cit. *Research Materials on the Cultural Revolution (Part 1)*, pp. 525–557.
92. Harry Harding, op. cit., p. 184.
93. "Call of the CCP Central Committee, the State Council, the Central Military Commission and the Central Cultural Revolution Group on the Launching of the Campaign to Support the Military and Cherish the People" (Zhonggong zhongyang, guowuyuan, zhongyangjunwei, zhongyangwengexiaozu guanyu kaizhan yongjun aimin huodong de haozhao) (25 August 1967); "Notice of the CCP Central Committee and the Central Cultural Revolution Group on the Launching of the Campaign to Support the Military and Cherish the People" (29 August 1967), op. cit. *"Cultural Revolution" Research Materials (Part 1)*, pp. 558, 559.
94. "Jiang Qing's Speech at the Founding and Celebration of the Beijing Revolutionary Committee (April 20, 1967)" (Jiang Qing zai beijingshi geming weiyuanhui chengli he qingzhu dahui shang de jianghua), op. cit., *Research Materials on the "Cultural Revolution" (Part 1)*", p. 435.
95. *Research Materials on the "Cultural Revolution" (Part 1)*, pp. 584, 585, 593.
96. "Instructions from the Central Committee of the CCP and the Central Cultural Revolution Group on the Restoration of Party Organizational Life in Units where Revolutionary Committees have been Established" (Zhonggong zhongyang, zhongyang-wengexiaozu guanyu yijing chengli le geming weiyuanhui de danwei huifu dangde zuzhi shenghuo de pishi) (27 October 1967), op. cit., *Research Materials on the Cultural Revolution (Part 1)*, p. 600.
97. William Hinton, op. cit., p. 177.
98. Ibid.

99. Wang Shaoguang, op. cit., p. 168.
100. Tang Shaojie, op. cit. "The Rise and Fall of the Tsinghua *Jinggangshan* Corps", p. 58.
101. Wang Shaoguang op. cit., pp. 173–174.
102. "Realizing the Great Revolutionary Union at the Climax of the Great Revolutionary Critique", *People's Daily*, September 14, 1967.
103. William Hinton, op. cit., p. 180.
104. According to Hinton's research, before the celebration of the Grand Coalition, there was a confrontation between the two over the name change, and the Grand Coalition was on the verge of bankruptcy, but in the end, Kuai Dafu gave in (William Hinton, op. cit., p. 181).
105. "Notice of the Central Committee of the Communist Party of China, the Central Cultural Revolution Group on the Consultation of Opinions on the Ninth Congress" (Zhonggong zhongyang zhongyangwenge guanyu zhengxun dui "jiuda" yijian de tongzhi (October 21, 1967), op. cit., *Research Materials on the "Cultural Revolution" (Part 1)*, pp. 598–600.
106. William Hinton, op. cit., p. 192.
107. Ibid., pp. 183–184.
108. Ibid., pp. 181–182.
109. *Jinggangshan*, op. cit., October 12, 1967.
110. "Who will be the headquarters member", (Zongbu weiyuan you shui laidang? *Jinggangshan*, op. cit., November 17, 1967.
111. William Hinton, op. cit., p. 185.
112. Ibid., p. 184.
113. Ibid., p. 187.
114. *Jinggangshan*, op. cit., December 28, 1967.
115. Chen Lining came from a landowning family and was a revolutionary cadre in Xiangtan City, Hunan Province. Before the Cultural Revolution, he wrote a long article criticizing Liu Shaoqi, for which he was dismissed from his post and finally imprisoned in a mental hospital. However, in January 1967, the Red Teachers' Union, a rebel group at Tsinghua University, demanded Chen's release in order to prove Chen's theory that "today's revolutionaries must emerge from among yesterday's opponents." The Beijing Public Security Bureau said that Chen's attacks on Mao exceeded even those on Liu Shaoqi and that

he would be prosecuted for counter-revolutionary crimes even after he was discharged from hospital. The Red Teachers' Union attacks the Ministry of Public Security. Wang Li and Qi Benyu of the Central Cultural Revolution Group considered Chen a "hero of the revolution" and a "modern madman" and supported the actions of the Red Teachers' Union, which demanded his release. At Tsinghua University, Kuai Dafu strongly advocated Chen Lining. In February 1968, six months after Wang Li stepped down, the Central Committee issued a statement that Chen Lining was a counter-revolutionary.

116. William Hinton, op. cit., p. 193.
117. Tang Shaojie, op. cit. "The Rise and Fall of the Tsinghua *Jinggangshan* Regiment", p. 59.
118. Hinton's and Tang Shaojie's studies provide a very detailed process of the Hundred Day War and will not be repeated here. William Hinton, *hundred Days of War: The Cultural Revolution at Tsinghua University*; Tang Shaojie, "The Death Knell of the Red Guard Movement: The Hundred Days of the Great Armed Conflict at Tsinghua University", ("Hongweibing yundong de sangzhong: Qinghua daxue bairi dawudou", collected in Liu Qingfeng, ed. op. cit., pp. 65–79.
119. Tang Shaojie, op. cit. "The Rise and Fall of the Tsinghua *Jinggangshan* Corps", p. 59.
120. Tang Shaojie, op. cit. "The Death Knell of the Red Guard Movement: The Hundred Days of the Great Armed Conflict" at Tsinghua University", p. 66.
121. Ibid., p. 66.
122. Ibid.
123. Uwe Richter, op. cit., pp. 76–86.
124. William Hinton, op. cit., pp. 178–179.
125. Ibid., p. 209.
126. Ibid., p. 217.

CHAPTER 7

The Reconstruction of the Regime and the Shanghai Model

Mao's support for the January Revolution in Shanghai allowed the Cultural Revolution to enter the third sub-movement period, which lasted from January 1967 to April 1969. The main objective for this sub-movement was building the Three-In-One Regime. During this period, Shanghai was taken as a model for the country's regime reconstruction, and the Shanghai Workers' Revolutionary Rebellion General Headquarters (hereinafter referred to as the "Workers' General HQ") commanded by Wang Hongwen replaced the student organizations as the main force of the Cultural Revolution. The workers' rebel groups, which were not among the vested interest groups in the pre-Cultural Revolutionary' times, were able to make their mark during this period. They gained the status of the highest privilege of revolution and thus the opportunity to seek power and benefits. The competition around various political rights and interests was more intense in the workers organizations because their involvement in interests at stake was more specific than that of the students, thus intensifying the rivalry between factions throughout the country. Armed conflict was a frequent occurrence. In this sense, it can be said that the January Revolution in Shanghai sparked national turmoil during the Cultural Revolution.

Y. Lijun, *Contest for Citizenship and Collective Violence during China's Cultural Revolution*, IPP Studies in the Frontiers of China's Public Policy, https://doi.org/10.1007/978-981-97-0906-9_7

This chapter will analyze how the "General Headquarters" evolved from its formation to the point of developing quasi-governmental functions. Specifically, by examining the characteristics of the three levels of actors—the central leadership, local party and government organs, and various social factions—involved in establishing the Three-In-One Regime in Shanghai, we arrive at a conclusion entirely different from previous studies. In summary, the "Three-in-One" political power structure in Shanghai resulted from the party-state power reconstruction process led by Zhang Chunqiao. It had fundamental differences from the power struggles that occurred in other regions. In other words, Shanghai was selected as a model for reconstructing political power because of its stable order. It provided a "governance" (*Zhi*) model to the rest of the country. This made Shanghai distinct from other cities that were experiencing "great turmoil" (*luan*).

To illustrate this argument, this chapter will focus on a series of questions that have been neglected in previous studies: Why did Mao shift his focus from the student organizations, which he had been supporting in the past, to the workers' rebellion in 1967 and use the workers' rebellion movement in Shanghai as a model for reestablishing order? What were the characteristics of the Three-In-One Regime in Shanghai? And what role did the workers' rebellion movement play in the stability of Shanghai at that time?

1 The Social Context in the Development of the Shanghai Model

Why was the rebel movement in Shanghai selected as a model for the reconstruction of political power throughout the country? To answer this question, we must first trace back to the effects of the Cultural Revolution on China's politics, economy, and society after the end of 1966. This is an extremely important issue for understanding the establishment of the Shanghai model and its position in the Cultural Revolution. The impact of the Cultural Revolution on Chinese society at that time was multifaceted, and we will mainly discuss it from the following three aspects.

1.1 *The Social Impact of the "Great Link-Up"*

The "Great Link-up" (*da chuanlian*) was, in Zhou Enlai's words, "to deepen the conduct of the Cultural Revolution," with "students from all over the country coming to Beijing to exchange experiences, and students from Beijing going out to deepen revolutionary ties."[1] The nationwide Great Link-up began in August 1966. On 16 November that year, the Central Committee of the CCP and the State Council instructed the suspension of "Link-up" using railways, shipping, and public transport. As a result, the nationwide student Link-up activities gradually subsided.[2] During these four months, Mao received students eight times in Tiananmen Square, and about 12 million students came to Beijing from the local area. In the first half of November, more than 250,000 students from all over the country came to Beijing daily, reaching a peak of 290,000.[3] In addition, while students from all over the country came to Beijing, many students from Beijing went to different parts of the country to exchange revolutionary experiences. Generally speaking, the Link-up activities in Beijing were called "going north," and the Link-up activities from Beijing to other places were called "going south." According to Zhou Enlai's speech, the "southward Link-up" peaked in October, when 40,000–50,000 students from Beijing went to various places every day.[4]

The original purpose of the Great Link-up was to share the Beijing experience of the Cultural Revolution with the whole country through organized visits to Beijing.[5] But since the party, which had long been the leading operator of the schools, had lost its command, it was difficult to achieve the organized Link-up that the central leadership had expected. In addition, the Great Link-up placed a considerable burden on the state treasury, as the government was responsible for food, lodging, transportation, and so on. To solve the problem of accommodation and food for a large number of arrivals, the central leadership mobilized the army, government agencies, factories, schools, and residents' committees, asking them to provide dormitories, classrooms, meeting halls, canteens, offices, and other public facilities, sometimes even by administrative order.[6] Moreover, as Zhou Enlai's comments and central documents show, providing medical care for the arrivals, transportation for meetings with Mao, and maintaining order were all urgent problems for Beijing, which became overwhelmed after November 1966.[7]

The southward Link-up began after Mao Zedong received the Red Guards in Tiananmen Square on August 18, 1966. Initially, the southward Link-up was a spontaneous action, not an organized activity, but in September, there was a tendency for more organization.[8] In the case of Shanghai, according to Li Xun's research, there were three large-scale southward movements, which took place in late August, mid-September, and early October 1966. The first was a spontaneous act of thousands, mostly Beijing students (1st HQ), with no unified name or leadership structure. The second time involved tens of thousands of people and was organized under the "Capital Red Guards Command Southward Troop," (2nd HQ) with a relatively more organized structure. The third southward movement was the smallest of the three. Still, it was centered on the 3rd HQ and other Rebel Red Guards organizations in Beijing, which were directly connected with the Central Cultural Revolution Group. This 3rd HQ-related southward movement directly influenced Shanghai's Cultural Revolution.[9]

Unlike the northward movement (which all went to Beijing), the southward movement was scattered in numbers and geography. As a result, the southward activity was less burdensome on local communities as in Beijing, but it did cause significant political disruption. Because the southward students brought the revolutionary style of the Cultural Revolution, which "doubted everything and overthrew everything," they shook the leadership of the local party and government organs to their core. The result upset the legitimacy of the local party and government in leading the Cultural Revolution. The southward students intensified the opposition between the masses and promoted the organization of factions and the rise of the worker rebellion movement. Using Shanghai as an example, the following are three aspects to illustrate this point.

First, the "southward" students contributed to the progress of the local Cultural Revolution. In contrast to locals, the students were characterized by the fact that their criticism was not directed at those in power within the schools and factories but directly at the heads of the provincial and municipal parties and government institutions. They did not have direct leadership relations with local party and government institutions as outsiders. They were not bound by the patron–client relations formed before the Cultural Revolution, nor were they worried that the cadres they had criticized would turn around and persecute them if the political situation changed. As such, their actions at the beginning of the Cultural Revolution were often more aggressive than those of the local rebels.

Most of the students who went south, whether rebels or conservatives before, became brave rebels once they arrived at their destination.[10]

According to Li Xun's research, Red Guard organizations existed in Shanghai before the first southward movement of Beijing students. However, unlike in Beijing, the Red Guards in Shanghai were supported and led by the city's party, government agencies, and school work teams. The leaders and members of these organizations were progressive students, similar to those of the Communist Youth League. The children of the municipal party and government cadres played a leading role in the Shanghai Red Guard organization. They acted under the leadership of the municipal party and government agencies. Also, there was no large-scale workers' organization in the factories during this period. The few workers dissatisfied with the status quo criticized the factory cadres and the "Four Clean-ups" Work Team cadres. Their actions were confined to the factories, and their influence hardly reached the municipal party and government institutions.[11]

This all changed when the southward students arrived in Shanghai. Shouting the slogan "bombard the Shanghai Municipal Party Committee," they directed their criticism directly at the party organizations.[12] On September 4, 1966, a group of students and workers in Shanghai issued an ultimatum to the then-mayor of Shanghai, Cao Diqiu, demanding that he acknowledge their complaints.[13] This became the turning point of Shanghai's Cultural Revolution.

Second, the southward students intensified the division among the local masses. Li Xun points out that the attacks of the students on the Shanghai party and government institutions made clear the previously unclear camps of Shanghai conservatives and rebels. Activists, "old workers," other progressive workers, cadre children, and progressive students who were close to the municipal party and government institutions rallied to protect them while resisting the actions of the southward students. In contrast, workers and students who were excluded from the circle of vested interests before the Cultural Revolution and were discontented with society supported the Beijing students' actions while attacking the city's party and government organizations.[14]

Third, the actions of the southward students contributed to the organization of the local masses, leading the Shanghai rebellion from within schools and factories to the general public. As a result, people with similar views deepened their communication and laid the groundwork for forming joint organizations at the municipal level. In this context, as shown in Table 1, several students and workers' unions were established in Shanghai between September and late November.

1.2 The Social Impact of the Workers' Rebellion

After October 1966, the Cultural Revolution, confined to cultural and educational circles, was extended to factories. The participation of industrial workers in the Cultural Revolution had a devastating impact on the nation's economy and social order in many ways.

First, the involvement of industrial workers in the rebel movement brought about the spread of economism. At the forefront of the financial demands were the mass organizations of the urban underclass.[15] Various rebel groups such as the half-worker/half-students(*bangong bandu xuesheng*), temporary-contract workers (*linshigong hetonggong*,[16] peasant workers (*guinong gongren*), and "support for inland regional workers" (*zhinei*),[17] As well as the trainee workers' (*shixi gongren*) rebel group[18] and the sent-down youth (*xiafang qingnian*) rebel group.[19] Among these, the National Red Workers' General Rebellion Group (hereinafter referred to as the "General All Reds") (*quanguo hongse gongren zaofan zongtuan, "quan hong zong"*), formed by temporary and contract workers, was the most active.[20] The demands of these mass organizations for economic rights were mainly focused on reforming the system and obtaining urban Shanghai hukou.[21] Beginning with mass organizations of the urban underclass seeking economic benefits, in December 1966, regular workers also launched activities nationwide to seek various economic benefits.[22]

The economic demands of the formal workers in Shanghai culminated in December 1966, launching a nationwide campaign for various economic rights.[23]

Their demands differed from those of the mass organizations at the low end of society and focused mainly on higher wages, allowances, and other monetary and material benefits.[24] According to recorded statements by Zhou Enlai and Jiang Qing, in January 1967, there was an indiscriminate payment of workers' wages and allowances in Shanghai, which resulted in a cost of more than 50 million yuan,[25] reaching between 10 and 20 million yuan in the province of Heilongjiang.[26] Faced with such a state of affairs, the government had to take military control measures on the banks and implement a policy of freezing corporate finances.[27]

Table 1 Composition of mass organizations in Shanghai in 1966

	Date established	*Conservative faction*	*Date established*	*Rebel faction*	*Notes*
United Student Organizations	12 September	Shanghai Red Guard Headquarters (Upper Red Headquarters)	12 October	Shanghai Revolutionary Committee of the Red Guards of Shanghai Universities and Colleges (Red Revolutionary Association)	
	26 September	Red Guard Shanghai University and College Headquarters (Shanghai Red University and College Headquarters)	22 November	Shanghai Red Guard Revolutionary Rebellion General Headquarters (Red Revolution Headquarters)	In December, the Red Revolutionary Headquarters split into the Shanghai Red Guard Revolutionary Rebellion Headquarters (Upper 3rd Headquarters) and the Shanghai Third Command of the Red Guards (Red 3rd Headquarters)
United Workers' Organization	26 November	Shanghai Workers' Scarlet Guards (the Scarlet Guards)	6 November	Shanghai Workers' Revolutionary Rebellion General Headquarters (Workers' General Headquarters)	

Note Abbreviations are given in brackets.
Source Chen Zu'en et al., A General History of Shanghai (vol. 11), p. 21; Shanghai People's Publishing House, 1999, 219; Li Xun, The Great Collapse—A History of the Rise and Fall of the Workers' Rebellion Headquarters in Shanghai, Times Publishing, 1996, p. 155

The disruption caused by the spread of economism was not only financial nature but also affected other dimensions. For example, the mass dislocation of workers from production sites caused rail and freight transport failures. The need for raw materials, electricity, and fuel supplies to sustain production led to a sharp drop in output. In addition, the surge in purchasing power caused by the indiscriminate disbursement of subsidies and other funds had a knock-on market effect, upsetting the balance between demand and supply.[28]

Second, the involvement of workers in politics caused significant disruption in society. While some workers' organizations joined the rebellion to achieve economic interests, others organized for political purposes. As a result, there were frequent strikes by many workers who broke away from production sites. This was especially seen in the big cities, which at one point caused disruptions in the railway operations in Beijing, Shanghai, Nanjing, and Bengbu for up to 12 days.[29]

In addition, while seeking organizational legitimacy, workers' rebels staged hunger strikes to complain about the repression of local party organizations and conservative workers during the early years of the Cultural Revolution, and there was a sharp increase in the number of forced train stops to Beijing by protestors lying on the tracks.[30] For example, at the Daqing oilfield, considered an industry model, more than 10,000 of the 70,000-plus workers went to Beijing, which interrupted production in part of the oilfield.[31] Some of the workers who went to Beijing tried to force their way into Zhongnanhai to tell Zhou Enlai and the Central Cultural Revolution Group about their experiences of repression and resorted to the extreme method of sit-ins in the Tiananmen Square and Wangfujing Street.[32] Under such circumstances, maintaining normal production activities took much effort.

Third, the emergence of national rebel groups posed a significant threat to the regime. Between the end of 1966 and the beginning of 1967, rebel groups of various names formed a nationwide coalition that sought political and economic rights. These organizations were not merely mass organizations mobilized to participate in the Cultural Revolution but took on the nature of interest groups. In particular, the nationwide rebel groups of military veterans posed a significant threat to the regime.[33] As a result, in February 1967, the Central Committee of the CCP reorganized the then 150 national organizations, banning nine of them, including the military veterans' organizations, and forcibly disbanding all those remaining.[34]

Fourth, the radicalization and disengagement of the mass organizations from government control led to massive social unrest. In addition to the violent storming of Zhongnanhai and the forced interception of trains mentioned above, mass organizations also occupied radio stations and other media in the name of the revolution and invaded grain silos and prisons. Acts like these that threatened social stability occurred in many places. Under these circumstances, in early 1967, the Central Committee of the CCP, the State Council, and the Central Military Commission jointly issued orders for the military to take control of banks, radio stations, grain warehouses, prisons, airlines, and other essential facilities.[35]

Society fell into complete disorder as the rebel movement expanded into the factories.

1.3 *The Destruction of the Bureaucracy and Its Effects*

In August 1966, the Central Committee of the CCP made "capitalist roaders within the party" the target of the Cultural Revolution, which significantly undermined the party's authority. In addition, the Cultural Revolution, which gave mass organizations the right to criticize party cadres and use mass mobilization to transform the party bureaucracy, effectively denied the legitimacy of the party leadership maintained since the country's founding. The result was that party cadres at all levels were challenged because actively leading the movement could be seen as suppressing the masses while ignoring them could lead to accusations of letting bad elements act against the party.

Before November 1966, local party cadres responded to the Cultural Revolution in three ways. Taking a proactive response, as the Shanghai example shows, the authorities responded by organizing progressives and using their followers in mass organizations to influence the mass movement. Pandering to the masses was also a more negative way of coping. For example, some lower-level cadres were pushed forward to be criticized by the masses in order to protect the top leadership or transferring criticized cadres to the military or other facilities to avoid directly confronting mass organizations.[36] A third type of response was to give up leadership completely. Some party cadres believed that, like the Hundred Flowers and Hundred Schools of Thought of the late 1950s, the Cultural Revolution was temporary and that order would be restored after some time.[37] In the spirit of "I would rather make mistakes sitting down than moving" (*ningyuan zuozhe fan cuowu, buyuan dongzhe fan cuowu*),[38] these party cadres gave up the leadership of the mass movement. They passed problems that needed to be solved to the central government. Sometimes they even ignored the central government's instructions to allow workers to go to Beijing to "appeal" (*shangshu*) to save themselves and maintain local order.[39]

By December 1966, with most cadres being criticized, more and more were relinquishing their leadership roles. There were many reasons for this, among which the most important three are as follows. First, due to the movement's expansion to criticize the "bourgeois reactionary line," most cadres were attacked by the mass organizations and lost their authority to lead the mass movement. As a result, local governments could not fulfill their functions of conveying and implementing policies. Second, the legalization of the workers' rebellion reduced the power of the local party and government organizations. Third, on December 9, 1966, the Central Committee of the CCP issued the *Ten Provisions on Grasping the Revolution and Promoting Production*, which explicitly forbade the obstruction of the Cultural Revolution under the pretext of promoting production.[40] For these reasons, only a few cadres could exercise their authority to lead the mass movement in late 1966. However, because these cadres held administrative power, their actions still greatly impacted society. Specifically, going into late 1966, local party cadres were the only means by which mass organizations could realize the demands of political and economic interests. Party cadres, who held the authority to review financial and material allocations, were forced by mass organizations to sign documents demanding various goods. Meanwhile, local cadres supported the mass organizations in going to Beijing by offering to pay for "fees" and "free tickets" to pass on to the central government problems that should be solved locally.[41] This resulted in the stranding of about 800,000 workers in Beijing who were separated from production sites between late 1966 and early 1967.[42]

As mentioned above, the expansion of the Cultural Revolution shook the legitimacy of the leadership of the Party Committees. In addition, the paralysis of government agencies responsible for mediating social conflicts and maintaining social order intensified factionalism and further exacerbated social unrest.

The central leadership attributed the harm caused by economism and many workers "appealing" to Beijing due to some local party cadres' resistance to the Cultural Revolution and blamed local party and government cadres.[43] However, in reality, the cause of the great chaos was not solely on the local officials, and it can be said that the central leadership also had a responsibility. This point can be explained in three ways.

First, the social upheavals can be seen as a reflection of the power struggles in the central leadership. For example, the general mobilization to criticize Liu Shaoqi allowed workers economically deprived by the state under the pre-Cultural Revolution policy to seek benefits in the name of revolution. During the October 1966 period, the rebellions of the masses to achieve economic benefits were also used to the maximum extent by the central leadership, including the Central Cultural Revolution Group, as a tool to criticize Liu Shaoqi.[44]

Second, the policy changes and policy differences among the central leadership led to confusion among local officials. In July 1967, the Central Committee of the CCP instructed that the Cultural Revolution should be carried out mainly in the cultural and educational sectors and the party and government organs. At the same time, the Cultural Revolution in the industrial and mining sectors should be carried out in a planned and organized manner during the Four Clean-ups campaign.[45] Furthermore, in September, to prevent the rebel movement from spreading to the industrial and mining sectors, the Central Committee of the CCP issued the *Circular on Grasping the Revolution and Promoting Production*, advocating the maintenance of production activities.[46] The priority given to maintaining production activities over participation in the revolution was the difference between this Circular and the aforementioned *Ten Provisions on Grasping the Revolution and Promoting Production* issued later in December. The September Circular, however, resulted from the CCP's concern that the Red Guards' autumn "Great Link-up" would significantly affect the crop harvest, while concerns about the impact on the industrial and mining sectors were only of secondary importance. Nevertheless, because of the CCP's clear instructions on maintaining production, local cadres could play a role in curbing the mass movement's development in a disorderly direction during this period. After November, however, the CCP Central Committee decided to carry out the Cultural Revolution in the industrial and mining sectors. It instructed local cadres to strictly prohibit the hindrance of the Cultural Revolution in the name of production.[47] This change reflected Mao's policy intentions and a confrontation of views within the central leadership.[48] It can be said that the chaos caused by the central leadership's policy changes greatly affected lower-level party and government organizations and mass movements.

Third, the social unrest resulted from the difference between the political and economic goals of the Cultural Revolution, namely "revolution" and "production," a point discussed later.

2 The Rise of the Workers' General HQ and the Shanghai Model

The previous section discussed the disruption of the political-economic-social order caused by the rebel movement. Though disruption resulted from Mao's Cultural Revolution, it differed from Mao's envisioned Cultural Revolution. In Mao's plan, revolution and production did not contradict each other, and the ultimate goal of the revolution was to develop productive forces.[49] This was explicitly stated in Article 14 of the *Sixteen Points*, which stipulated the relationship between revolution and production: "The Great Proletarian Cultural Revolution is designed to revolutionize people's thinking so that all work can be done better, faster, and more economically. As long as the masses are fully mobilized and proper arrangements are made, it is possible to ensure that the cultural revolution and production are not lost and that all work is completed with high quality. The Great Proletarian Cultural Revolution is a powerful driving force for the development of the productive forces of our society. It is wrong to regard the Cultural Revolution as opposed to the development of production." Wang Li's memoirs also contain a record of Mao's remarks on the harmonious relationship between "revolution" and "production." Indeed, the six-month period from the beginning of the Cultural Revolution to the end of 1966 shows Mao's heavy emphasis on grasping the revolution and promoting production.

However, in reality, the revolution had an extremely negative impact on production. This situation prompted Mao Zedong to search for a new approach to maintain economic order while carrying out political revolution and promoting the simultaneous development of revolution and production. It can be said that Mao's support for the Workers' General HQ in Shanghai was closely related to his idea of integrating revolution and production. To understand why Mao Zedong chose Shanghai's rebellion as a model for rebuilding political power, we must first examine his conception of the Cultural Revolution.

2.1 *Mao's Vision of the Cultural Revolution*

In Mao's view, the Cultural Revolution was a revolution in the ideological sphere to solve the problem of cadres' detachment from the masses, bureaucratization, and privilege. This was similar to the Communist Party's previous "rectification campaigns." However, the Cultural Revolution was different from the rectification movement because the right to monitor and evaluate cadres was given to the masses. Mao saw the use of popular power to transform the existing bureaucratic system to achieve the ultimate goal of integrating the government and the people into a single political system. However, Mao did not have a concrete plan for achieving this goal but adopted a trial-and-error approach that adjusted to changing circumstances. Throughout the Cultural Revolution, Mao constantly explored the most effective ways to express his ideas. When he found a method that worked, he immediately supported it. When a method failed to align with his hopes, he abandoned it and focused on discovering and implementing new approaches. As the examples of the Old Red Guards and the Tsinghua University *Jinggangshan* Corps in the previous chapter show, the alternation between the two groups in the political arena in 1966, their rise and decline can be said to be Mao's process of discovery and abandonment. From this perspective, the end of 1966 was when it was necessary to discover an order to control the aforementioned social upheaval. Against this backdrop, it can be argued that Mao prioritized stabilizing the economic order and rebuilding the party-state power system as the central task at the end of 1966 to control the social chaos that had arisen.

In reading Mao's speech in late December 1966, one will find that he initially expected party and government cadres to take the lead in revising the trajectory of the Cultural Revolution. According to Wang Li's records, Mao said, "The cadres must actively participate in the Cultural Revolution. Go deep into the masses, join the workers, peasants, and students in the Cultural Revolution, and realize the revolutionization of the cadres."[50] This can be said to be a manifestation of his "coming from the masses and going to the masses" ideology. Concerning the conduct of the Cultural Revolution, Mao instructed: "In this movement of class struggle, the Central Committee of the CCP should lead it in a focused manner, first in Beijing, then in Shanghai, Tianjin, and the Northeast."[51] Coincidentally, during the same period, in January 1967, rebel groups in Shanghai jointly published the "*Letter to the People of Shanghai*" (*gao shanghai quanshi renmin shu*) and the "*Urgent Notice*" (*jinji tonggao*). This can be considered the inspiration for the idea of a "three-in-one" regime in which revolution and production would go hand in hand. With the publication of these two important circulars, Shanghai was positioned as a template for regime reconstruction during the Cultural Revolution.

What exactly is the Shanghai model? In short, it is the unification of mass organizations and the establishment of the Three-In-One Regime consisting of revolutionary cadres, revolutionary soldiers, and revolutionary masses. In February 1967, this regime was formed in Shanghai. Still, it was the third in the country regarding time sequence, with earlier iterations in Heilongjiang and Shandong provinces.[52] Nonetheless, Shanghai was chosen as the template for regime reconstruction, not only because of its economic position in the country but more importantly because the Cultural Revolution in Shanghai in late 1966-early 1967 contained elements consistent with Mao's vision. With this in mind, we must return to the Workers' General HQ formation to illustrate this.

2.2 The Rise of the Workers' General HQ

On November 6, 1966, the "Shanghai Workers' Revolutionary Rebel General Headquarters" (also known as the "Workers' General HQ") was established in Shanghai. Led by Wang Hongwen (who later became the vice chairman of the Central Committee of the Communist Party of China), it was the earliest worker rebel organization in China. It had thousands of members in the early stages, basically young workers with low political and economic status before the Cultural Revolution.[53] Since the *Sixteen Points* did not contain any provision allowing workers to form mass organizations, the first action of the Workers' General HQ after its establishment was to seek its legitimacy from the party and government agencies, specifically, (1) to recognize the legitimacy of the Workers' General HQ; (2) to request the attendance of the Mayor of Shanghai, Cao Diqiu, in a mass meeting in the afternoon of November 9th to accept the criticism; and (3) to provide the Workers' General HQ with propaganda and transportation facilities. However, after obtaining the Central Cultural Revolution Group's approval, the Shanghai Municipal Committee refused to support the actions of the Workers' General HQ, nor did it attend the November 9th meeting. In response, thousands of Workers' General HQ members forcibly boarded a train to Beijing to petition the Shanghai Municipal Committee's decision. On instructions from the State Council, the planned train to Beijing stopped at the Anting Station near Shanghai. This led to a confrontation at Anting Station between members of the Workers' General HQ, led by Wang Hongwen, and the conservative workers who tried to prevent them from going to Beijing. Members of the Workers' General HQ laid on the tracks to prevent trains from passing, suspending trains for more than 31 hours. This disruption led to suspending more than 100 passenger and freight trains, affecting the entire Nanjing-Shanghai-Nanjing route, and obstructing north–south traffic.[54]

The incident in Shanghai, China's largest industrial city, was a significant issue for the CCP Central Committee in Beijing. Zhou Enlai and Chen Boda ordered the Workers' General HQ to "return to Shanghai with a correct attitude" immediately, but to no avail. On November 12, 1966, Zhang Chunqiao was sent to Shanghai by the Central Cultural Revolution Group to solve the problem. On his first day in Shanghai, Zhang adopted the same approach as the Central Cultural Revolution Group and the Shanghai Municipal Committee, stressing the seriousness of the problem to the leaders of the Workers' General HQ and putting pressure on them to return to Shanghai immediately, but still to no avail. The following day Zhang changed tactics and began to make concessions. He negotiated with the Workers' General HQ. He used the signing of the "Five Demands," which recognized the organization's legitimacy, as a condition for them to immediately return to Shanghai, resolving the issue.[55]

In settling the Anting Incident, Zhang Chunqiao violated the organizational principle of the Shanghai Municipal Committee, which had decided not to recognize the legitimacy of the Workers' General HQ organization, and Tao Zhu supported this decision. However, at the negotiation scene, Zhang Chunqiao did not implement the decision of Tao and the Shanghai Municipal Committee.[56] In previous studies of the Cultural Revolution, many researchers have pointed out that Zhang Chunqiao used mass organizations to exert pressure on the central leadership and the Shanghai Municipal Committee to seize control of the committee.[57] But considering the situation at the time, preventing the situation from worsening and restoring railway operations was a more urgent issue than recognizing the legitimacy of the mass organization. This was also what Mao Zedong expected. From this point of view, Zhang Chunqiao's approach is more pragmatic than the decision of Tao and the municipal committee, and this is probably why Zhang Chunqiao had Mao's support.[58] This way, after the Anting Incident, the Workers' General HQ became active in Shanghai's political arena as a legitimate rebel group.

2.3 *The Confrontation Between the Workers' General HQ and the Scarlet Guards*

In late November 1966, the Shanghai General Command Provisional Committee of Workers' Scarlet Guards to Defend Mao Zedong Thought (hereinafter referred to as the Scarlet Guards) emerged as the rival organization of the Workers' General HQ. Most of the members of the Scarlet Guards were Communist Party members, members of the Communist Youth League, activists, and other workers called "progressives."[59] "The Scarlet Guards was founded with the support of the majority of workers and was an overwhelming majority compared to the Workers' General HQ."[60] They focused their criticism on Zhang Chunqiao and the Workers' General HQ. They received support and cooperation from the major leaders of the party and government institutions, such as Shanghai Mayor Cao Diqiu and Municipal Party Secretary Chen Pixian.[61]

In late 1966 there were frequent clashes between the Workers' General HQ and the Scarlet Guards. The Liberation Daily Incident and the Kangping Road Incident further escalated the clashes into armed conflict.

Liberation Daily Incident

The *Liberation Daily* Incident stemmed from the Shanghai Rebel Red Guard's demands that their propaganda newsletter, the *Red Flag Battle News (hongqi zhanbao)*, be jointly distributed with the Shanghai Municipal Party Committee's party newspaper, the *Liberation Daily*. In November 1966, under Jiang Qing's instigation, Nie Yuanzi, a lecturer at Peking University and known nationwide as the first to criticize the university Party Committee at the beginning of the Cultural Revolution in May 1966, went to Shanghai and mobilized the student rebel groups for a week or so. She intended to force Deng Xiaoping, a central government leader, to step down by toppling and criticizing the principal leaders of Shanghai's party and government bodies, Cao Diqiu and Chen Pixian. As a result, on 22 November and the 25th, the student rebels in Shanghai held a general meeting in the name of the Shanghai Red Guard Revolutionary Rebellion General Headquarters and launched a criticism of Cao Diqiu. After the conference on the 25th, the Shanghai Red Guard Revolutionary Rebellion General Headquarters requested that the Shanghai Municipal Party Committee's ninth issue of their newsletter *Red Flag Battle News*, which contained the criticism, be distributed as an insert in the party newspaper *Liberation Daily*. However, Tao Zhu instructed that the mass organizations' propaganda newspaper was completely different from the party newspaper and could not be jointly distributed. The Shanghai Municipal Committee accepted Tao's instructions and refused their request. Subsequently, the student rebels demanded cooperation from the newspaper distribution department of the Shanghai Post Office and the head of the distribution department of the *Liberation Daily* but were again refused. After this refusal, the student rebels blocked the *Liberation Daily* by force, preventing the normal distribution of the newspaper.[62] The situation gradually escalated, and workers' organizations got involved, eventually creating a confrontation between the Workers' General HQ, which rebelled against the Shanghai Municipal Committee, and the Scarlet Guards, which supported the Shanghai Municipal Committee. The chaos caused by the *Liberation Daily* Incident lasted for seven days. After several injuries, the Central Cultural Revolution Group issued a permit for the joint publication, and the incident gradually subsided.[63]

The expansion of the *Liberation Daily* Incident was not only due to competing interests among mass organizations but also to the response of the Shanghai party and government organizations. Specifically, after the Workers' General HQ intervened in the incident to support the student rebellion, the Shanghai Municipal Committee contacted the Scarlet Guards to inquire about whether the Scarlet Guards intended to interfere. According to Li Xun's research, at this stage, the Shanghai Municipal Committee tended to agree to the demands of the student rebels and the Workers' General HQ, hoping that the Scarlet Guards would not intervene. That is to say, once the Scarlet Guards intervened, the fact that the Shanghai Municipal Committee agreed to the demands of the Workers' General HQ would put the Scarlet Guards, which had always maintained a cooperative relationship with the Shanghai Municipal Committee, in a difficult position. It would be challenging for the Shanghai Municipal Committee to gain future support from the Scarlet Guards. Out of this concern, the Shanghai Municipal Committee inquired to the Scarlet Guards in advance. However, the actions of the Shanghai Municipal Party Committee were misunderstood by the leaders of the

Scarlet Guards. When asked whether they would intervene in the incident, they thought the committee wanted them to intervene but could not give a clear directive, so they used this method to imply their intention.[64] The Scarlet Guards' mistaken intervention escalated the conflict between conservative and rebel forces, resulting in armed confrontation.

"Kangping Road Incident"

The Kangping Road Incident was closely related to the change in policy and the divergence of views within Shanghai's party and government institutions. In the Anting Station and the *Liberation Daily* incidents, the Central Cultural Revolution Group supported the Workers' General HQ, which brought about a split and a change in policy within the Shanghai party and government. First, Ma Tianshui, Secretary of the Shanghai Municipal Committee, Wang Shaoyong, alternate secretary, and other cadres changed their stance to support the rebels. At the same time, the "Writing Group" (*xiezuo ban*) of the Shanghai government, which was responsible for propaganda work, also accepted the instructions of Zhang Chunqiao and Yao Wenyuan and set up a Revolutionary Rebellion Liaison Station of the Municipal Committee Organs. The support of Mao Zedong and the Central Cultural Revolution Group for the Workers' General HQ and the rebellion within the municipal party and government agencies brought about a change in the policy of the Shanghai party and government agencies, specifically that the Shanghai party and government agencies led by Cao Diqiu and Chen Pixian withdrew the support they had been giving to the Scarlet Guards and began to support the Workers' General HQ. The Scarlet Guards, outraged by these actions, changed their position and began criticizing the Shanghai party and government organizations.[65]

On December 23, 1967, the Scarlet Guards held a criticism meeting with the Shanghai Municipal Committee. They put forward the Eight Demands, insisting that the Shanghai Municipal Committee recognize the legitimacy and revolutionary nature of the Scarlet Guards organization and restrict the actions of the Workers' General HQ. Cao Diqiu was forced to agree to these terms.[66] Still, at a meeting of the Shanghai Municipal Committee the next day, Chen Pixian and other members of the Committee decided to support the Workers' General HQ in the future, and Cao Diqiu's signature was considered not representative of the will of the Party Committee. Having no choice, Cao Diqiu had to comply with the decision of the Municipal Committee.[67]

On December 28th, two days after Cao Diqiu's statement was made public, over 20,000 Scarlet Guards gathered on Kangping Road, where the East China Bureau and the Shanghai Municipal Committee were located, demanding to see Cao Diqiu and requesting a statement that their "Eight Demands" were still valid. To monopolize the right to criticize the Shanghai Municipal Committee, the Workers' General HQ mobilized about 100,000 members to surround the Scarlet Guards on Kangping Road on the 29th. Before dawn the next day, they fabricated the accusation that the Scarlet Guards had "searched and looted Zhang Chunqiao's residence" and began attacking the Scarlet Guards. The armed conflict lasted several hours and ended with the Scarlet Guards' defeat.[68]

The Kangping Road Incident was the most significant armed struggle in 1966; over 150,000 were involved,[69] and it resulted in the injuries of 91 people.[70] The incident seriously undermined Shanghai's administration, production, transportation, and social order, and its effects lasted long after the end of the struggle. For example, more than 70 of the 80 Shanghai Railway Dispatch Office employees were Scarlet Guard members who left their workplace after the incident to "appeal" to Beijing. This resulted in the repeated interruption of railway services. In the 26 hours between dawn on the 30th and the 31st, nearly 100 passenger trains and freight trains were suspended. Coupled with the prevailing trend of economism, Shanghai's economy and security were on the verge of collapse by the end of 1966.[71]

It is important to note here the role played by the Workers' General HQ in ending the chaos. At that time, Cao Diqiu and Chen Pixian had utterly lost all authority over Shanghai. On December 31, 1966, the Workers' General HQ issued an "Emergency Order," announcing that the heads of the Scarlet Guards would be brought to the Public Security Bureau.[72] This "Emergency Order" received the support of Zhang Chunqiao and the cooperation of the Public Security Bureau, and 240 leaders of the Scarlet Guards were detained.[73] The Scarlet Guards, which had lost the support of the municipal party and government agencies and its faction leaders, began to disintegrate and disappeared from the stage of the Cultural Revolution. By January 1967, most of the Scarlet Guards members had been absorbed by the Workers' General HQ, which had defeated its rivals and became Shanghai's most influential mass organization.[74] Additionally, in January 1967, the rebel group seized power of the newspaper *Wen Hui Bao*. After the seizure, it published "A Letter to the People of Shanghai," signed by 12 rebel groups on 4 January, and "An Urgent Notice," signed by 32 rebel groups on 9 January. These two articles aimed to advocate grasping the revolution and promoting production and to stop the spread of economism. On 10 January, cadre–worker–student rebel groups spontaneously set up the Shanghai Anti-Economism Liaison Department, which was responsible for managing the outflow of irregular funds from banks, disbanding organizations seeking economic benefits, and arranging for the return-to-agriculture workers and support-the-interior workers to their original places of residence. The Shanghai Anti-Economism Liaison Department had secretarial, organizational, investigative, reception, writing, and policy research groups, which essentially performed the functions of the government.[75]

2.4 *The Establishment of the Shanghai Model and Mao's Concept of the Cultural Revolution*

The above series of events illustrates the following points, and from them, we can see why Mao Zedong chose Shanghai as the model of the reconstruction of political order.

First, in 1966 the party and government institutions in Shanghai could no longer resolve the chaos. On many occasions, rather than solving problems, their actions and intervention could even be said to have exacerbated the conflict. Of course, the inconsistency of the political views of the CCP Central Committee was a decisive factor in the social chaos. However, the bureaucratic approach, which could not respond flexibly to changing situations and only knew how to rigidly carry out orders from above, was what Mao disliked most and wanted to reform through the Cultural Revolution.

Second, during the Kangping Road Incident, the Workers' General HQ employed forceful intervention to resolve the situation where the Scarlet Guards besieged the city's party and government agencies. This action dismantled the Scarlet Guard's organization, effectively ending the ongoing confrontation between conservatives and rebels that had lasted for months and eliminating the resulting social chaos. In the aftermath, particularly regarding resuming production, stabilizing order, and combating the trend of economism, the Workers' General HQ took the most decisive actions. These measures objectively stabilized the precarious public order in Shanghai, which aligned with Mao Zedong's desire to maintain social stability.

Third, the rebel organizations in Shanghai were the earliest in the country to advocate for "grasping revolution and promoting production" and were self-organized groups that opposed the economic trend. This is of great significance. During the chaotic period of the Kangping Road Incident, the industrial workers' strike caused the "Three Stops" event of power outages, water supply stoppages, and train stoppages, causing enormous economic losses and great social disorder in Shanghai. The Workers' General HQ claimed that the responsibility rested with the Shanghai Municipal Party Committee and the "Scarlet Guards." Still, in reality, the Workers' General HQ and Zhang Chunqiao, who incited the violence, can be said to bear greater responsibility. However, considering the situation at the time, preventing the situation from worsening was more important than pursuing responsibility. Therefore, after publishing the "Notice to the People of Shanghai" and the "Urgent Notice," Mao Zedong immediately supported them.

Fourth, apart from the above, the most crucial reason for Mao's appreciation of the "Letter to the People of Shanghai" and the "Urgent Notice" was probably the fact that the voice of the people advocating grasping the revolution and promoting production was spread through the joint organization of the cadres, workers, and student rebels. In particular, the formation of rebel groups by the cadres in Shanghai to participate in the movement was of great importance in Mao's view. As mentioned above, in late 1966, Mao expected party and government cadres to play a role in stabilizing social order and reestablishing power and wanted them to participate in the Cultural Revolution by reaching out to the masses. Mao Zedong said: "Revolutionary students in Shanghai have risen, revolutionary workers have risen, and cadres of revolutionary organs have risen. Therefore, there is hope for the Cultural Revolution in Shanghai."[76] He also instructed Wang Li and other relevant cadres that party and government cadres should not confront the mass movement and give up leadership but should mingle with the masses and actively participate.[77] From this point of view, the Shanghai joint statement provides an example of Mao's Cultural Revolution concept, and it can be said that the Shanghai joint statement is a realistic version of Mao's Three-In-One Regime idea. As discussed in the *People's Daily* commentator article "Shanghai Revolutionary Rebels Launch Total Attack on the Bourgeois Reactionary Line," "The second important indicator of the favorable situation of the proletarian cultural revolution in the Shanghai region is the emergence of revolutionary rebels within the city-level party and government agencies. They broke through various constraints imposed by the differentiation between insiders and outsiders, joining forces with workers and revolutionary students to criticize the bourgeois reactionary line and struggle against the handful of authorities within the party who pursued a bourgeois reactionary path."[78]

For these reasons, establishing the Three-In-One Regime became the primary goal of the Cultural Revolution from 1967 onwards, and Shanghai was taken as a model for achieving this goal. As such, the development of the Cultural Revolution in Shanghai after 1967 was different from that of other regions.

3 The Establishment of the Three-In-One Regime: National Chaos and Shanghai's Stability

In January 1967, Mao Zedong praised the seizure of power by the Shanghai *Wen Hui Bao* newspaper. Subsequently, incidents of mass organizations seizing power from party and government agencies occurred nationwide, and the Cultural Revolution entered the "comprehensive seizure of power" phase. Mao Zedong's original goal in forming a Three-In-One Regime government through "great unity" was to revolutionize the political sphere and the orderly development of the economic sphere. However, in the context of a paralyzed bureaucratic and social control system, the unrestricted "seizure of power" transformed into power struggles among mass organizations, further disrupting social order throughout China and leading to armed conflicts between factions in various places. Given the "seizure of power" situation nationwide at the time, despite small-scale conflicts in Shanghai, the city maintained relative stability compared to regions like Wuhan, Chengdu, Chongqing, and Guangzhou, where armed conflicts persisted for an extended period. So, why could Shanghai maintain relative stability when most cities and regions experienced chaos? To understand this, it is necessary to examine Zhang Chunqiao's unique approach to handling factional interest competition while forming the Three-In-One Regime.

3.1 Organized Power "Grabs"

In 1967, Zhang Chunqiao's role in maintaining order in Shanghai was vital. This, of course, also included his motive of self-interest. Zhang Chunqiao was the Secretary of the Shanghai Municipal Party Committee for Culture and Education before the Cultural Revolution. In May 1966, he was elected deputy head of the Central Cultural Revolution Group. Still, he had the lowest political status among the five: Tao Zhu, Kang Sheng, Chen Boda, and Jiang Qing. At the beginning of the Cultural Revolution, Zhang Chunqiao's activities were not remarkable, but with the resolution of the Anting Incident, he began to gain Mao Zedong's attention. In this context, Zhang Chunqiao and Yao Wenyuan were appointed by Mao Zedong to lead the Cultural Revolution in Shanghai at the end of 1966 and returned to Shanghai as agents of the Central Cultural Revolution Group. For Zhang Chunqiao, whose political position in the central leadership was still unstable, whether he could gain Mao's favor through his performance in Shanghai was essential for his future political position. Moreover, Mao had already designated Shanghai as a model for rebuilding the regime. Hence, Zhang Chunqiao needed to grasp the reins of the Cultural Revolution in Shanghai and turn it into a model to Mao's liking. He planned every step of the progress of the Cultural Revolution in Shanghai very carefully after his return.

For example, on January 8, 1967, Wang Hongwen was instructed by Zhang Chunqiao to set up the Shanghai "Grasp Revolution and Promote Production" Command in the name of a mass organization. Zhang Chunqiao called it the "Economic Soviet" and expected it to exercise the functions of the People's Congress. In addition, on 15 January, the Shanghai Revolutionary Rebel Organization Liaison Headquarters was established to take over the functions of the municipal government. After seizing power from the Shanghai party and government organs in February, it was named the Shanghai People's Commune. The above measures were reported to Mao for support but failed to do so. From the above steps, we can see that Zhang Chunqiao was trying hard to push the Cultural Revolution in Shanghai in the direction Mao wanted.

It can be said that it was thanks to the efforts of Zhang Chunqiao that a vertical command system similar to the party's administrative leadership, consisting of Mao Zedong → Central Cultural Revolution Group → Zhang Chunqiao and Yao Wenyuan → Workers' General HQ, was constructed in Shanghai after the party and government institutions became paralyzed. This helped to minimize the harm caused by factionalism among the masses. This is the biggest difference compared to other areas where the party and government institutions became paralyzed and failed to produce individuals or groups capable of performing their functions. To explain this, constructing the Three-In-One Regime power must be examined.

Curbing Chaotic Power Seizures

After the collapse of the Scarlet Guards, a three-pronged system emerged in Shanghai: the Cadres' Rebellion, the Workers' Rebellion, and the Students' Red Guard Rebellion.[79] Not only were there rivalries between these three organizations but there were also factional rivalries within each. Since the Anting Incident, the largest workers' organization, the Workers' General HQ, had been divided into the Workers' General HQ, the 1st Corps, the 2nd Corps, and the 3rd Corps.[80] The 1st Corps, 2nd Corps, and 3rd Corps were nominally subordinate organizations of the Workers' General HQ, but in practice, they were independent. The Red Guards groups were not originally a collaborative organization; their most active groups, the Red Revolutionary Association, the Red 3rd Headquarters, and the Upper 3rd Headquarters, were all independent student rebel groups.[81] In January 1967, like other regions, various rebel organizations in Shanghai also engaged in power struggles for control of the party and government. Before establishing the Shanghai People's Commune, there were four power struggles to control the city's party and government organs.

The first joint operation by the 2nd Corps and the Upper 3rd Headquarters was established on January 15, 1967. Before the "seizure of power," the head of the 2nd Corps, Geng Jinzhang, reported the action to Zhang Chunqiao, who instructed him to "report to the central authorities and then take action." But Geng Jinzhang and Zhao Quanguo, the head of the Upper 3rd Headquarters, started their operations without waiting for instructions from the Central Committee. Although Mao Zedong supported this "seizure of power," Zhang Chunqiao believed that other mass organizations did not participate in the action. He convinced Geng Jinzhang to relinquish power and withdraw from the city's government facilities to prevent internal chaos among the rebel organizations.[82]

The second was the Upper 3rd Headquarters-led seizure of power on 22 January by the Red Guards rebel group at the university. After the seizure of power, the leader, Zhao Quanguo, falsely claimed to have been instructed by Zhou Enlai to legitimize his action. Still, when this was revealed, Wang Hongwen turned him over to the public security authorities, and the seizure of power was declared invalid.[83]

The third occurred on January 25th, when the Red Guards rebel organization, the Red Revolutionary Association, and Pan Guoping, one of the Workers' General HQ leaders, jointly attempted a seizure of power. Zhang Chunqiao persuaded Pan to return the Party Committee's official seal, rendering the "seizure of power" invalid.[84]

The fourth occurred when the 2nd Corps joined forces with the 1st Corps, the 3rd Corps, and other rebel groups to confront the Shanghai People's Commune under Zhang Chunqiao and Yao Wenyuan. On the eve of establishing the Shanghai People's Commune, the Shanghai Revolutionary Rebel Committee was formed to confront them and announced that it had taken over the power of the Shanghai Municipal Committee. However, Zhang Chunqiao persuaded Geng Jinzhang to dissolve the Shanghai Revolutionary Rebel Committee, invalidating the seizure of power.[85]

In summary, like other regions, Shanghai experienced struggles for power among mass organizations in January 1967. Some of the "seizure of power" actions received praise from Mao Zedong. However, Zhang Chunqiao, to prevent chaos within the rebel organizations, persuaded leaders of various factions to abandon the power they had seized and suppress actions by not supporting the "seizure of power" efforts. At the same time, Zhang Chunqiao and Yao Wenyuan actively promoted the government's reconstruction. This point will be elaborated upon in the following section.

Zhang Chunqiao's Role in the Establishment of the Three-In-One Regime

On January 5, 1967, Wang Hongwen was instructed by Zhang Chunqiao to set up a Shanghai Revolutionary Rebel Organization Liaison Station to prepare for the seizure of power from the Shanghai Municipal Committee. Through this organization, Zhang Chunqiao and Yao Wenyuan led the elections and the distribution of representative seats in the Shanghai People's Commune. Thus, the Shanghai People's Commune was formally established on 5 February as a Three-In-One Regime and was later renamed the Shanghai Revolutionary Committee at Mao's suggestion. In addition to Zhang Chunqiao and Yao Wenyuan, the Shanghai Revolutionary Committee consisted of three workers' representatives (all members of the Workers' General HQ), two peasant representatives, one student representative, one representative of the Revolutionary Rebellion Liaison Station of the Municipal Committee Organs, and a representative of each of the seven departments under the committee, and the heads of the three armies in Shanghai.[86] Shortly after that, Zhang arranged for Ma Tianshui, the former secretary of the Municipal Committee, Wang Shaoyong, the alternate secretary, and other cadres who had supported the rebels during the Anting Incident to assume critical leadership positions in the Shanghai Revolutionary Committee.[87] In this way, under the active leadership of Zhang Chunqiao and Yao Wenyuan, the struggle for power among the mass organizations in Shanghai and the confusion over the handling of former cadres in the Three-In-One Regime were kept to a minimum, and the reconstruction of the regime was completed through an organized power seizure.

Another critical question needs to be explained here: How was Zhang Chunqiao able to control the mass organizations in Shanghai in a political situation where anarchism was rife and "overturned everything? (*dadao yiqie*)" There are three main reasons for this.

First, Zhang Chunqiao had the legitimacy and absolute authority to lead the mass movement, as granted by the supreme leader Mao Zedong and the Central Cultural Revolution Group. For example, in late January 1967, dissatisfied with the distribution of political power in the Shanghai People's Commune, university rebels organized a massive demonstration to criticize Zhang Chunqiao, launching the "28 January bombarding of Zhang Chunqiao." But the Central Cultural Revolution Group issued an urgent telegram supporting Zhang Chunqiao and Yao Wenyuan, and the Red Guards rebel group's action was aborted halfway.[88]

Second, Zhang Chunqiao was supported by the PLA, which Mao Zedong instructed to intervene in the Cultural Revolution in January 1967. According to Harding's research, the PLA's support of conservative mass organizations as "leftists" in many areas brought about rebel attacks on the PLA. It intensified the factional struggle between the conservatives and the rebels. Not only that, but conflicts also arose between the PLA and the Central Cultural Revolution Group, which supported the rebels. Thus, the antagonistic structure was further complicated, and the chaos gradually expanded and escalated.[89] However, the situation in Shanghai was entirely different. The leadership of the military gave Zhang Chunqiao full support and cooperation. For example, during the January 28 incident in 1967, the PLA accepted Zhang Chunqiao's instructions and rushed into Fudan University to rescue the head of the Revolutionary Rebellion Liaison Station of the Municipal Committee Organs, Xu Jingxian, whom the student rebels had detained. When the operation failed, the rebel Red Guards detained even the military-political commissar Xu Haitao. To legitimize their attack on Zhang Chunqiao, the Red Guards forced Xu Haitao to reveal the behind-the-scenes figures who sent the troops. Still, Xu Haitao refused to implicate Zhang Chunqiao. Establishing the City Revolutionary Committee in Shanxi, Heilongjiang, Shandong, and Guizhou also showed the same results.[90] This demonstrates that the army's support and cooperation with local cadres were essential in stabilizing order.

Third, Zhang Chunqiao had the cooperation of the largest mass organization in Shanghai, the Workers' General HQ. This point will be explained in detail in the next section.

3.2 *The Role of the Workers' General HQ in the Stabilization of Shanghai*

The cooperation between Zhang Chunqiao and the Workers' General HQ was crucial in stabilizing the social order in Shanghai. Wang Hongwen and the Workers' General HQ served as a de facto substitute for the government in conveying and implementing policies for Zhang Chunqiao and Yao Wenyuan.

With the party and government organizations in Shanghai paralyzed, Zhang Chunqiao and Yao Wenyuan were entrusted to lead the city's Cultural Revolution. Under these circumstances, relying solely on the cooperation of the cadres from the Municipal Committee Revolutionary Rebel Liaison Station formed after the *Liberation Daily* Incident was not enough. To implement Zhang Chunqiao's instructions among the masses, it was essential to secure the collaboration of mass organizations. Moreover, since Mao Zedong also emphasized the combination of revolutionary cadres and the masses, relying solely on the cadres from the Revolutionary Rebellion Liaison Station of the Municipal Committee Organs could potentially provoke rebellion from mass organizations and even lose Mao Zedong's support.

More importantly, in the context of factional struggles for party and government power, it was difficult for Zhang Chunqiao to handle conflicts between factions independently without relying on the strength of mass organizations. The outcome would be clear—if they failed, they might face criticism from the central leadership and mass organizations, just like the Shanghai Municipal Committee. Under these circumstances, the Workers' General HQ became Zhang Chunqiao's tool for resolving factional conflicts. On the other hand, from Wang Hongwen's perspective, Zhang Chunqiao's support was an excellent opportunity to undermine the opposing factions within the Workers' General HQ. The alignment of interests between the two played a crucial role in stabilizing Shanghai and resolving factional conflicts. The following examples can help illustrate this point.

Weakening of the Red Guard Rebel Organization

As mentioned earlier, in 1967, Shanghai saw the emergence of three groups: the worker rebellion, the cadre rebellion, and the Rebel Red Guards in universities. The rebel group of cadres, the Revolutionary Rebellion Liaison Station of the Municipal Committee Organs, was originally a rebel group under the direction of Zhang Chunqiao and Yao Wenyuan.[91] For this organization, Zhang Chunqiao and Yao Wenyuan could exercise direct influence. Various factions were within the workers' rebellion, and the rivalry between the factions and the leaders was fierce. Nevertheless, Zhang Chunqiao could still exert his influence through Wang Hongwen and Geng Jinzhang, the leaders of two major organizations, the Workers' General HQ and the 2nd Corps. However, the power struggle between Wang Hongwen and Geng Jinzhang also created instability for Zhang Chunqiao, especially since Geng Jinzhang often disregarded Zhang Chunqiao's instructions and acted independently. Yet, from the perspective of the unsuccessful "seizure of power" cases, it was not difficult for Zhang Chunqiao to control them, as their primary goal was to seize power.

By contrast, Zhang Chunqiao had no mechanism to control the Rebel Red Guards groups, which practiced anarchism and did not have the means to absorb them into his leadership circle. Therefore, Zhang could only remind the heads of the workers' rebel groups to avoid being swayed by the Rebel Red Guards.[92] In other words, it can be argued that Zhang Chunqiao intended to cut the organizational ties between the worker and the student rebellion groups and indirectly weaken the power of the students by supporting the workers.

Against this backdrop, a subtle rivalry arose between the Rebel Red Guards groups, which Zhang Chunqiao did not support, and the worker and cadre rebel groups under his leadership.[93] Thus, this potential rivalry became increasingly apparent in early 1967 around the struggle for power during the establishment of the Three-In-One Regime and took on a complex shape under the influence of two significant developments. First, the workers' rebel groups had become the main body of the movement since 1967, mainly since the Workers' General HQ played an overwhelming leadership role in Shanghai compared to the student rebel groups. Without the cooperation of the worker rebel groups, it was highly unlikely that the student rebel groups alone could seize power in the municipal party and government institutions. Second, the power struggle between faction leaders within the worker rebel groups, especially between Wang Hongwen and Geng Jinzhang, weakened the cohesion and eased the competition between the remaining worker rebel groups and the Rebel Red Guards groups that confronted the Workers' General HQ.

As a result, the following interests were formed between Zhang Chunqiao, the Workers' General HQ, the 2nd Corps, and the Rebel Red Guards groups. First, Geng Jinzhang, the head of the 2nd Corps, took the lead in seizing power through his links with the Red Guards, trying to establish a superior position to Wang Hongwen. To this end, he carried out the first power seizure against the Shanghai Municipal Committee to confront Wang Hongwen, who had monopolized the leadership of the workers' rebellion group. In this case, the power struggle between Wang Hongwen and Geng Jinzhang and the fact that Zhang Chunqiao showed a greater preference for Wang Hongwen than Geng Jinzhang contributed to the Red Guard rebellion and the 2nd Corps joining forces for the sake of their joint political interests. On the other hand, the fight against the Red Guard rebellion's arbitrary seizure of power killed two birds with one stone for Wang Hongwen. It was an opportunity to eliminate a powerful rival in the power hierarchy while showing loyalty to Zhang Chunqiao, who disliked the Rebel Red Guards group, and indirectly to gain a favorable position in authority.

In this context, between January and February 1967, the Workers' General HQ took the following measures against the Rebel Red Guards and their leaders. In late January, following the failure of the second seizure of power by the Upper 3rd Headquarters, its leader, Zhao Quanguo, was handed over by Wang Hongwen to the Municipal Public Security Bureau on the grounds of forging instructions from Zhou Enlai. Thus, the power of the Upper 3rd Headquarters declined. In addition, shortly after the "28 January bombarding of Zhang Chunqiao," the Workers' General HQ began investigating the incident. As a result, more than 2500 people from the Rebel Red Guards group were investigated, and about 440 were sent to study classes (*xuexi ban*).[94] In addition, when the Shanghai People's Commune was established, Wang Hongwen excluded the Rebel Red Guards group from power distribution because the Red Guard rebels who had participated in the "28 January bombarding of Zhang Chunqiao" were ineligible to participate in the Shanghai People's Commune.[95] In this way, the Rebel Red Guards groups were weakened, and one of the critical factors leading to social unrest was eliminated. However, the rivalry within the Workers' General HQ became more pronounced.

Dissolution of the Three Corps

After the collapse of the Red Guards, the struggle for power within the worker rebellion began to surface, mainly between the Workers' General HQ and the 2nd Corps. After the Kangping Road Incident, the 2nd Corps, under Geng Jinzhang's leadership, absorbed many of the Scarlet Guards and expanded its power dramatically, becoming an organization almost as powerful as the Workers' General HQ.[96] Since 1967, there had been several conflicts between Wang Hongwen's Workers' General HQ and Geng Jinzhang's 2nd Corps. On the one hand, Wang Hongwen tried to exclude Geng Jinzhang and other leaders of worker rebel groups from the power distribution structure. While on the other hand, Geng Jinzhang tried to replace Wang Hongwen at the top of the worker rebel groups. The struggle between the two intensified on the eve of establishing the Shanghai People's Commune. At the Shanghai People's Commune preparatory committee, Wang Hongwen stipulated that "only municipal organizations could participate in the Shanghai People's Commune." As a result, the 1st Corps, 2nd Corps, 3rd Corps, and 3rd Division of Workers (formally known as the Third Command of the Shanghai Workers' Revolutionary Rebellion) were excluded from leadership roles in the Shanghai People's Commune. Subsequently, the excluded groups formed the Shanghai Workers' Revolutionary Rebellion Joint Command (*Shanghai gongren geming zaofan lianhe zhihuibu*). They attempted to establish the Second Shanghai People's Commune, which opposed the Shanghai People's Commune. However, under the persuasion of Zhang Chunqiao, the organization voluntarily disbanded, and the plan was aborted.[97]

Even if Zhang Chunqiao could control to some extent the factional rivalry within the workers' rebellion, as long as the factional rivalry still existed, it would significantly affect the stability of Shanghai. Moreover, the attempt of the 2nd Corps to establish the Second Shanghai People's Commune posed a significant threat to the original Shanghai People's Commune. Against this background, Zhang Chunqiao gave Wang Hongwen the full authority to resolve the factional rivalry within the worker rebellion to settle the disputes.

Two weeks after establishing the Shanghai People's Commune, Wang Hongwen lured Geng Jinzhang to the No. 31 Shanghai Cotton Factory on the pretext of discussing the "Grand Alliance" and had him arrested and sent to the Public Security Bureau. After that, Dai Zuxiang, the head of the 1st Corps, was handed over to the Public Security Bureau by Wang Hongwen. After losing their leaders, the 2nd Corps and 1st Corps were automatically disbanded. Under these circumstances, the opposing 3rd Corps and 3rd Division of Workers suddenly found themselves isolated and soon dissolved as their members gradually left.[98]

This outcome not only strengthened Wang Hongwen's political position but also achieved Zhang Chunqiao's goal of using factional interest competition to resolve disputes and strengthen the power of the Shanghai People's Commune.

The Destruction of the United Headquarters

After the disintegration of the 2nd Corps and other organizations, the Workers' General HQ absorbed some of the members of these groups and grew into an organization with more than 10,000 members. No other organization in Shanghai could match its size and strength. Because of this, Shanghai could maintain a more stable social order than other regions. But there was still a hidden element of instability, and it came from the Shanghai Diesel Engine Plant.

Since the beginning of the Cultural Revolution, there were two opposing factions in the Shanghai Diesel Engine Factory, the United Headquarters (*lian si*—officially named the "Revolutionary Rebel United Headquarters") and the Eastern Red Headquarters (*dongfanghong zongbu*). Because the Eastern Red Headquarters was a subordinate organization of the Workers' General HQ, the Workers' General HQ intervened in the conflict between the two factions. In June 1967, the Workers' General HQ declared the United Headquarters a reactionary organization, leading to dissatisfied workers' rebel factions and weakened Red Guard factions to rally together at the "Support Station of the Revolutionary Rebel United Headquarters," under the leadership of Quan

Xiangdong, to launch actions against the Workers' General HQ while also criticizing Zhang Chunqiao. In July 1967, a violent conflict erupted between the United Headquarters and the Eastern Red Headquarters, leading to the death of Xie Fuxi, a member of the Eastern Red Headquarters. Using this event as an excuse in August, Wang Hongwen sent members of the Workers' General HQ to launch a violent attack on the United Headquarters, ultimately destroying it. After the event, as many as 663 people were arrested by Wang Hongwen and sent to the public security bureau. At the same time, Wang Hongwen took on the role of Chairman of the Revolutionary Committee of the Diesel Engine Factory and eliminated the remnants of the United Headquarters. Thus, another factor affecting the stability of Shanghai was eradicated.[99]

According to Li Xun's research, Mao Zedong happened to be in Shanghai when the Workers' General HQ launched its forceful attack on the United Headquarters. In September 1970, Mao, who happened to be in Shanghai then, said to Wang Hongwen, "I have seen your "movie," you have achieved a great victory. The situation in Shanghai is stable."[100] In other words, although the actions of the Workers' General HQ were highly radical, adopting a violence-against-violence approach, including the selfish aim of expanding the power of the faction, the result settled the factional disputes and played a role in stabilizing the order in Shanghai, which Mao greatly appreciated.

In all three cases, the Workers' General HQ took the lead. Still, Zhang Chunqiao's tolerance, support, and work behind the scenes played a vital role. Wang's actions would not have been successful without Zhang's help and the power given to him to suppress other factions and their leaders. However, it can be argued that without the cooperation of Wang Hongwen and the Workers' General HQ, which understood and carried out Zhang Chunqiao's intentions, Zhang would not have been able to control the chaotic situation in Shanghai. Therefore, the three incidents described above can be seen as both a result of factional competition and Zhang Chunqiao's use of competition between factions to control the situation in Shanghai.

3.3 *The Quasi-Governmentalization of the Workers' General HQ*

Another important factor in maintaining stability in Shanghai was the quasi-governmental role played by the Workers' General HQ. Starting from a massive grassroots organization with significant mobilizing power in Shanghai, the Workers' General HQ gradually evolved through four stages to assume the functions of a de facto government.

During the first stage, the "Wen Gong Wu Wei Command" was established, monopolizing the authority to legitimately use violence. After September 1967, to prevent the rise of opposing forces like the United Headquarters, Zhang Chunqiao and Wang Hongwen responded to Jiang Qing's call for "attack with words and defend with force" (*Wen Gong Wu Wei*) and established the "Wen Gong Wu Wei Command," which effectively militarized the Workers' General HQ. The Wen Gong Wu Wei Command played the militia role and was led by Zhang Chunqiao and Wang Hongwen. In addition, in January 1967, Wang Hongwen had already seized control of the city's public security bureau and held it in his own hands. Specifically, after the Third Red Headquarters seized control of the Shanghai Public Security Bureau in January 1967, the rebel organizations, including the Public Revolutionary Association, the Workers' General HQ, and the Red 3rd Headquarters and others, established the "Proletarian Cultural Defense Committee" (hereinafter referred to as "Wen Bao Hui"), which integrated the functions of the Public Security Bureau, Procuratorate, and Court. Wang Hongwen served as its director.[101]

As a result of this series of actions, by September 1967, Wang and Zhang were in control of the whole law enforcement and legal apparatus in Shanghai. They monopolized the legal use of violence in Shanghai by arming the Workers' General HQ organization. In 1970, the Wen Gong Wu Wei Command was renamed the "Shanghai Militia," but its essence remained unchanged and continued until the collapse of the Gang of Four.[102]

The second stage was the institutionalization of the organization through the establishment of the Shanghai Workers' Revolutionary Rebel Congress (hereinafter referred to as the "Workers' Congress"). In late 1967, the Workers' Congress was held. Five members, Wang Hongwen, Pan Guoping, Chen Ada, Ye Changming, and Jiang Zhoufa, were appointed as standing committee members at this meeting. Since these five were all leaders of the Workers' General HQ, it could be said that the Workers' General HQ had a monopoly on the leadership of the Congress. As a result, the Workers' General HQ, which was initially a mass organization, was transformed into a workers' command structure.[103]

During the third stage, by dispatching the Workers' Mao Zedong Thought Propaganda Teams, the Workers' General HQ gained control of

Shanghai's leadership. In the summer of 1968, Mao Zedong's directive to dispatch these propaganda teams to schools allowed Workers' General HQ to expand its influence. The influence of the Workers' General HQ originally existed only among industrial workers. Zhang Chunqiao and Yao Wenyuan advocated that the Worker Mao Zedong Thought Propaganda Teams should lead all areas, including party and government institutions, industrial and mining sectors, commercial departments, schools, and neighborhood committees. As a result, the propaganda teams were dispatched to all these departments.[104] As a result, the Workers' General HQ gained control of Shanghai's leadership.

The fourth stage was to increase the voice of the party by bringing the leaders of the Workers' General HQ into the party. In 1968, the Central Committee of the CCP, while conducting a campaign of criticism of factionalism and anarchism, began to rebuild the party organization, reorganize the class ranks, and resume production activities. These became the main tasks of the newly established revolutionary committees. In many cities and regions, the power of the rebel groups was weakened since most party members had joined the conservatives, and the reactivation of the party meant that the conservatives were politically superior to the rebels. Moreover, the "reorganizing of the class ranks" was unfavorable to the rebels because of the large number of non-red five class members within the rebels. However, the reorganization and reconstruction of the party in Shanghai allowed the Workers' General HQ leaders to join the party. In February 1967, the Central Committee and the Central Cultural Revolution Group issued a document entitled Opinions and Problems Concerning the Reorganization, Restoration, and Reconstruction of Party Organizations, which not only stipulated that former party members, such as counter-revolutionary revisionists, should be expelled, it also specified that activists from the workers, peasants, and Red Guards should be selected to join.[105] In this context, Zhang Chunqiao instructed Wang Hongwen and others to admit Chen Ada, Dai Liqing, Huang Jinhai, and Ye Changming, the leaders of the Workers' General HQ into the Communist Party and nominated them as representatives to the Ninth National Congress of the CCP.[106] Not only that, after joining the party, Chen Ada also absorbed many of his people into the Communist Party and appointed them as cadres of the party and other political organizations. According to Li Xun's research, more than 200 rebels in the No. 31 Shanghai Cotton Factory, where Huang Jinhai worked, joined the party. Additionally, in 1968, 16 rebels from the No. 17 Cotton Factory were inducted into the party when the first reorganization occurred. Having been appointed by Wang Hongwen, a previous cotton factory employee, these individuals served in many positions in the party and other political organizations.[107] The Workers' General HQ used this method to increase its voice within the party.

4 Concluding Remarks

Zhang Chunqiao made use of the rivalry and competition among Shanghai's mass organizations, employing methods such as manipulation, merging, weakening, and disintegration to control these organizations, expand the influence of organizations cooperating with him, and help them improve their political status. Specifically, he first granted the Workers' General HQ the dominant position among Shanghai's mass organizations, allowing this organization to control the actions of other organizations. Then, by taking advantage of opportunities such as the establishment of the Shanghai Revolutionary Committee and dispatching Worker Mao Zedong Thought Propaganda Teams, he achieved the quasi-governmentalization of the Workers' General HQ. It can be said that Zhang Chunqiao successfully controlled the competition of interests among various organizations through these methods, avoiding the frequent occurrence of large-scale violence seen in other regions. This is the fundamental reason why Shanghai was able to maintain stability.

The power struggle that began with the January Revolution in Shanghai intensified factional competition in many places, leading to large-scale armed conflicts. However, in Shanghai, Zhang Chunqiao adopted a method of controlling factions with factions and using violence to curb violence. He established a "three-in-one" regime of revolutionary cadres, revolutionary military personnel, and revolutionary masses, realizing Mao Zedong's vision of simultaneous revolution and production. Zhang's method of stabilizing order was different from traditional administrative measures, and its methods also had violent characteristics. However, this can be considered an effective method for stabilizing social order in a situation where party and government institutions were paralyzed, and the police and military institutions responsible for maintaining social order were all involved in the Cultural Revolution. Furthermore, this method was not only approved by Mao Zedong, but Zhang Chunqiao and Wang Hongwen were also promoted to central leadership.

Zhang Chunqiao joined the Central Politburo at the Ninth National Congress of the Party in 1969. After the Lin Biao incident, his ranking in the central leadership rose to fifth, only behind Mao Zedong, Zhou Enlai, Kang Sheng, and Jiang Qing. Wang Hongwen, as a worker representative, was nominated as a Central Committee member at the Ninth National Congress of the Party and entered the Central Politburo at the Tenth National Congress of the Party in 1973, becoming the party's vice chairman. At this time, Wang Hongwen's position within the party surpassed Zhang Chunqiao and Jiang Qing, ranking third after Mao Zedong and Zhou Enlai. Jiang Qing, Yao Wenyuan, and Kang Sheng, who actively promoted the radical line of the Cultural Revolution, all entered the Central Politburo at the Ninth National Congress of the Party, and their positions within the party were elevated.[108] As a result, their radical ideas began to be incorporated into national policies, which in turn affected the entire country. Although Zhang Chunqiao and Wang Hongwen entered the central leadership, their political foundation remained in Shanghai until the downfall of the Gang of Four. The Shanghai model had an impact on the national political movement. In 1976, shortly after Mao Zedong's death, Zhang Chunqiao faltered in the power struggle. In the subsequent political reconstruction, the radical line he promoted became an obstacle to rebuilding the government and was abandoned and criticized. In addition, his achievements, once highly praised by Mao Zedong, were completely negated as the root cause of chaos in the *Resolution on Certain Historical Issues of the Party since the Founding of the People's Republic* adopted by the Sixth Plenary Session of the Eleventh Central Committee of the Communist Party in 1981.

Notes

1. "Zhou Enlai's Speech at the Conference to Receive Revolutionary Teachers and Students from Abroad in Beijing" (Zhou Enlai zai jiejian waidi laijing geming shisheng dahuishang de jianghua) (August 3, 1966), op. cit., *Research Materials on the "Cultural Revolution" (Part 1)*, p. 109.
2. "Notice of the Central Committee of the Communist Party of China and the State Council on the Revolutionary Link-up of Teachers and Students" (zhonggongzhongyang, guowuyuan guanyu geming shisheng jinxing geming chuanlian wenti de

tongzhi) (November 16, 1966), op. cit., *Research Materials on the Cultural Revolution (Part 1)*, p. 161.

3. "Speech by Zhou Enlai at an Emergency Meeting on the Reception Work at the Workers' Gymnasium in Beijing on November 16" (unknown source) (Zhou Enlai 11yue 16ri zai Beijing gongren tiyuguan jiedai gongzuo jinji huiyi shang de jianghua), in *Speech by the Comrade in Charge of the Central Committee on the Great Proletarian Cultural Revolution (sequel) (Zhongyang fuze tongzhi guanyu wuchanjieji wenhua da geming de jianghua, xubian)*, January 1967, p. 54.
4. "Zhou Enlai's Latest Oral Instructions" (Zhou Enlai zuijin koutou zhishi), (October 24, 1966), op. cit., *Compilation of Materials on the Great Proletarian Cultural Revolution* (wuchan jieji wenhua da geming ziliao huibian), p. 280.
5. In June 1966, Mao Zedong instructed: "Support students from all over the country to come to Beijing and let them enter for free". See Wang Nianyi, *The Age of Great Turmoil*, p. 77. After August, a few students from all over the country came to Beijing to complain to the central government about persecution by local Work Teams and Party Committees. This action was praised by Chen Boda and reported in the Party's theoretical magazine, *Red Flag*, and thus began the nationalization of Beijing student action throughout the country. On September 5 of the same year, the CCP Central Committee and the State Council issued a notice on organizing revolutionary students from non-Beijing higher schools, representatives of revolutionary students from secondary schools, and representatives of revolutionary staff to come to Beijing to observe the Cultural Revolutionary Movement, which was a general mobilization in the name of the Party and the government. The Circular made the following provisions.

   > Revolutionary students from non-Beijing institutions of higher education (including full-time and Part-Worker-Part-Peasant institutions of higher education), except for those who are ill, have already been here, or have other reasons for not being able to come, may organize visits to Beijing. The staff of higher schools may elect one representative of the revolutionary staff for every fifty students.
   >
   > Secondary schools in the field (including full-time and Part-Worker-Part-Peasant, semi-working secondary specialized schools,

> and general secondary schools) shall elect one representative of the revolutionary students for every ten students; the teaching staff shall elect one representative of the revolutionary staff for every one hundred students.
>
> All visits to Beijing will be free of charge by train. …… The subsistence allowance and transportation costs for revolutionary students and revolutionary staff visiting Beijing will be paid by the state public finance.
>
> Upon arrival in Beijing, meals and accommodation shall be arranged by Beijing municipal government. Meals while in Beijing were paid for by the state public finance. (*Research Materials on the Cultural Revolution (Part 1)*, p. 112)

The above central document on the "Link-up" shows that the original purpose of the Great Link-up was to popularize the Beijing experience of the Cultural Revolution to the whole country by organizing visits to Beijing.

6. On November 16, 1966, to solve the problem of receiving 1.1 million visitors to Beijing, Zhou Enlai gave the following instructions: For the Beijing party and government organs, factories, enterprises, residents' committees, and police stations, in addition to the number of people already received, the task of receiving 500,000 more people was added. For the Central Party and government organs, in addition to the number of people already received, 300,000 more people were to be received. Universities and Colleges received an additional 200,000 people. Concerned about the shortage of manpower in schools because of the "southward Link-up", 10,000 PLA troops were dispatched to schools to support the food supply. The PLA headquarters and other organs increased their reception duties by 100,000 people ("Zhou Enlai's Speech at the Emergency Meeting on Reception Work at the Workers' Gymnasium in Beijing on November 16", op. cit., "*Speech of the Comrade in Charge of the Central Committee on the Great Proletarian Cultural Revolution*" (sequel), January 1967, p. 54).
7. On November 2 and 14, 1966 Zhou Enlai twice summoned the heads of the various Red Guard organizations in Beijing to an emergency meeting on the reception of those who had entered the capital. At the meeting on the 2nd, Zhou Enlai

gave orders to the heads of the various Red Guard organizations in Beijing not to attend the Tiananmen reception and to persuade the non-Beijing students who had already been received to leave the capital. At this meeting, Zhou Enlai told the heads of the Red Guard organizations, "There are 240,000 people entering the capital every day, and less than 20,000 people leaving the capital. If this continues, Beijing will not be able to bear it in a few days". On the 16th, Zhou Enlai summoned the heads of party and government organs, factories, enterprises, residents' committees, and police stations to an emergency meeting on the distribution of reception tasks. In addition, the CCP Central Committee and the State Council also issued three documents ordering the temporary suspension of the Link-up activities during the period from October 31 to December 1, 1966 ("Comrade Zhou Enlai's Speech at the Emergency Meeting of the Comrades Responsible for the Red Guards in the Capital on November 2, 1966" (Zhou Enlai tongzhi 11yue 2ri zai shoudu ge hongweibing zeren tongzhi jingji huiyi shang de jianghua), op. cit. *Speech by the Comrade in Charge of the Central Committee on the Great Proletarian Cultural Revolution (sequel)* pp. 6–7; "Emergency Circular of the Central Committee of the Communist Party of China and the State Council on Maintaining Order in Railway Transport" (Zhonggong Zhongyang, guowuyuan guanyu weihu tielu yunshu zhixu de jinji tongzhi) (October 31, 1966); "Circular of the Central Committee of the CCP and the State Council on the issue of Revolutionary Students and Teachers Conducting Revolutionary Link-ups" (Zhonggong Zhongyang,guowuyuan guanyu geming shisheng jinxing geming chuanlian wenti de tongzhi) (November 16, 1966); "Supplementary Circular of the Central Committee of the CCP and the State Council on the Issue of Revolutionary Students and Teachers Conducting Revolutionary Link-ups in Universities and Secondary Schools" (Zhonggong Zhongyang, guowuyuan guanyu dazhong xuexiao geming shisheng jinxing geming chuanlian wenti de buchong tongzhi), op. cit. *Research Materials on the Cultural Revolution (Part 1)*, pp. 152, 162, 180).

8. Li Xun, *The Great Collapse—A History of the Rise and Fall of the Shanghai Workers' Rebellion*, op. cit., pp. 30–31.

9. Ibid.
10. "Speech by Comrades Qi Benyu and Guan Feng to Some Red Guard members of the Capital 3rd HQ of at the *People's Daily* News Agency on 25 October" (QI Benyu, Guan Feng tongzhi 10yue 25ri zai Renmin Ribao she dui shoudu hongweibing disan silingbu bufen hongweibing zhanshi de jianghua), op. cit., *Compilation of Materials on the Great Proletarian Cultural Revolution*, p. 286.
11. Li Xun, op. cit. *The Great Collapse—A History of the Rise and Fall of the Shanghai Workers' Rebellion*, p. 33.
12. Chen Zu'en et al., *A General History of Shanghai (Vol. 11)* (Shanghai tongshi), Shanghai People's Publishing House, 1999, p. 216.
13. Li Xun, op. cit. *The Great Collapse—A History of the Rise and Fall of the Shanghai Workers' Rebellion*, p. 37.
14. Ibid., pp. 33–40.
15. On the rebellion of temporary and contract workers, "half-work" students, and sent-down youths, see the following studies: Tsuneto Yamamoto, "The Policy Trinity of Labor, Education, and Deportation in 1960 and Its Destruction", in Mitsuyuki Kagami, ed. *Frustration in Contemporary China*, op. cit., pp. 147–226; Li Xun, "The "Economist Wind" in Shanghai during the Cultural Revolution", in Luo Jinyi et al., eds, *Beyond the Catastrophe: Rethinking of the Cultural Revolution*, pp. 30–49; Fang Yuan, "Pioneer of the Free Workers Movement—Commemorating the 30th anniversary of the establishment of the independent trade union General All Red"—*Beijing Spring*, March, April, and May 1997.
16. In the early 1960s, when Mao Zedong withdraw to the second line and Liu Shaoqi was active in the political arena as the first-line leader, Liu Shaoqi vigorously promoted a "two-legged labor system" and a "two-legged education system" in order to achieve efficiency and rationalization in urban management. As a result, a large number of temporary workers, contract workers, "half-worker" students and sent-down youths emerged throughout the country. They are at a disadvantage compared to regular workers and full-time school students in terms of labor insurance, medical insurance, and promotion. In October 1966, in the context of the criticism of Liu Shaoqi's line, temporary and contract workers,

"half-worker" students and sent-down youths at the bottom of the city in the name of criticizing Liu Shaoqi's line, rose up and put forward various economic demands. Their actions bring to the surface the problems that had previously been hidden at the bottom of Chinese society.

17. In 1965, the State put forward the policy of the Third Five-Year Plan to "accelerate the construction of the Third Line". The "first line" of the "three lines" was "the old developed coastal industrial bases of Shanghai, Tianjin and Guangzhou", the "second line" is "the initially developed economic areas between the first and third lines", and the "third line" is "Sichuan, Gansu, and Xinjiang, which are located deep inland and have less developed industrial, transportation and infrastructure areas". The "Third Line" was built by decentralizing an industrial system centered on military industry to the interior areas. The implementation of this policy led to the relocation of many workers from major cities such as Shanghai to the interior. These workers, having experienced the disparities between the regions, started a rebellion after the beginning of the Cultural Revolution to return to their original cities.
18. According to the State Council, new regular workers must undergo an internship period before becoming skilled, during which they are not paid a salary but only a living wage of 18 yuan per month. In addition, the internship period varied according to the type of work and the degree of difficulty and was not uniformly regulated by the State Council, except for a maximum of three years. During the Cultural Revolution, some internship workers demanded to be converted to regular workers and paid back wages on the grounds that their internship had been made permanent. For a study of the rebellion of internship workers, see Li Xun's article "The 'Economist Wind' in Shanghai during the Cultural Revolution", p. 37.
19. In the 1960s, to curb the burden of the overpopulation of the cities and the surge in employment, the massive return of the urban population to agriculture was accompanied by a vigorous campaign to mobilize and organize urban youths to go to the countryside. Young people who had been deported to the countryside during the Cultural Revolution organized themselves to demand their return to the cities.

20. The General All Reds is a national federation with subordinate organizations everywhere and is the largest of the mass organizations demanding economic benefits. The demands of this organization can be summarized in three main points. First, the factories must not unilaterally dismiss temporary-contracted-outsourced workers, abolish the employment system of temporary-contracted-outsourced workers, and demand job security for the above-mentioned occupations. Second, they are entitled to the same political and economic treatment as workers under state ownership. Third, reinstatement of temporary, contract, and outsourced workers who were dismissed during and on the eve of the Cultural Revolution (Li Xun, "The "Economist Wind" in Shanghai during the Cultural Revolution", pp. 30–38). In February 1967, when the "Economist Wind" was opposed, the "General All Reds" were banned as a reactionary organization. The following eight other national rebel organizations were banned during the same period: The National Headquarters of the Rebel Group for the Elimination of the Capitalist Army, The National Red Rebel Corps of State Farms, The National Revolutionary Rebel Group of the Sent-Down Youth for the Defending the Truth, The National Revolutionary Rebel Group for the Soldiers of the Military reclamation, The International Red Guards' Chinese Detachment, The National Sent-Down Youth Red Revolutionary Rebel Corps, The Red Front Combat Team of the National Sent-Down Youth, and The United Headquarters of the National Deaf Revolutionary Rebellion ("Comrade Fu Chongbi's Speech when Receiving Representatives of the Leftist Team" (fuchongbi tongzhi jiejian zuopai duiwu daibiao shi de jianghua) (February 2, 1967), *Beijing Reference Materials on the Great Proletarian Cultural Revolution*, edited by The Congress of the Red Guards of the Capital University and College, Political and Legal Commun of the Beijing Institute of Political Science and Law), 1967, No. 32, pp. 507–508.
21. Li Xun, op. cit. "The 'Economist Wind' in Shanghai during the Revolution", p. 37.
22. Ibid.
23. Ibid., p. 39.
24. Ibid., pp. 39–42.

25. "Minutes of the Talks of Comrades Zhou Enlai, Chen Boda, Kang Sheng, Jiang Qing, Wang Li, Qi Benyu and Others" ("Zhou Enlai, Chen Boda, Kang Sheng, Jiang Qing, Wang Li, QI Benyu deng tongzhi de zuotan jiangzuo" (January 10, 1967), collated by the Eastern Red Commune of the Beijing Mining Institute, printed by The Xicheng Branch of the Capital Workers' Revolutionary Rebellion Headquarters (Beijing kuangye xueyuan dongfanghong gongshe zhengli, shoudu zhigong geming zaofan zongbu xibu fenbu yinshua), January 11, 1967, 3 pages; "The speech of comrades Zhou Enlai, Chen Boda, Jiang Qing, Kang Sheng and Others at the Great Hall of the People when they received representatives of the revolutionary rebels, students, workers, representatives of military academies and other parties"("Zhou Enlai, Chen Boda, Jiang Qing, Kang Sheng deng tongzhi zai renmin dahuitang jiejian geming zaofan xuesheng, gongren, junshi yuanxiao daibiao he qita fangmian daibiao de jianghua") (January 10, 1967), *The collection of Reference Materials on the Great Proletarian Cultural Revolution* (*youguan wuchan jieji wenhuadageming de cankao cailiao*), 1967, No. 3, p. 59.
26. "Comrade Zhou Enlai's Address to Representatives of the PLA's Ministry of Culture Cultural Revolution Task Force Rebellion Liaison Station and Some Representatives of the Four Clean-up Task Force" (Zhou Enlai tongzhi jiejian jiefangjun canjia wenhuabu wenhua geming gongzuodui zaofan lianluozhan daibiao he bufen siqing gongzuodui daibiao de jianghua) (January 11, 1967), *A Collection of Speeches by the Head of the Great Proletarian Cultural Revolution*, printed by *Red Flag*, The Beihang Institute, April 1967, p. 159.
27. "Minutes of the Talks of Comrades Zhou Enlai, Chen Boda, Kang Sheng, Jiang Qing, Wang Li, Guan Feng, and Qi Benyu" (January 10, 1967), op. cit., collated by the Eastern Red Commune of the Beijing Mining Institute, printed by the Western Component Department of the Capital Workers' Revolutionary Rebellion Headquarters, January 11, 1967, p. 3.
28. For specific examples, see Li Xun's article, "The 'Economist Wind' in Shanghai during the Cultural Revolution", pp. 44–45.
29. "Comrade Zhou Enlai's Speech at Zhongnanhai when Receiving Some members of the August 1st Combat Regiment of the

Beijing Construction and Engineering Institute and *Jinggangshan* Corps zhouenlai tongzhi zai Zhongnanhai jiejian Beijing jiangong xueyuan 8–1 zhandoutuan, jinggangshan bufen zhanshi shi de jianghua) (January 7, 1967)", op. cit. *in The collection of Reference Materials on the Great Proletarian Cultural Revolution*, 1967, No. 3, p. 35.

30. "Speech by Zhou Enlai When Receiving Representatives of the Chengdu Workers' Rebel Corps on 4, December" (Zhou Enlai 12yue 4ri jiejian chengdu gongren zaofantuan daibiao shi de jianghua), op. cit., *Compilation of Materials on the Great Proletarian Cultural Revolution*, p. 501; "Speech by Zhou Enlai When Receiving Representatives of the Chengdu Workers' Rebel Corps on the Evening of December 14" (Zhou Enlai 12yue 14ri wan jiejian chengdu gongren daibiaotuan daibiao shi de jianghua), op. cit., *Speech of Comrade in Charge of the Central Committee on the Great Proletarian Cultural Revolution (sequel)*, January 1967, p. 172.
31. Comrade Zhou Enlai's Speech on Reception and Hospitality at the Great Hall of the People (11 January 1967) (Zhou Enlai tongzhi zai renmin dahuitang guanyu dui jiedai, zhaodai gongzuo de jianghua, op. cit., *Collection of Speeches on the Proletarian Cultural Revolution by the Chiefs of Central Committee*, p. 166.
32. Comrade Zhou Enlai's Speech at the Workers' Gymnasium to Receive Revolutionary Teachers and Students and Representatives of Oil Industry and Mining Enterprises (January 8, 1966) (Zhouenlai tongzhi zai gongren tiyuguan jiejian geming shisheng he shiyou gongkuang qiye daibiao dahuishang de jianghua), op. cit., *The collection of Reference Materials on the Great Proletarian Cultural Revolution* 1967, No. 3, p. 40; "Comrade Zhou Enlai's Speech on November 26 at a Discussion with Forty Delegates of the Southwest Joint Complaint Group to Beijing" (Zhou Enlai tongzhi 11yue 26ri tong xinan fujing lianhe gaozhuangtuan 40 ming daibiao zuotanshi de jianghua), op. cit., *The speech of Comrade in Charge of the Central Committee on the Great Proletarian Cultural Revolution (sequel)*, January 1967, p. 100.
33. "Speech by Comrades Zhou Enlai and Li Xiannian on Receiving Cadres of the Directorate of Finance and Trade (February 18, 1967)" (Zhou Enlai, Li Xiannian tongzhi jiejian caimaokou sijuzhang yishang ganbu de jianghua), op. cit., *The collection of*

Reference Materials on the Great Proletarian Cultural Revolution 1967, No. 29, p. 456.

34. "Speech by Comrade Fu Chongbi when Receiving Representatives of the Left Procession (February 24, 1967)" (Fu Chongbi tongzhi jiejian zuopai duiwu daibiao shi de jianghua), *Reference Materials on the Great Proletarian Cultural Revolution* 1967, No. 32, pp. 507–508.
35. *Research Materials on the "Cultural Revolution" (Part 1)*, pp. 245, 254, 259, 261.
36. Harry Harding, op. cit. *Organizing China*, p. 153.
37. Bi Xia, "Leaving the Material of Memory to History" ("Ba huiyi de cailiao liugei lishi"), edited by Xu Youyu, *1966: Memories of Our Generation (1966: women nayidai de huiyi)*, The China Federation of Literary Press (Zhongguo wenlian chubanshe), 1998, p. 194.
38. "Speech by Tao Zhu when Receiving the to-Beijing Combat Group of the East China Engineering College on November 12", (Tao Zhu 1yue 12ri jiejian Huadong gongcheng xueyuan fujing zhandoutuan shi de jianghua) op. cit., *Compilation of Materials on the Great Proletarian Cultural Revolution*, p. 353.
39. "Comrade Pan Fusheng's Communication of the Central Working Conference on November 1", op. cit., *Compilation of Materials on the Great Proletarian Cultural Revolution*, p. 316.
40. "Tao Zhu's Speech at the Meeting of State Ministers and Cadres on 3 December" (Tao Zhu 12yue 3ri zai guojia buzhang ganbuhui shang de jiang hua), op. cit., *Compilation of Materials on the Great Proletarian Cultural Revolution*, p. 500.
41. "Comrade Zhou Enlai's Speech at the Meeting with the Anhui Revolutionary Rebels (January 20, 1967)" (Zhou Enlai tongzhi jiejian anhui geming zaofanpai shi de jianghua); "Comrades Zhou Enlai, Chen Boda, Kang Sheng and Jiang Qing's Speech at the Meeting with the Masses of to-Beijing Link-up Groups. (January 22, 1967)" (Zhou Enlai, Chen Boda, Kang Sheng, Jiang Qing tongzhi zai jiejian waidi laijing qunzhong dahuishang de jianghua), op. cit., *The collection of Reference Materials on the Great Proletarian Cultural Revolution*, 1967, No. 11, pp. 166, 168.

42. "Speech by Comrades Zhou Enlai, Chen Boda, Jiang Qing and Kang Sheng at the People's Great Hall when they received representatives of the revolutionary rebels, students, workers, representatives of military academies and other parties (January 10, 1967)" (Zhou Enlai, Chen Boda, Jiang Qing, Kang Sheng deng tongzhi zai renmin dahuitang jiejian geming zaofanpai xuesheng, gongren, junshi yuanxiao daibiao he qita fangmian de daibiao shi de jianghua) op. cit., *The collection of Reference Materials on the Great Proletarian Cultural Revolution*, No. 3, 1967, p. 61.
43. Ibid., pp. 57–64.
44. "Zhou Enlai's Speech at the Meeting of Teachers and Students of Half-Working and Half-Study Schools in the Capital and Abroad to Fire Fiercely at the Bourgeois Reactionary Line on 31 October" (Zhou Enlai 10yue 31ri zai shoudu ji waidi bangongbandu xuexiao shisheng xiang zichanjieji fandong luxian menglie kaihuo dahuishang de jianghua); "Transcript of Comrades Jiang Qing, Chen Boda and Others' Speech to Some Representatives of the National Red Labour Rebellion General League on 26 February", (Jiang Qing, Chen Boda deng tongzhi 12yue 26ri jiejian quanguo hongse laodong zaofan zongtuan bufen daibiao jianghua jilu), *Compilation of Materials on the Proletarian Cultural Revolution*, pp. 304–306, 620–622.
45. "Notice of the Central Committee of the CCP and the State Council on How to Launch the Cultural Revolution Campaign in Industrial and Transportation Enterprises and Infrastructure Sectors (2 July 1966)" (Zhonggong Zhongyang, guowuyuan guanyu gongye jiaotong qiyan he jiben jianshe danwei ruhe kaizhan wenhuadageming yundong de tongzhi), op. cit., *"Cultural Revolution" Research Materials (Part 1)*, p. 54.
46. "Notice of the Central Committee of the Communist Party of China on Grasping the Revolution and Promoting Production (September 14, 1966)" (Zhonggong Zhongyang guanyu zhuageming cushengchan de tongzhi), op. cit., *"Cultural Revolution" Research Materials (Part 1)*, p. 114.
47. "Ten Provisions of the Central Committee of the CCP on Grasping the Revolution and Promoting Production (9 December 1966)" (Zhonggong Zhongyang guanyu zhuageming cushengchan de shitiao guiding), op. cit. *"Cultural Revolution" Research Materials (Part 1)*, p. 182.

48. Su Caiqing, "Three Rounds of Struggle in the Early Years of the Cultural Revolution" (wenge chuqi sange huihe de douzheng), in Zhang Hua et al., eds., *Looking Back at the Cultural Revolution (part 2) (huishou wenge xia)*, pp. 735–763.
49. Wang Li, *Live History: A Chronicle of the Cultural Revolution* (Xianchang lishi: wenhuadageming jishi), Oxford University Press, 1993, p. 102.
50. Ibid., p. 104.
51. "Speech by Comrades Zhang Chunqiao and Yao Wenyuan at the Meeting with the Heads of the Red Revolution Societies of the Schools in Shanghai (January 9, 1967)" (Zhang Chunqiao, Yao Wenyuan tongzhi zai shanghai jiejian gexiao honggehui fuzeren shi de jianghua), in *A Collection of Speeches by the Heads of the Great Proletarian Cultural Revolution*, printed by Red Flag of Beihang, April 1967, p. 117.
52. Jin Chunming, *A History of the "Cultural Revolution"* (*Wenhua dageming shigao),* Sichuan People's Publishing House, 1995, p. 217.
53. Li Xun, op. cit. *The Great Collapse—A History of the Rise and Fall of the Shanghai Workers' Rebellion*, pp. 65–69.
54. Ibid., pp. 91–108.
55. Wang Nianyi, "Some Thoughts on the 'January Revolution' in Shanghai"; Chen Zu'en, op. cit.
56. Li Xun, op. cit. *The Great Collapse—A History of the Rise and Fall of the Shanghai Workers' Rebellion*, pp. 118–121.
57. The following studies hold such views: Li Xun, op. cit., *The Great Collapse—A History of the Rise and Fall of the Shanghai Workers' Rebellion*, p. 117; Jin Chunming, op. cit., *A History of the "Cultural Revolution"*, p. 194; Wang Nianyi, op. cit., "Some Thoughts on the 'January Revolution' in Shanghai", p. 130.
58. After hearing Zhang Chunqiao's report on the handling of the Anting incident from the Central Political Bureau, Mao Zedong expressed his attitude of defending Zhang Chunqiao's actions in violation of the Party Committee's instructions: "First there are facts, then there are concepts".
59. Li Xun, op. cit. *The Great Collapse—A History of the Rise and Fall of the Shanghai Workers' Rebellion*, pp. 150–152.

60. "Speech by Comrades Zhang Chunqiao and Yao Wenyuan when they received the to-Beijing delegation of the Shanghai Workers' Revolutionary Rebellion Headquarters in the third floor lounge of the auditorium of the Chinese People's Political Consultative Conference on 6 December" (Zhang Chunqiao, Yao Wenyuan tongzhi 12yue 6ri zai zhengxie litang 3lou xiuxishi jiejian shanghai gongren geming zaofan silingbu fujing daibiaotuan shi de jianghua), op. cit., *Speeches by Comrades in Charge of the Central Committee on the Great Proletarian Cultural Revolution" (sequel)*, January 1967, p. 149. According to Yao Wenyuan's speech, the Red Guards had 400,000 members by early February.
61. Li Xun, op. cit. *The Great Collapse—A History of the Rise and Fall of the Shanghai Workers' Rebellion*, p. 151.
62. Chen Zu'en, op. cit., pp. 219–220.
63. Wang Nianyi, op. cit., "Some Thoughts on the 'January Revolution' in Shanghai" pp. 131–132.
64. Ibid., pp. 171–172.
65. Ibid., pp. 177–209.
66. Elizabeth J. Perry and Li Xun, *Proletarian Power: Shanghai in the Cultural Revolution* (Boulder: Westview Press, 1997), p. 86.
67. Chen Zu'en, op. cit., pp. 220–221.
68. Ibid., p. 222.
69. Ibid., p. 230.
70. Wang Nianyi, op. cit. "Some Thoughts on the 'January Revolution' in Shanghai", p. 91.
71. Jin Chunming, op. cit., p. 198.
72. Li Xun, op. cit. *The Great Collapse—A History of the Rise and Fall of the Shanghai Workers' Rebellion*, p. 231.
73. Wang Nianyi, op. cit., "Some Thoughts on the 'January Revolution' in Shanghai", p. 170.
74. Li Xun, op. cit. *The Great Collapse—A History of the Rise and Fall of the Shanghai Workers' Rebellion*, pp. 234–239.
75. Ibid., p. 259.
76. "Speech by Comrades Zhang Chunqiao and Yao Wenyuan at the Meeting with the Heads of the Red Revolution Association of the Schools in Shanghai (9 January 1967)" (Zhang Chunqiao, Yao Wenyuan tongzhi zai shanghai jiejian gexiao honggehui fuzeren shi de jianghua), op. cit., *Collection of Speeches by the Heads of the Great Proletarian Cultural Revolution*, p. 117.

77. Wang Li, op. cit., p. 104.
78. "Shanghai Revolutionary Rebels Launch General Attack on Bourgeois Reactionary Line" (Shanghai geming zaofanpai xiang zichanjieji fandong luxian faqi zong jingong) *People's Daily*, January 12, 1967.
79. An Wenjiang, "I Have No Regrets", edited by Xu Youyu, op. cit., *1966: The memories of our generation*, p. 100.
80. Li Xun, op. cit. *The Great Collapse—A History of the Rise and Fall of the Shanghai Workers' Rebellion*, pp. 127–128.
81. An Wenjiang, op. cit., p. 93.
82. Li Xun, op. cit. *The Great Collapse—A History of the Rise and Fall of the Shanghai Workers' Rebellion*, pp. 319–323.
83. Ibid., pp. 326–331.
84. Ibid., pp. 331–335.
85. Ibid., pp. 356–358.
86. Ibid., p. 359.
87. Ibid., p. 317.
88. Jin Chunming, op. cit., *A History of the "Cultural Revolution"*, pp. 205–207.
89. Harry Harding, op. cit., pp. 169–170.
90. Liu Geping, a cadre of the Shanxi Provincial Committee, was able to establish the "Municipal Revolutionary Committee" at an early stage because of the support of Zhang Riqing, the second political commissar of the provincial military district. Pan Fusheng, the first secretary of the Heilongjiang Provincial Committee, not only served as the first political commissar of the Heilongjiang Military Region himself, but also received strong support from Wang Jiadao, the commander of the provincial military region. Wang Xiaoyu, the vice mayor of Qingdao in Shandong Province, was able to control the power struggle among the masses and set up the "Shandong Provincial Revolutionary Committee" thanks to the support of Tong Guogui, the commander of the provincial military district. In Guizhou Province, the "Provincial Revolutionary Committee" was established under the direct leadership of Li Zaihan, the second political commissar of the provincial military district (Jin Chunming, op. cit., *A History of the "Cultural Revolution"*, pp. 213–214; Liu Guokai, ed., The *history that cannot be blocked*, published by the Editorial Board of Re-evaluation of the Cultural Revolution Series, pp. 347–348).

91. Li Xun, op. cit. *The Great Collapse—A History of the Rise and Fall of the Shanghai Workers' Rebellion*, pp. 186–202.
92. Ibid., p. 321.
93. An Wenjiang, the head of the "Red Third Headquarters", once stated in his memoirs that the reason why the "Red Third Headquarters" took over the power of the Municipal Public Security Bureau in January 1967 was that in the "power seizure" campaign, by taking the lead, they would be able to build up more credit than the workers' and cadres' rebel groups. In other words, by building credit, it was possible to gain the support of the political leaders and thus regain the political status that was being lost (An Wenjiang, op. cit., p. 100).
94. Jin Chunming, op. cit., *A History of the "Cultural Revolution"*, p. 207.
95. Li Xun, op. cit. *The Great Collapse—A History of the Rise and Fall of the Shanghai Workers' Rebellion*, p. 355.
96. Ibid., p. 292.
97. Ibid., pp. 348, 356–358.
98. Ye Yonglie, *The Rise and Fall of Wang Hongwen*, Times Literary Publishing House, 1989, p. 258.
99. Chen Zu'en et al., op. cit. General History of Shanghai (vol. 11), pp. 232–235.
100. Li Xun, op. cit. *The Great Collapse—A History of the Rise and Fall of the Shanghai Workers' Rebellion*, p. 384.
101. Ibid., p. 317.
102. Chen Zu'en et al., op. cit. General History of Shanghai (vol. 11), p. 236.
103. Li Xun, op. cit. *The Great Collapse—A History of the Rise and Fall of the Shanghai Workers' Rebellion*, p. 408.
104. Ibid., pp. 348, 421.
105. "Opinions and Issues of the Central Committee of the CCP and the Central Cultural Revolution on Rectifying, Restoring and Rebuilding Party Organizations (2 February 1967)", op. cit., *"Cultural Revolution" Research Materials (Part 1)*, pp. 640–642.
106. Li Xun, op. cit. *The Great Collapse—A History of the Rise and Fall of the Shanghai Workers' Rebellion*, pp. 438–444.
107. Ibid., p. 444.
108. Masashi Ando et al., op. cit. *The Cultural Revolution and China*, pp. 117, 140.

CHAPTER 8

Competing Strategies for Citizenship—The Cultural Revolution in Small Cities

The first three chapters of this work delve into the behaviors of the three primary factions of the three sub-movements that took place during the Cultural Revolution. The focus is on analyzing how the interaction between the state and society influences and regulates factional behavior and how this behavior, in turn, affects political competition among top political leaders. It is worth noting that the interaction between the state and society played a significant role in pushing the Cultural Revolution from one sub-movement to another, leading to changes in the goals and main factions of the movement. Additionally, this work addresses the equally important issue of how non-mainstream factions in non-political central cities are formed, how they participate in the movement, and how they compete for citizenship during the movement.

This chapter examines factional behavior during the Cultural Revolution, focusing on the factional struggle at the Dongfanghong Textile Factory (DFH) in Yuci City, Shanxi Province. The study will explore factional behavior at the periphery and micro level, as opposed to the political center. By analyzing the behavior of factional leaders rather than ordinary participants, we can gain insight into the sustained enthusiasm, motivation, access to resources, mobilization ability, and competitive strategies crucial to any collective action. Comprehending the behavior of factional leaders in the Cultural Revolution is key to understanding

Y. Lijun, *Contest for Citizenship and Collective Violence during China's Cultural Revolution*, IPP Studies in the Frontiers of China's Public Policy, https://doi.org/10.1007/978-981-97-0906-9_8

the analysis of factional competition. This chapter aims to answer a series of questions by studying factional leader behavior: How did the Cultural Revolution unfold in cities far from the political center? How were ordinary people drawn into the movement? Who were the first to rebel, and why did they actively participate? What kind of people became factional leaders, and how did they organize and mobilize collective behavior? Finally, how did factional leaders respond when the movement's goals changed or when other political opportunities arose?

The choice of the DFH Textile Factory in Yuci City, Shanxi Province, as the subject of this study is due to the availability of historical data. Micro-level research on the Cultural Revolution is rare, and research on factional behavior is typically limited to the city level. This is primarily because obtaining micro-level data is often challenging. Fortunately, during the author's collection of Cultural Revolution materials in Shanxi in 2001, relevant materials on the Cultural Revolution movement in DFH Textile Factory in Yuci City were collected. These materials include investigation records of the Four Cleans Movement in the factory, personal files of employees, and materials on the exposure and criticism during the Cultural Revolution. With access to these materials, the author could focus the analysis of participant behavior in the Cultural Revolution on factional leaders and trace their motivations back to before the Cultural Revolution. This approach has enabled the author to explore the Cultural Revolution in Shanxi beyond the limitations of previous research that only studied the Cultural Revolution as a single movement. By considering how political movements such as the Four Cleans before the Cultural Revolution influenced factional behavior during the Cultural Revolution, this study provides a more comprehensive understanding of the dynamics of the Cultural Revolution.

Besides the availability of historical materials, the analysis of the Cultural Revolution-related records from DFH (Dongfanghong Textile Factory) offers valuable insights into the dynamics of the Cultural Revolution at the local level. Yuci City, at that time, was a typical local urban center with a population of approximately 500,000, primarily focused on light industry. Surprisingly, it witnessed intense factional struggles during the Cultural Revolution comparable to those in major cities. Therefore, examining the factional struggles within DFH contributes to our understanding of the localized nature of the Cultural Revolution.

Moreover, from a political perspective, Yuci City served as the administrative center for party and government organs in the Jinzhong area, located near Taiyuan, the political hub of Shanxi Province. It was also geographically near Dazhai, which was celebrated as a national model for agricultural production. Within Yuci City, there were three prominent state-owned enterprises, including DFH. The power struggles between party and government officials at the levels of the CCP Central Committee, Shanxi Province, Jinzhong area, and Yuci City, as well as the conflicts between party and government officials and military personnel, and among military officers themselves, were intertwined with the opposition between Chen Yonggui, a highly acclaimed model worker from Dazhai, endorsed by Mao Zedong, and the party, government, and military officials. This intricate political landscape of turmoil provided a fertile ground for studying the factional struggles within DFH, offering valuable material for comprehending the Cultural Revolution from a micro-level perspective.

However, examining factional behavior through the lens of complex political structures is not the focus of this book, and the author will explore it in future research. This chapter will only focus on the period from the launch of the Four Cleans Movement in early November 1965 to the temporary dissolution of the Defend the Army faction at the end of March 1967. This chapter consists of the following five sections. The first section delves into the formation of factional leaders. The second section explores how the Four Cleans Movement before the Cultural Revolution influenced collective behavior during the Cultural Revolution. The third section focuses on how factional leaders maintain collective behavior amid political uncertainty. This includes examining the mobilization strategies and methods used to attract new members to their faction and prevent the loss of existing members. Additionally, section three explores the methods that factional leaders employ to salvage the legitimacy of their faction when the political situation takes a turn for the worse. The fourth section examines the mobilization methods and strategic means that factional leaders typically use to organize collective behavior in competition with opposing factions. This section sheds light on how factional leaders ensure their faction maintained a competitive advantage. Finally, the last section provides a concise summary of the main findings of this study.

1 The Rise of the Faction Leaders

Factional conflict in DFH arose between two groups: the DFH Textile Factory Revolutionary Organization Alliances Headquarters (hereinafter referred to as the East Alliance) and the Defending Mao Zedong Thought Revolutionary Rebel Army (hereinafter referred to as the Defending Army). The Eastern Alliance was established on January 5, 1967, with about 500 members at the beginning and a rapid increase in membership one month later, reaching a peak of approximately 4000 members (out of over 6000 employees in the factory). The core members of the Eastern Alliance were those who were removed from their positions during the Four Cleans campaign and those who were persecuted during that time.

The Defending Army was established in February 1967 with around 1200 to 1300 members, most of whom were Four Cleans cadres, representatives, and active participants. Although the Defending Army was established later than the Eastern Alliance, it was reorganized from the Red Guards Headquarters formed in September 1966 and the "Defend Station" (officially known as the "Defend Mao Zedong Thought Liaison Station") formed on January 7, 1967, and its organizational foundation was established earlier than that of the Eastern Alliance.

Looking at the factional conflict structure within the East Alliance, it becomes clear that the formation of factional leaders and the conflict between factions are closely tied to the Four Cleans Movement. The factional conflict at DFH originated from the conflict between four cleans cadres and four uncleans cadres. To thoroughly explain this point, this chapter will analyze a one-year period, from the expansion of the Four Cleans Movement in the factory in November 1965 to the rise of the factory mass rebellion in January 1967, in four stages. By examining the political changes that occurred during these four stages, we can better understand how the Four Cleans Movement influenced and shaped the factional conflict within the DFH during the Cultural Revolution.

1.1 Selection of the Four Cleans Movement Delegates

The first stage of the factional conflict at DFH occurred from November 1965 to the end of January 1966, during which four cleans representatives were selected. This stage laid the groundwork for the factional conflict unfolding during the Cultural Revolution. The Four Cleans Movement at DFH was initiated in November 1965 when a work group of approximately 100 individuals was dispatched to the factory, with Bu Hongyun, the deputy secretary of the Jinzhong Regional Party Committee, serving as the group leader.[1, 2] The initial task of this group was to select representatives for the Four Cleans campaign in the factory. To achieve this goal, the working group focused on all party members and active elements in the factory. After about three months of checking personal files and conducting oral inquiries, 345 representatives were selected from the party members and active elements in February 1966.[3]

While it was an honor to be selected as a Four Cleans Movement delegate, it became a liability when the movement led to the Cultural Revolution. Indeed, the destiny of the four cleans delegates during the Cultural Revolution was determined at the outset when they were selected. The delegates had to cooperate closely with the Work Team. They were required to provide information to the Work Team to investigate other people in the factory, particularly party cadres who became main targets during the Four Cleans Movement. Furthermore, the delegates had to write very detailed political investigation records, which were added as individual files as a part of personnel history. This sort of political investigation record was called the "black material" (*heicailiao)* during the Cultural Revolution. From this point of view, one can argue that when they were selected as delegates, they also became targets of opposition by others. Factory Party Secretary Song Mingyuan and deputy chief manager of the political department Li Gengyang from the factory Party Committee were examples of four cleans delegates facing antagonism. At the advent of the Cultural Revolution and as the Work Team withdrew from the factory in October 1966, Song and Li and the Work Team leader Bo Hongyun were denounced as "capitalist roaders within the party" and were seriously criticized by mass rebels. Meanwhile, the other four cleans delegates were transferred to the lowest work sections, or the weaving rooms, to be reformed through labor after the East Alliance seized power of the factory in end-January 1967.[4] Most of the four cleans delegates joined pro-local authorities' mass groups, the Red Guards Headquarters, and the Defending Station in late 1966 and some of them played leadership roles. As described in the next section, their positions and actions during the Four Cleans Movement significantly constrained their political opportunities and political mobilization resources.

1.2 Cadre Evaluation

The political investigation was the second stage of the cadre evaluation process. It created a deep cleft among cadres and activists and provided faction leaders with network resources to mobilize their supporters and followers. "Clean politics, economy, organization, and ideology" were the objectives of the Four Cleans Movement. The Work Team and delegates started to evaluate party cadres and activists in February 1966 based on these objectives. The cadre evaluation process included the reviewing of individual files, soliciting confessions, exposing others, and so on. Party members and activists were not only obliged to confess their own wrongdoings but also had a duty to expose the wrongdoing of others; at the same time, they were also concerned about the possibility of being exposed by others. When the evaluation process was over, cadres were placed according to their evaluation of four types: (1) Good, (2) Relatively Good, (3) Problematic, and (4) Significantly Problematic. The first two groups were assessed as "clean cadres and party members," and the remaining two were "unclean cadres and party members." Also, cadres classified in the fourth group, the so-called "problematic" group, were designated as "class enemies" and became the target of the Four Cleans Movement.[5]

The evaluation and categorization processes led to splits among party cadres and members. Forced confessions and the prevalent practice of encouraging people to expose one another had produced irreparable cracks in relationships and sown resentments among party cadres, members, and political activists. It was an ugly and cruel period. All party cadres, members, and political activists were placed in a situation whereby they were asked to expose their colleagues, friends, neighbors, and even immediate family members, including husbands, wives, or parents. If refusing to expose others, they would be punished for not making a clean break with those under investigation and could be seen as not demonstrating loyalty to the party. In the report documents, the author found many cases of cadres' evaluation categories having been changed from clean to unclean simply because they were unable to provide useful information on others.[6] By contrast, if a party cadre and member acted enthusiastically to expose someone else, he/she would gain political credit, which might lead to a promotion opportunity. Party cadres and members would often find themselves in a situation where the person they tried to shield and refused to expose jumped the gun to expose them instead to protect themself. This was a typical "prisoner's dilemma" situation that thrives on the logic that, if having been exposed, they might have to expose someone else who had committed more serious political wrongdoing in order to elevate their ranking in the evaluation; otherwise, they might be placed in the four uncleans categories. This, therefore, became an endless cycle. Indeed, since people continually exposed one another during the process, the Work Team had to change the evaluation ranking several times. Without a doubt, these exposures involved tremendous false accusations and exaggeration of reality. As a result, several counter-revolutionary groups, such as the "counter-revolutionary group" in the 2nd factory section and the "corruption group" in the south factory canteen, were "discovered." These so-called counter-revolutionary groups involved over 200 party cadres. Factory Party Committee members Wang Zhaobin, Guo Ji, Guo Junshan, and Shi Zunren, accused of engaging in past political activities and acts of corruption, were targeted as "class enemies" and "power-holders of capitalist roaders within the party."[7] About 180 party cadres were denounced for hiding their past, which included "spy activities" and participation in secret organizations such as Yi Guan Dao, and 50 of them were targeted as class enemies for hiding their counter-revolutionary history, such as being the heads of secret organizations. Most of those marked as class enemies became the first rebels a year and a half later. Some of them became leaders of the rebel group East Alliance from the early beginnings of this organization until the end of the factional conflict.[8]

As mentioned earlier, categorizing party cadres into clean and unclean groups and identifying activists and enemies had led to two opposing camps. This deeply impacted the structure of factions and bred antagonism during the period of the Cultural Revolution. The evaluation process of party cadres, especially the denunciation of counter-revolutionary groups, was also highly divisive as individuals were divided along faction lines, splintering into several factions. While these individuals were not related to each other, they started to bond, sharing the same political destiny during the process of denunciation. Therefore, the common political destiny and experiences shared between these unclean groups provided faction leaders with mobilization in networks and political resources to achieve their goals later.

1.3 *Establishment of the Red Guards Headquarters and Cultural Revolution Preparation Committee*

The third stage spans from August 1966 to the end of November, marking the period when factions began to take shape. During this time, with the establishment of the Red Guards Headquarters and the Cultural Revolution Preparatory Committee and the adjustment of middle-level cadre leadership positions, the promotions and demotions within the Four Cleans Movement became clear, and the divisions between the two camps began to emerge. From mid-August 1966, the factory's Four Cleans Movement transitioned from the "investigation of cadres" phase to the Great Four Cleans phase, known as the "struggle against enemies." This period was characterized by frequent struggle sessions targeting the counter-revolutionaries exposed during the "investigation of cadres" phase.[9] In August 1966, following the central government's instructions, the Cultural Revolution Preparation Committee was established in each organization across the country to replace the original Party Committee and lead the Cultural Revolution. On 14 August, the Work Team of DFH established the Cultural Revolution Preparation Committee, whose 214 members were all four cleans cadres and four cleans delegates and activists.[10] Zhi Feng, a member of the Standing Committee of the party bureau of the factory, became the head of the Cultural Revolution Preparation Committee, and Li Gengyang and Wang Fugui, two four cleans delegates, were appointed as deputy directors. Apparently, Zhi, Li, and Wang belonged to the first category of cadre evaluation. Before the Four Cleans Movement, Wang was a party member with a working-class background. As he hailed from the proletarian class, had an untainted political and historical record, and was identified as an activist based on his political attitudes in the past years, he was chosen as a four cleans delegate by the Work Team. After a year and a half, he was promoted to deputy director of the Cultural Revolution Preparation Committee. In October, further to his promotion, Wang had concurrent appointments as deputy director and the party Secretary of the Temporary Leading Group of the Maintenance Section. In December 1966, acting on the instruction from the factory party authority, Wang appointed himself as team leader of the Eastwind Team (*Dongfeng dui*), a mass organization set up to oppose the linkup students attempting to enter the factory.[11] In January 1967, the Cultural Revolution Preparatory Committee was forced to cease its activities due to criticism from the rebel faction for not being democratically elected. However, the representatives of the Cultural Revolution Preparatory Committee continued to form mass organizations under the leadership of the Party Committee, using party general branches and party branches as units to counter the linkup students and worker rebel organizations. On January 6, 1967, the aforementioned mass organizations formed the Defense Station.[12]

In the early summer of 1966, the student rebellion organization, Red Guards, proliferated the capital city, Beijing. After Mao's Red Guard rally in Tiananmen Square on 18 August, the Red Guards as a new type of political organization spread widely across the country. At the time, the right to take part in the Cultural Revolution was limited only to the cultural and education sector, such as schools, propaganda departments, and their party and government organizations, and it did not expand into factories and other production sectors. Moreover, the Work Team still held absolute leadership in DFH. Even though the Red Guards' influence expanded in Yuci through the mass media, similar mass rebellion organizations had also started to sprout up in schools, factories, and other organizations. Under this scenario, DFH established its own Red Guards organization consisting of about 200 members, in mid-September. It was named the Red Guards Headquarters in order to distinguish it from the student Red Guards organizations and was headed by Li Funian, assistant chief of the factory security section.

The Red Guards Headquarters, unlike the student Red Guard groups in Beijing, did not mobilize spontaneously but in a top-down approach through the party power structure by the Work Team and factory party authorities. In other words, members of the Red Guards Headquarters were chosen by the Work Team based on two criteria. First, the candidate must belong to a red-class background. Second, the number of Red Guards must be less than three percent of the total number of each organization, such as *chejian* (a unit of a production department), *banzu* (a subsidiary unit under chejian), and *keshi* (a unit of an administrative department). The percentage of Red Guards in any organization was much lower compared to that of party members at 13.7 percent on average. Hence, based on the two criteria, only four cleans party cadres, members, and activists had the potential to be selected. Since most members were party cadres, the party power structure was automatically inducted into the Red Guards Headquarters.[13] Therefore, as long as the legitimacy of the party organization's leadership was maintained, it was impossible for the structure of the Red Guards Headquarters to deviate from the party organization. The overlapping structure and coordination of actions had determined and limited how this mass group acted during the Cultural Revolution.

In October 1966, the Work Team withdrew from the factory. Just before its withdrawal, the Work Team reformed the factory leadership based on the cadre evaluation result. About 99 cadres were appointed to be in charge of the factory and more than half of the middle-ranking cadres were dismissed or demoted. In addition, about 28 Party members with working-class backgrounds and lower-ranking cadres were promoted to middle-ranking cadres. Liu Guishan, the second-generation head of the Red Guards Headquarters to be discussed in a later section, was one of the newly promoted cadres. Soon after the mass organizations formed, 74 out of the 99 cadres participated in the Defending Army while 25 participated in the Eastern Alliance.[14] As documented in the "Four Cleans Movement" investigation record archives, out of the 25 persons who participated in the Defending Army, 24 were demoted or not able to be promoted because of their social status: they did not belong to the "red" class background, or they or their family members or relatives had been denounced for committing some political wrongdoing or had "complicated" histories, such as being members of secret organizations or had worked for the Nationalist Party government.[15] On the other hand, among the 28 who were promoted, 25 joined the Defending Army. Wang Fugui, Li Jian and Wang Youshan became faction leaders and played a significant role throughout the faction conflict.[16] From this stage onwards, it became evident that the differentiation between the demoted groups and the promoted ones directly resulted in factional conflict, the so-called fighting between the rebel and conservative groups. Most of the promoted cadres joined the Defending Army which was labelled as the conservative groups, and most of the demoted cadres joined the Eastern Alliance, widely known as the rebel group, during the Cultural Revolution.

1.4 The Formation of the Factions

The fourth stage, from December 1966 to January of the following year, was the period of factional formation. By the end of 1966, the political environment in Yuci changed completely, as was the case in DFH. In the process of factional formation, several significant events that occurred nationwide at the end of 1966, as mentioned in previous sections, were crucial. First, the public criticism of the "capitalist reactionary line" of Liu Shaoqi and Deng Xiaoping; second, central directives explicitly instructing the redress of grievances for the masses oppressed by the Work Group; third, the legitimization of the Anting Incident in Shanghai and the legalization of rebel organizations formed by Shanghai workers; fourth, the widespread dissemination and influence of new revolutionary methods brought by linkup Red Guards. The changing political environment gave new and timely opportunities to those who experienced dissatisfaction during the Four Cleans Movement. In December 1966, several four unclean cadres posted big-character posters that accused the factory Work Team of wrongful persecution and demanded that their own names be cleared, and their positions be reinstated. Most of these were written by several people and signed off as the combat team (*zhandou dui*) instead of their real names. On 26 December, the *People's Daily* published an editorial to advocate that workers should organize themselves to join the Cultural Revolution. This led to a proliferation of worker rebellion groups across the country.[17]

On January 5, 1967, nearly 20 rebel groups united to form a coalition named the East Alliance, headed by Li Jinlong and Peng Yuexuan, both having been removed from their positions by the Work Team a few months earlier. The East Alliance Standing Committee members, Zhou Hanqing and Chang Liantong, were four uncleans cadres. Zhou was especially identified as counter-revolutionary by the Work Team. Guo Ji, Guo Junshan, Shi Zunren, and Wang Zhaobin were members of the factory Party Committee before the Four Cleans Movement but had been dismissed from their positions by the Work Team. The four dismissed cadres did not participate in the East Alliance on the surface but were shadow leaders and played very important roles in the East Alliance.[18] In other words, when the East Alliance was formed, most of its leaders and members had already been previously accused of political wrongdoing or corruption. For example, the Duozhuangzhi Fighting Team (*Duozhuangzhi zhantoudui*) and the Changying Combating Team (*Changying zhandoudui*) were two of the most active teams in the East Alliance. Eight of the 11 members in the two teams had previously been accused of political wrongdoing or having "complicated" histories.[19] In February 1967, the East Alliance experienced a large expansion, with its membership base reaching 4000 members and 300 of them holding leading positions. However, over 100 of these 300 people had been previously accused of committing political wrongdoing, corruption, or having "problematic backgrounds."[20]

On January 6, 1967, in order to compete with the East Alliance, four cleans cadres and activists, under the factory Party Committee's guidance, united 14 small like-minded groups, such as the East Wind, which were established by the Cultural Revolution Preparation Committee in December 1966, into one single alliance named the Defending Station. The East Wind was one of the most active groups in the Defending Station.[21] At the same time, the Red Guards Headquarters was having difficulty surviving because most of its members, especially the leaders, worked closely with the Work Team and were associated with the Four Cleans Movement investigation. Li Funian, head of the Red Guards Headquarters, is one such example. As assistant chief of the security section and "good"-category party cadre, Li was relied on by the Work Team and deeply involved in the four cleans cadre investigation and other political activities. When the Work Team withdrew, Li was denounced for having participated in "suppressing masses" and for having collected black materials of four unclean cadres and others. Therefore, for ideological legitimacy reasons, Li had to step down. On 7 January, the Red Guards Headquarters changed its leaders. The new leadership comprised 18 people; most were delegates of the Cultural Revolution Preparation Committee, the Four Cleans Movement, and newly promoted middle-ranking cadres. On the surface, the new head, Zhang Zhixiang, was an ordinary worker, party member, and activist. However, he was only a puppet. The real authority was held by the newly promoted middle-ranking cadre Liu Guishen, who was a demobilized soldier.[22] Nevertheless, the changed leadership strategy did not help the Red Guards Headquarters escape from its disadvantaged situation. The newly established Defending Station also faced the same problem. These two groups started to lose their members. On 3 February, the leaders of the Red Guards Headquarters and the Defending Station announced the dissolution of these two organizations and integrated their remaining members to form a new organization named the Defending Army.[23] Since there were no sharp differences between the Defending Station and the Defending Army in terms of members' backgrounds and political action, and the Defending Station had merely existed for almost a month, the author will, for the convenience of analysis, use the Defending Army to describe these two groups.

2 Solidarity and Sectarianism

Taking into consideration the discussion above, the author explores the dynamics of factional conflict from the following five perspectives. First, from the Four Cleans Movement to the early stage of the Cultural Revolution in DFH, there was a complete reversal of the power structure. Those delegates evaluated as good cadres and promoted by the Work Team were later criticized for following Liu and Deng's "bourgeois counter-reactionary line" at the start of the Cultural Revolution. On the other hand, those who were evaluated as four unclean cadres or counter-revolutionaries and were demoted or dismissed by the Work Team became ideologically correct since they were suppressed by Liu and Deng's "bourgeois counter-reactionary line." For students who were persecuted by the Work Team, rehabilitation was not too complicated. However, in the factory, rehabilitation was associated with power struggles. Demoted or dismissed cadres not only demanded rehabilitation but also the reinstatement of their previous positions. Restoring reputation also implied restoring positions that newly appointed cadres had already occupied. Therefore, the demand for rehabilitating deposed cadres threatened the power acquired by the ascending cadres during the Four Cleans Movement. From this point and the subsequent factional struggles during the Cultural Revolution, it becomes clear that the fundamental source of opposition between factions lies in faction leaders, specifically the power struggle between the four cleans cadres and the four unclean cadres within the factory. The competition for political power is the most crucial and intense aspect of competing for citizenship. Therefore, the power struggles among the factory elites revealed in this chapter reflect the most intense aspect of competition for citizenship.

Second, the polarization of the formal administrative structure resulted in the formation of factions. Both the East Alliance and the Defending Army were composed of several subsidiary organizations, namely, "combat teams." For example, the East Wind was one of the subsidiary groups under the Defending Army, and Duozhuangzhi was one of the subsidiary groups under the East Alliance. Most of these subsidiary groups were based on the factory's formal administrative structure: chejian (a unit of the production department), banzu (a subsidiary unit under chejian), and keshi (a unit of the administrative department). This phenomenon was attributed to two factors. First, the Four Cleans

Movement mobilization was based on the formal administrative structure, and the Red Guards Headquarters and the Defending Station were mobilized by the Work Team and were based on this structure. The administrative structure had a significant impact on the formation and development of factions. The formal network had helped facilitate faction leaders to mobilize faction activities. Second, due to such operations as exposure, enforced confession, and position usurpation between acquaintances among small units, the administrative structure had blurred the boundary between political divergence and personal grievance, thus forging strong social solidarity and breeding antagonism simultaneously. The coexistence of solidarity and antagonism, which Mao had emphasized and was even deemed contradictory, in fact, provided faction leaders with great political resources to mobilize their followers. The fact that political mobilization processes deployed a top-down approach targeting various small units implies that these small-scale like-minded groups could be amalgamated into a larger one more easily and quickly when political opportunities arose. In other words, the formal administrative network was an effective tool to expand factions.

Third, the changing political environment was a key factor for splintering group solidarity into factions. Although the Four Cleans Movement, which developed a consciousness of solidarity, paradoxically laid the foundation for faction formation, it was the changing political environment during the Cultural Revolution—such as political mobilization by the central government and top leaders, demonstrations of the student linkup groups, and loss of political legitimacy of local and factory party organizations—that had thrown up a galvanizing force for the awareness of solidarity and pursuit of collective interests. In short, the changing political environment induced the awareness of group solidarity and provided a mechanism for collective action. For example, when the Red Guards Headquarters was established, it lacked a clear aim and a sense of solidarity. However, these later emerged because their political interests were challenged by opponents.

Fourth, initial faction conflicts already emerged between promoted cadres and demoted or dismissed cadres during the Four Cleans Movement. Both sides were former or present factory party cadres, members, and activists; their family backgrounds or socio-political statuses were not the main factors contributing to antagonism. Of course, it is undeniable that the Defending Army was dominated by party cadres, members, activists, and workers with red family backgrounds, while the East Alliance was dominated by four unclean party cadres, members, and workers who belonged to non-red family backgrounds and other discriminated categories. This characterization was, however, a consequence of mobilization by faction leaders and changing political opportunities. In fact, the beginning stage of the Cultural Revolution only saw conflicting interest groups, such as promoted cadres and the demoted ones, who had a strong motivation to participate. Those who were marginalized from the power structure, such as ordinary workers or workers from non-red family backgrounds, did not harbor such a motivation because there was no possibility of promotion. Regardless of which side held power, they could not benefit politically or economically.

Another reason that contributed to the differences in the constitution of these two factions was exclusionism. When the Red Guards Headquarters was established in the summer of 1966, it was an extremely exclusive organization that permitted the admission of just three percent of selected four cleans party cadres and members. Although there was keen interest from many people of various family backgrounds in joining, they were rejected. In January 1967, when it was legitimate for workers to organize their own group, the Red Guards Headquarters was no longer the only dominant mass group in the factory. To compete with the East Alliance, the Red Guards Headquarters opened membership to workers with different family backgrounds, but due to its conservative color, it lost its popularity and appeal. In early 1967, the Defending Army, in emphasizing its political legitimacy, claimed that the East Alliance was dominated by "monsters and ghosts" (*niugui sheshen*). However, the birth bloodline theory (*xuetonglun*) was strongly criticized then. Evidently, having a prominent family background and political status would not lead to any political benefits and would instead generate strong opposition from ordinary workers. The Defending Army's emphasis on its own political legitimacy had thus strengthened the sense of solidarity among people who had been persecuted by the Work Team. It is also evident that how the structure of factions took shape was closely related to the development process of political movements.

Fifth, despite the obviously conflicting political interests and power struggles between the Defending Army and the East Alliance, both sides had to achieve the objectives of the Cultural Revolution when the movement began. This explains why opposing factions, even though at loggerheads with one another, employed the same political ideology and political slogans. In this respect, seizing the same political opportunity to pursue faction aims had obscured the boundary between group (private) interests and common political ideology. This had indeed facilitated faction leaders to mobilize people to be involved in their political activities.

3 Resource Mobilization by Faction Leaders

As revealed earlier, the formation of factions within the DFH results from political construction. The political processes from the Four Cleans Movement to the Cultural Revolution constructed two opposing factions and the competition between them for political power within the factory. This competition went beyond the usual power struggles. First, as mentioned in previous chapters, this competition occurred within the institutional space of the state and society without exit options. Second, the gain or loss of power signified one's political status and the quality of one's citizenship. Being labeled as one of the Four Unclean cadres would mean being relegated as a political untouchable. If not allowed redress, this would affect one's entire life and even extend to one's family. Thus, the competition between the two camps was irreconcilable from the start.

Under these circumstances, faction leaders with no way out had to rely on resources and smarts to manage their factions, ensuring their survival and development. Generally speaking, there were two main factors influencing the survival and development of factions: the resource mobilization capabilities and competitive strategies of faction leaders and the creation of political opportunities. The two are interconnected. The creation of political opportunities led to changes in the flow of political resources, affecting the resource mobilization methods and competitive strategies of faction leaders. These could also contribute to the growth of factional power, thereby influencing the construction of political opportunities through pressure, although this influence is limited in Mao-era China.

This section mainly discusses the mobilization methods employed by faction leaders to secure political resources, ensuring the survival and development of their factions in a volatile political environment. As such, the survival and development of factions were largely dependent on the results of resource mobilization and strategic competition by faction leaders.

3.1 *Mobilization of Political Resources*

Political resources are a determining factor for the support that faction leaders have. There are multiple ways to draw political resources, such as getting support from political leaders or Party Committees. Of course, attaining ideological legitimacy is the most essential. Since leaders of the East Alliance and Defending Army had undergone different political situations in 1966, their previous experiences strongly affected the methods (and costs) of acquiring political resources. More specifically, the majority of the Defending Army leaders were holding positions in party organizations—such as factory Party Committees, the general party branch, or chejian party branches—when the mass rebellion movement spread to the factory. This implied that mass group leaders could easily garner political support from the factory party authorities initially but found it difficult to act autonomously since the Defending Army leaders, being party cadres or members, had to enforce the administrative order of the Party Committee. On the other hand, since all interests were determined by the hierarchical structure of political power, the authority to appoint personnel was held by superiors. Under such institutional circumstances, it was in the interest of the lower-level party branch members to protect their superiors' leadership in order to protect themselves. The overlap between party organizations and the Defending Army suggested that Defending Army faction leaders faced constraints in their attempts to take advantage of political opportunities. That also determined how they deployed mobilization strategies.

From the establishment of the factions on January 6, 1967, to the announcement of their dissolution on March 24, there were two and a half months during which the Defending Army exhibited highly divergent actions, with the end of January serving as a dividing line. Before the end of January, the Defending Army followed the Party Committee's instructions to oppose external rebel groups, such as the linkup students and internal rebel workers, in order to maintain the existing top-down political order. However, when the factory Party Committee lost control at the end of January, the Defending Army completely reversed its attitude. It began to violently attack the top leaders of the Party Committee, like Song Mingyuan, whom the Defending Army had previously supported. The Defending Army shifted its position from supporting the Party Committee to challenging it; its behavior seemed inconsistent on the surface. As a matter of fact, the inconsistent attitude only indicated that the Defending Army leaders had changed their strategies in order to maintain the faction's legitimacy.

In contrast with the Defending Army leaders, the majority of the East Alliance leaders experienced persecution during the Four Cleans Movement. After the Work Team withdrew, their experience of persecution became advantageous as a political resource. Furthermore, the unfolding of the "criticizing bourgeois counter-reactionary line" movement provided the East Alliance leaders with strong ideological support. In January 1967, the East Alliance held several mass rallies to criticize Song Mingyuan and Bo Hongyun, leader of the factory Work Team. Through these mass rallies, the East Alliance expanded quickly and soon became the largest mass group of DFH. Also, in order to seize power from the factory Party Committee, the East Alliance made efforts to connect to external rebel groups in the city. It successfully united others to form an alliance called the Jinzhong Prefectural Revolutionary Rebel Headquarters (*Jinzhong diqu geming zaofan zongsilingbu*). On 18 January, the East Alliance declared that it had seized power from the factory Party Committee. By the end of January, the East Alliance held the vital power of the factory since the alliance had triumphed over the power struggle of the Jinzhong Prefectural Party and government organizations. On 4 March, the East Alliance officially seized power when the faction leaders cooperated with Party Standing Committee member Zhi Feng and established the three-in-one revolutionary committee. From then onwards, the power of the factory was officially transferred from the four cleans cadre group to the four uncleans cadre group.

To acquire political resources, one of the important methods was to establish connections with ideologically supported political leaders and organizations. If mass faction groups could procure support from them, it would be a booster in helping the faction win the battle. For example, on January 26, 1967, the East Alliance held a struggle session to criticize the "bourgeois counter-revolutionary line" and successfully invited cadres of the Jinzhong Prefectural Military Sub-area #4558 Troops to participate in the session and deliver a keynote speech. This had a significant impact on the confrontation between the two factions. Many workers, who did not take sides with either the Defending Army or the East Alliance, chose to join the East Alliance, with many members leaving the Defending Army. Moreover, three days later, the East Alliance again successfully invited the 4558 Troops to march together in celebration of the "Assisting the Left by People's Liberation Army." This political event accelerated the exit of

Defending Army members from their organization. Of course, joining the opponent faction was not easy for active faction members, especially the leaders. A case in point was Song Mingwang. Song was one of the lowest-level team leaders of the Red Guards Headquarters. At the end of January, he wanted to defect to join the East Alliance. The East Alliance accepted him, but with a precondition—he had to write a "letter of surrender." Song's letter was distributed widely in the city in order to split the opposing faction and attract more workers who remained undecided about joining the East Alliance. For ordinary Defending Army members, they were welcome to join the East Alliance but had to prove their allegiance. One of the common practices required those who surrendered to spit on their original armbands in a mass rally to show their determination.[24] It is noteworthy that changing sides was very common during the period of faction conflict, although this was only prevalent among ordinary members and some activist members; most faction leaders maintained their stand until the end of the Cultural Revolution. According to an East Alliance report, only four Defending Army leaders, namely Song Mingwang, Yan Xiaoping, Song Xitao, and Wang Tongsheng changed sides to the East Alliance between January and March 1967 when faction members actively changed their positions.[25]

Establishing connections with political powers could benefit a faction, but for faction leaders, it was not easy to garner support from political leaders or organizations. Faction leaders had to demonstrate skillfulness in employing tactics, propaganda, and psychological strategies. In this perspective, a faction conflict could be interpreted as a competition of mobilization strategies to acquire political resources. The incident in early 1967, when Mao Zedong relied on the military to restore political order, could well illustrate this. The 4558 troops were sent to control DFH. The army arrived at the factory on 11 March. As the army maintained its neutrality in the first few weeks, both the Defending Army and the East Alliance tried to take this political opportunity to benefit their group. Factions applied strategies at times to create the guise that their group was supported by the army. The Defending Army distributed leaflets to induce the public into believing that "the PLA's involvement in the Cultural Revolution of the factory indicates that now it is the right time to launch a general offensive to oppose monsters and demons" and attempted to rally all the like-minded forces to fight back. The East Alliance, on the other hand, tried to produce false evidence showing that its group had the support of the PLA. Some of the strategies were very interesting. For example, when the army finished its military training and paraded back to the factory, the East Alliance leaders would line their members up behind the army and shout political slogans along with the army in order to stress that they threw their support behind the army. The purpose was to consolidate the group's cohesiveness and weaken the opponent's forces.

Similarly, connecting with political leaders was beneficial to factions in acquiring political resources. However, the downfall of political leaders could very well affect their coalition with factions. The Defending Army's alliance with and support for Song Mingyuan, who was the factory party secretary, and Wang Xiujin, who was Jinzhong prefectural party secretary, is a case in point. By the end of January, Wang Xiujin had lost the power struggle against the prefectural Party Committee members Ren Jingfu and Wang Zhengguo. As a result, the Defending Army was questioned about its legitimacy and survival. In order to overcome this disadvantaged situation, the Defending Army announced a withdrawal statement to break up the alliance with the Jinzhong prefectural party and the government organization rebel group—Headquarters Station, namely, the Jinzhong Prefectural Proletariat Revolutionary Faction Coalitions Headquarters Station, which was established on 15 January and reported directly to Jinzhong prefectural party secretary, Wang Xiujin. However, such a move had little impact; numerous Defending Army members had left, and only leaders and active members remained by the end of January. By contrast, the East Alliance scored a landslide victory by fighting alongside their alliance group, General Headquarters, namely, the Jinzhong Prefectural Proletariat Revolutionary Faction General Headquarters, established on 18 January and reported directly to Ren Jingfu and Wang Zhengguo. Ren and Wang's victory in the power struggle against Wang Xiujin brought tremendous political resources to the East Alliance. The East Alliance expanded rapidly and became a dominant mass faction organization in the factory.

In addition, utilizing politically charged events is also a primary method for mobilizing political resources. He Wenhua, a demobilized soldier who worked in the security section of the factory without a title, was a good example.[26] He believed he should be promoted to section chief, or at least to the deputy chief level, because of his military background. But that was not to be the case. He rallied other demobilized soldiers and sent a letter to the Jinzhong prefectural Party Committee to expose the factory Party Secretary Song Mingyuan, accusing him of not implementing the Maoist class policy and suppressing demobilized soldiers in the mid-1950s. He Wenhua and his followers were convicted as counter-revolutionary group members by the factory Party Committee in 1958. In 1967, his previous persecution experience allowed him and his followers to express their grievances. He Wenhua himself and his personal experience had become political resources for either local political leaders or mass group leaders who were enemies of Song Mingyuan. They had tried to build good relationships with He Wenhua and frequently invited and encouraged him to disclose Song Mingyuan's offenses. For these local political leaders and mass group leaders, He Wenhua and his previous persecution experience presented a useful political weapon to attack their enemy Song Mingyuan.[27] The Zhang Tiebao counter-revolution incident is another example. Zhang was a party member and ordinary worker in the factory. He developed a liking for a widow, Zhou, whose late husband was a counter-revolutionary. Zhang often helped Zhou with her heavy housework and was exposed for departing from the correct party line and having immoral sexual relations with Zhou during the Four Cleans Movement. Zhang summoned other four uncleans party cadres and members to expose the factory Party Secretary Song Mingyuan's wrongdoing in order to escape from the unfortunate circumstances he was facing. Zhang and 30 other people were convicted as a counter-revolutionary group by the Work Team. At the end of December 1966, Zhang and the others formed the Proletariat Unyielding Fighting Force (*Wuchan jieji yinggutou zhandoudui*) and held a mass struggle session to denounce the Work Team and party leaders. When the coalition mass organization East Alliance was formed, the resistance experience of Zhang and his followers earned them considerable political credits. Zhang became one of the faction leaders of the East Alliance, and his group became one of the core groups in the East Alliance.[28]

3.2 Mobilization of Human Resources

In addition to the discussion on mobilizing political resources mentioned above, let's briefly explore the methods and effects of mobilizing human resources by faction leaders. Due to the collective nature of factionalism, acquiring human resources is equally important as acquiring political resources. Just as factional structures exhibit heterogeneity, the human resources that faction leaders can mobilize also possess heterogeneity, resulting in varying mobilization methods. Simply put, human resources can be divided into three categories. First, ordinary social members;

second, talented individuals (for example, those able to establish connections with political leaders or organizations and with unique capabilities such as mobilizing support and resources, generating publicity and public speaking, and so on); and third, party and government officials in significant positions, as well as individuals who have experienced political persecution.

Although the three categories of resources mentioned above are important in the context of factional leaders, the mobilization methods may vary depending on the targets. For general participants, the most common method was to attract and entice more people to join the organization through material incentives such as distributing movie tickets, Mao Zedong badges, and other tangible means because collective actions require the strength of individuals. Demonstrations, denunciation meetings, and other activities require a large presence to create a revolutionary atmosphere. Sometimes, coercion and threats were also used to achieve the desired goals, such as using one's authority to assign those who refuse to participate or support opposing factions to undesirable work positions. Another commonly used method is to create facts. For example, signing the names of opponents under one's own faction's big-character posters, leaving them with no way out.[29]

The most commonly used method for the third type of human resources was the combination of conciliation and criticism. As evidenced in the cases of He Wenhua and Zhang Tiebao, people who had gone through persecution previously were still a force to be reckoned with because their experiences were great weapons to attack political leaders. They were important evidence to "accuse the bourgeois reactionary line" and effectively mobilize political resources. Party and government officials were also targeted in the competition for resources primarily due to the establishment of the Three-In-One Regime structure. Cooperation between factions and party/government officials was crucial in the political landscape following the establishing the Three-In-One Regime structure in 1967. If a faction can obtain support from party and government officials who were potentially recognized by higher authorities, the leader of that faction had a chance to participate in the Three-In-One Regime. However, it was challenging for party and government officials caught between the struggles of two factions. Not expressing a clear stance will invite criticism from both sides, supporting one faction will result in retaliatory denunciation from the other side, and not supporting either side would lead to the worst outcome as they will face severe criticism from both factions.

Furthermore, as mentioned earlier, denunciation meetings serve as an arena to demonstrate revolutionary fervor. Both factions strove to bring the most prominent "capitalist roaders within the party" to the forefront for denunciation to heighten the revolutionary fervor or rebellion and attract more people to join their respective factions. However, top local leaders were considered scarce resources compared to the abundance of mass organizations. Therefore, in the factional struggles at the DFH, there were instances of fierce competition to target the capitalist roaders within the party. To monopolize the opportunity to criticize the capitalist roaders within the party, factions would control and hide the target of denunciation. In contrast, the opposing faction would rush to the hiding place to seize them. Alternatively, they would ambush and intercept the target while escorted to the denunciation venue. Another commonly used method is sabotaging or disrupting the opponent's meeting venue before or during the denunciation meeting, preventing the meeting from occurring.

Personal historical files and political movement investigation records (*diaochajilu*), or the so-called black materials, were significant political tools for faction leaders to mobilize human resources. At the beginning of 1967, the Defending Army leaders tried to prevent the black materials from falling into the hands of the East Alliance and other mass faction groups that wanted to seize and destroy them because there were negative records. However, by the end of February, the Defending Army leaders changed their strategies. They selectively released and promulgated the contents of the black materials, especially relating to personal and family political histories and records of political wrongdoing and corruption. By doing so, the Defending Army attempted to convince others that the East Alliance leaders were not truly leftists but were "monsters and demons" instead. Instead, their rebellion and seizure of power were acts of vengeance and conducted solely for their self-interests. At the beginning of 1967, on the contrary, the East Alliance leaders tried to destroy the records and asked the authorities to rehabilitate their names. However, they soon realized that the black materials were useful political sources. The East Alliance leaders also began to use black materials to denounce the Defending Army leaders. The East Alliance leaders accused the factory's political leaders of implementing the bourgeois counter-revolutionary line because they were involved in cadre investigation in the Four Cleans Movement. Given that the exposed secret record was an effective tool to breed hatred, both factions used black materials extensively to motivate people to join their groups, especially in attracting talented individuals.

4 Faction Conflict and Competition for Political Resources

The competition between factions is essentially a competition for resource mobilization, with the ultimate goal of acquiring more or, one might say, higher-quality citizenship. Moreover, since this competition takes place within a national and societal system that does not allow for withdrawal, it becomes an ongoing process that cannot automatically come to a halt. If one side gains an advantage, the other will make even greater efforts to mobilize resources and devise strategies for stronger collective action. This, in turn, prompts the former to change strategies to gain an advantage in the next round of competition in an endless cycle. In this context, we will briefly explain the three common methods of resource competition among factions during the Cultural Revolution, focusing on the case of the DFH.

First, conflicting factions contended with each other for the same political resources. Most of the competition for political resources during the Cultural Revolution belonged to this category. The factional formation process was also a political process based on the ideology of power struggles. Although the two conflicting factions aimed at attaining the same goal, they had to act competitively and confrontationally since there could only be one winner. Therefore, both factions competed to secure ideological legitimacy each time political instructions were issued by the central party-state. Both factions held spectacular mass rallies to claim that their faction was the first and only one that had thoroughly implemented the political instructions of Mao and the central party-state. In this competition, antagonistic factions were constantly attacking one another. The two common ploys used to interrupt the opposing faction's political performance were to occupy a venue before the opposing faction held its rally there and kidnap political leaders denounced as "bourgeois elements inside the party." The purpose was to block the opponents' ability to find a place or outlet to express their political statements. The competition between factions appeared to be filled with the violence of revolution. Still, precisely this revolutionary violence served as the widely utilized means of achieving victory and the most popular "glamorous packaging" used to disguise personal or factional interests at that time. The competition for expressing factional political attitudes was deemed necessary to gain support from the authorities, it also enhanced collective cohesion building and sought to attract new members from the opposing faction. In this sense, the faction leaders needed political performance to reinforce their ideological justification.

Furthermore, the Cultural Revolution was not only a policy confrontation between Mao Zedong and Liu Shaoqi but also a power struggle between the two. Therefore, the flow of political resources constantly changed with the policy adjustments of the central leadership and the decline of the top leaders, which affected the goals, methods, and means of resource mobilization and the structure of the factions. As long as the unstable coordination mechanism beyond factions existed, competition and confrontation between factions would become ongoing.

For example, in April 1967, a power struggle erupted between members of the Jinzhong prefectural Cultural Revolution Core Group (*Wenge hexin xiaozu*, or the equivalent to the Jinzhong prefectural Party Committee) including Wang Zhenguo, Ren Jingfu, and Zhang Huaiying, and Jinzhong military sub-area leader Cui Bing that took place in April 1967. The East Alliance supported Cui Bing initially. However, in July of the same year, the central party held a meeting to discuss issues in Shanxi, where Cui Bing received criticism while Wang Zhenguo gained support. As a result, the legitimacy of the East Alliance came under fierce criticism from the opposition, leading to self-criticism within the faction. On the contrary, the Defending Army initially supported Cui Bing. Still, after receiving internal information at the end of June, it shifted their support to Wang Zhenguo and others, thereby reclaiming their lost political legitimacy. The acquisition of political legitimacy led many wavering participants who had been siding with the prevailing faction to return to the "Defending Army." In this situation, the membership of the Defending Army began to increase, and the dwindling organization was able to rebuild itself. Changes greatly influenced the opposition between the two factions and the fluctuation of their power in the political environment. The competition persisted until the open confrontation between factions was completely prohibited in 1969.

The second category was the competition between opposing factions for different resources. Because political resources were constantly changing and fluid, the political resources possessed by the two factions were not equal during a specific period. In other words, the changing political situation can favor one faction while disadvantaging another. In such a situation, under the same political opportunity structure, if one faction lacks resources, they either have to supplement them with alternative resources by exploring new avenues or resort to seizing resources from the opposing faction.

For example, in mid-January 1967, when the confrontation between the two factions reached a stalemate, the leader of the East Alliance took advantage of the visit of the leader of the Defending Army to the command headquarters. He pretended to receive a phone call personally from Zhou Enlai, spreading false information that the East Alliance had received direct support from Zhou Enlai. The purpose was to disrupt the morale of the Defending Army and cause the loss of their resources.

The third category was self-resource protection undertaken to avoid the weakening of one's faction. Whether faction leaders had abundant resources depended on the political opportunity structure. Changes in the political opportunity structure caused by adjustments in the policies of the central leadership and the decline of party and government officials could lead to changes in the flow of political resources. To prevent the loss of political resources, faction leaders often adjusted their faction's goals and relationships with relevant organizations. They underwent thorough internal reforms to maintain the political legitimacy of their faction.

For example, in early 1967, the East Alliance and the Defending Army fiercely opposed each other in establishing the Three-In-One Regime in a factory. As a minority faction, the Defending Army announced its dissolution on March 4th to gain an advantage in this round of competition. The so-called dissolution of the organization was not an abandonment of the competition but rather a means of winning in the competition, just like the leadership change and dissolution and restructuring of the Red Guards Headquarters and Defending Station into the Defending Army. Because the construction of the Three-In-One Regime required factions to dissolve their organizations, cease opposition, and achieve a grand alliance of mass organizations. On the one hand, dissolving the organization was the politically correct action at the time. On the other hand, returning to the traditional "workshop" and "team" structure after dissolving the organization meant that members of the Defending Army who occupied advantageous positions in the pre-Cultural Revolution power structure would gain more resources. When examined in detail regarding factional behavior, the seemingly passionate revolution is revealed to be a mobilization and strategic competition for political resources, with the ultimate goal of acquiring citizenship and avoiding becoming "inferior people"—class enemies.

5 Conclusion

To deepen our understanding of the Cultural Revolution, we must pay greater attention to major movements in the pre-Cultural Revolution history of the People's Republic of China, particularly the Four Cleans Movement. The Cultural Revolution was not a discrete political movement. To a large degree, it was a continuation of previous movements, all of which had a cumulative effect on the Cultural Revolution. What the literature of social movement calls "repertoire" had developed in the Cultural Revolution, meaning that similar discourses, tactics, strategies, and many other aspects of a movement that appeared during the Four Cleans Movement reappeared during the Cultural Revolution.

Some main findings of this case study can be summarized here. First, the origin of factionalism during the Cultural Revolution was closely related to the Four Cleans Movement. The initial mass conflict appeared between the four cleans cadres and the four uncleans cadres. The legacies of this conflict continued and persisted throughout the entire period of the Cultural Revolution. From the Four Cleans Movement to the early stage of the Cultural Revolution in DFH, the promoted and demoted groups in the power structure were reversed completely, and factionalism in DFH was closely related with personal struggle for political interests from the very beginning. Because of the conflicts, it was difficult to coordinate the two groups. Based on the DFH case study, we find that family background and interest group factors are the only consequences; political ambiguity and constant change at the beginning of the Cultural Revolution spawned uncertainty for the conflict groups and limited their political opportunities and mobilization strategies, but these did not originate from faction conflicts. In other words, faction conflicts had emerged before the advent of the Cultural Revolution.

Second, faction leaders played a significant role in faction organization and mobilization during the Cultural Revolution. It is impossible to describe faction conflicts without understanding the role of faction leaders. From this perspective, the political and social turmoil during the Cultural Revolution was not only caused by the political leaders and their political mobilization but also by the competition for political opportunities and mobilization strategies between opponent faction leaders. With changes in the power structures, mass groups experienced alliances, splits, reunions, dissolutions, re-establishments, and changes in their leadership. All these actions, however, were often tactics for leaders of the mass groups to maintain or gain political justification for their existence.

Third, the mobilization of political resources took place at both the state and societal levels. There was political mobilization by political leaders from the top and resource mobilization by faction leaders from the bottom. Political leaders often provided both opportunities and constraints for faction leaders. In contrast, the faction leaders maintained a degree of autonomy in mobilizing available resources at hand as well as in deciding which political leaders they should support. Faction leaders played very significant roles during the formation of factions and in engendering competition with other factions. In other words, mass conflicts and violence arose not only due to political mobilization from above but, more importantly, as a consequence of resource mobilization from the bottom. Political mobilization from above during the pre-Cultural Revolution period, especially the Four Cleans Movement, and mobilization from the bottom during the Cultural Revolution produced enormous tensions among social groups. These tensions turned out to be important tools during faction formation and in factional struggles.

Fourth, the formation of factions is a political construction process, the result of switching the top-down mobilization of the Four Cleans Movement and the bottom-up political mobilization of the Cultural Revolution in opposite directions in a short period; at the same time, it is also the result of a major reversal of the Four Cleans Movement between the beneficiaries and the losers. It was also the result of a major reversal between the beneficiary groups and the losing groups in the Cultural Revolution. The formation of factions in the DFH Textile Factory shows that the core of the faction rivalry was the struggle for power between the cadres of the four cleans and the cadres of the four uncleans. In Mao-era China, the gain or loss of power would mean the quality of citizenship—whether to be the most progressive citizen or to be reduced to a "pariah" of the class enemy. Thus, the confrontation between factions is irreconcilable. Moreover, it is clear from this analysis that the class differences and the "patron–client relationship" emphasized in the earlier studies as elements of resource mobilization prior to the Cultural Revolution played a role in forming factions. Still, they were not the direct cause of factional antagonism. The uncertain political environment at the beginning of the Cultural Revolution, which Andrew Walder emphasizes as an influence on the participants' behavior,[30] also does not explain the faction formation at the DFH Textile Factory. Factional rivalry in the DFH Factory was apparent before the Cultural Revolution began at the factory, and the behavioral goals of both were quite clear from the beginning—to try not to be at a disadvantage in the uncertain and changing political environment. In the process of competition between the two, the organizational and mobilization capacity of the faction leaders, their competitive strategies, and means played a large role in the outcome of the competition. However, the key to victory or defeat was still high-level politics, that is, the creation of political opportunities.

Notes

1. This chapter was originally published in Japanese: "Resource Mobilization and the Rise of Factions during the Chinese Cultural Revolution", *The Japanese Association of Asian Studies*, vol. 51, no. 4 (2005), pp. 32–49, and revised and published in English: "From the 'Four Cleans' Movement to the Cultural Revolution: The Origins of Factional Conflicts", *China: An International Journal*, vol. 13, no. 3 (2015), pp. 1–24. The author thanks both journals for the copyright permission.
2. Dongfanghong Textile Factory Archives, Report of Expose by Seven Work Team Delegates: Regarding Bo Hongyun's Heinous Crimes Committed on the "Four Cleans Movement", 15 March 1967.
3. Dongfanghong Textile Factory Archives, The Categories of the Factory Party Members in the First Quarter of 1965.
4. Dongfanghong Textile Factory Alliance Headquarters, The Formation of the East Alliance and its Activities in Jinzhong before the "April Black Wind", 21 November 1967.
5. Dongfanghong Textile Factory "Four Cleans Movement" Archives, Reinvestigating the Party's Grassroots Organizations and Party Members, 28 April 1966.
6. Dongfanghong Textile Factory "Four Cleans Movement" Archives, Reinvestigation of the Party Organizations and Party Members after the Struggle against the Enemy, 23 October 1966.

7. The Dongfanghong Textile Factory Revolutionary Organization Alliance Station, False "Four Cleans," True Restoration, 26 October 1967.
8. The 10th Team of Defending the Mao Zedong Thought Revolutionary Rebellion Army of the Dongfanghong Textile Factory, Reveal the Inside Story of a Handful of People with Ulterior Motives in Dongfanghong Textile Factory "Revolutionary Organization Alliance," 27 February 1967.
9. Report of Exposé and Criticism by Wu Jinzeng, 5 July 1967.
10. Dongfanghong Textile Factory "Four Cleans Movement" Archives, The List of the Cultural Revolution Delegates in the Dongfanghong Textile Factory, August 1966.
11. Dongfanghong Textile Factory "Four Cleans Movement" Archives, The List of the Cultural Revolution Delegates in the Dongfanghong Textile Factory, August 1966; and the Dongfanghong Textile Factory Chinese Communist Party Committee, The List of Names of Responsible Persons in Each Division, 14 October 1966.
12. Compiled by the Dongfanghong Textile Factory Revolutionary Organization Alliance Station Propaganda Department, Dongfanghong Textile Factory Proletarian Cultural Revolution Chronicle of Events, December 1967.
13. The Dongfanghong Textile Factory Revolutionary Organization Alliance Station, Reviewing the Five Magic Weapons of the Red Guard Headquarters in Our Factory, January 1967.
14. Zhonggong dongfanghong fangzhichang dangwei (Dongfanghonh Textile Factory Chinese Communist Party Committee), *Dangwei xinbanzi chengyuan mingdan (The list of Names of New Team Leaders in Each Division)*, 12 October 1966.
15. Dongfanghong fangzhichang siqing yundong dangan ziliao (Dongfanghong Textile Factory "Four Cleans Movement" Archives), *Dongfanghong fangzhichang wenhua geming daibiao mingce (The list of the Cultural Revolution Delegates in the Dongfanghong Textile Factory)*, August 1966.
16. Dongfanghong fangzhichang dangan ziliao (Dongfanghong Textile Factory Archives), 1965–1967.
17. The Dongfanghong Textile Factory Revolutionary Organization Alliance Station Propaganda Department, The Compilation of Documents, 1966–1968.

18. The 10th Team of Defending the Mao Zedong Thought Revolutionary Rebellion Army of the Dongfanghong Textile Factory, Reveal the Inside Story of a Handful of People with Ulterior Motive in Dongfanghong Textile Factory "Revolutionary Organization Alliance", 27 February 1967.
19. Ibid.
20. The Dongfanghong Textile Factory Revolutionary Organization Alliance Station, Cui Bing, and His Like were Responsible for Repression of Revolutionary Organizations, The Part Two of Exposing the Crimes of Cui Bing and His Like in the Dongfanghong Textile Factory and Hard Lesson, *Compilation Documents of Turning Against One's Own Side*, Part Two, 1967, p. 11.
21. Compiled by the Dongfanghong Textile Factory Revolutionary Organization Alliance Station Propaganda Department, Dongfanghong Textile Factory Proletarian Cultural Revolution Chronicle of Events, December 1967, pp. 12–13.
22. "Four Cleans Movement" Archives, The List of Names of the Temporary Leaders of The Red Guards Headquarters, 10 January 1967; Dongfanghong Textile Factory "Four Cleans Movement" Archives, The List of the Cultural Revolution Delegates in the Dongfanghong Textile Factory, August 1966; and the Dongfanghong Textile Factory Chinese Communist Party Committee, The List of Names of Responsible Persons in Each Division, 14 October 1966.
23. Compiled by the Dongfanghong Textile Factory Revolutionary Organization Alliance Station Propaganda Department, Dongfanghong Textile Factory Proletarian Cultural Revolution Chronicle of Events, December 1967.
24. The Dongfanghong Textile Factory Revolutionary Organization/ Student Station, News Express, 26 January 1967.
25. Compiled by the Dongfanghong Textile Factory Revolutionary Organization Alliance Station Propaganda Department, Dongfanghong Textile Factory Proletarian Cultural Revolution Chronicle of Events, December 1967, pp. 1–33.
26. The Dongfanghong Textile Factory Revolutionary Organization Alliance Station, Dongfanghong Textile Factory "May Eighth Counter-revolutionary Incident: The Crime of Suppression of

Revolutionary Cadre He Wenhua and His Peers as Well as Revolutionary Masses", 1967.
27. Reprinted by the Dongfanghong Textile Factory Revolutionary Organization Alliance Station, It Was an Appalling Unjust Case of Political Persecution, 3 September 1967; and the Dongfanghong Textile Factory Revolutionary Organization Alliance Station, False "Four Cleans", True Restoration, 26 October 1967.
28. Dongfanghong Textile Factory "Four Cleans Movement" Archives, Reinvestigating the Party's Grassroots Organizations and Party Members, 28 April 1966.
29. The 10th Team of Defending the Mao Zedong Thought Revolutionary Rebellion Army of the Dongfanghong Textile Factory, Reveal the Inside Story of a Handful of People with Ulterior Motives in Dongfanghong Textile Factory "Revolutionary Organization Alliance," 27 February 1967.
30. Andrew G. Walder, "Beijing Red Guard Factionalism: Social Interpretations Reconsidered", *The Journal of Asian Studies*, vol. 61, no. 2 (May 2002), pp. 437–471.

CHAPTER 9

Radicalization of Factional Competition and Collective Violence

This book views the violence of the Cultural Revolution as a competition between factions for citizenship. Factional rivalries during the Cultural Revolution, like other collective behaviors, began to develop their autonomy once they occurred and were characterized by formations of political opportunity that affected the movement itself. If there were no institutional constraints, factional competition would have taken place in anarchy, resulting in the radicalization of competition. The factional competition during the Cultural Revolution can be divided into three types: (1) ideological break with orthodoxy; (2) mass withdrawal of participants; and (3) collective violence. During the various sub-movements of the Cultural Revolution, inter-factional competition led to the radicalization of the movement's subjects in terms of both speech and action. Particularly during the third sub-movement of the Cultural Revolution, these three types of radicalizations peaked simultaneously with the onset of national unrest. Of these, collective violence was the most extreme manifestation of factional conflict. The frequent occurrence of collective violence eventually led to the intensification of the means of controlling factional rivalry by the CCP Central Committee, represented by Mao Zedong, and the eventual decline of factional rivalry.

Chapters 4 to 6 of this book have analyzed the processes and causes of the three sub-movements of the Cultural Revolution that resulted

Y. Lijun, *Contest for Citizenship and Collective Violence during China's Cultural Revolution*, IPP Studies in the Frontiers of China's Public Policy, https://doi.org/10.1007/978-981-97-0906-9_9

from factional rivalries. In comparison, Chapter 7 focused on unfolding factional struggles in a small local city in the context of these three sub-movements. As the analysis in the previous chapters has shown, the radicalization of factional competition during the Cultural Revolution, especially collective violence, occurred during all three sub-movements. This chapter will therefore build on the previous empirical analysis to categorize factional conflicts during the Cultural Revolution and draw on additional empirical material to further analyze their causes, focusing on collective violence. The analysis will elaborate on the argument that the radicalization of factional conflict results from competition for factional interests and is also closely related to state policy and the words and actions of state agents, especially at the policy decision-making level. The containment of factional conflict is also the result of the interaction between the decline of social movements and the strengthening of the state's means of controlling them.

This chapter consists of four sections. The first section reviews the various types of collective violence from 1966 to 1969. The second section uses specific examples to examine the causes of collective violence from the perspective of the interaction between the state and society. The third section analyzes the relationship between the spread of heretical thinking and the emergence of dropouts, two types of radicalizations other than collective violence, and the competition for access to citizenship. Finally, the fourth section analyzes the state's role in the decline of sectarian conflict.

1 Types of Collective Violence

As discussed in the preamble, collective violence during the Cultural Revolution was mainly (1) between different factions; (2) within factions; (3) between Red five groups against Black five and other social groups that suffer from political discrimination; and (4) factions against state agents. These four main types of collective violence are analyzed below.

1.1 Inter-Factional Collective Violence

Collective violence between different factions refers mainly to violence between conservatives and rebels. For three years, from 1966 to 1969, armed struggles between the conservatives and the rebels culminated in four main periods of collective violence.

The first period was from November to December 1966. During this period, the rebel groups gained support from the central leadership and expanded their power, threatening the survival of the conservatives. In the main cities of the Cultural Revolution, such as Beijing and Shanghai, some members of the conservative faction attacked the Central Cultural Revolution Group or local party organizations that supported the rebel groups to preserve their losing political status, resulting in counter-attacks from rebel groups. For example, in late 1966, armed conflicts in Beijing universities existed between the conservative Red Guards and the rebel Red Guards, the 3rd HQ. Additionally, in Shanghai, there were incidents such as the *Jiefang Daily* Incident and the Kangping Road Incident between the conservative workers' organization, the Scarlet Guards, and the rebel organization, the Workers' General HQ.[1] Both incidents ended with the conservatives' defeat as the central leadership supported the rebels. Nationwide, the mainstream of factional struggle during this period was in the form of big-character posters and debates, and by 1967, the armed conflict was only happening in a few places.

The second period lasted from February to April 1967. During this period, armed conflict occurred between conservative and rebel groups around Mao Zedong's instructions for the PLA to "support the left."[2] Since the central leadership had not clearly defined the criteria for judging "leftists," the PLA tended to support conservative organizations with a majority of advanced elements, such as party-youth League members. This allowed the conservative organizations in a weak position in late 1966 to expand their power. Simultaneously, rebel groups felt encouraged to attack the army. As a result, the army directly attacked the rebel groups that attacked military installations while manipulating conservative groups into confronting rebel groups. Conservative organizations proactively assaulted rebel groups with the help of the military. Armed conflict between conservative organizations supported by the army and rebel organizations supported by the Central Cultural Revolution Group frequently occurred throughout the country. Since the PLA had received authorization from Mao Zedong to judge which organizations were leftist, the armed conflict between the two was an excellent opportunity for the PLA to suppress rebel groups, with some being banned by the PLA in cities such as Wuhan.[3]

The third period lasted from July to September 1967. In the first half of 1967, with the support of the PLA, the conservative organizations expanded in varying degrees and resurrected themselves on the Cultural

Revolution stage. The rebel groups, suppressed by the army and weakened during the February crackdown, were revived in April 1967 by Mao's restriction of the PLA's authority. Against this backdrop, the rebel groups responded to the call of the central leadership to "expose the small group of people in power within the army who have taken the capitalist road" and attacked the army and the conservative groups supported by the army. On the other hand, the PLA, which had been stripped of its power by Mao, could no longer openly suppress the rebel groups, so it could only use various methods to get weapons to conservative groups to support their actions against the rebels. For example, the conservative groups were informed in advance of weapons stockpile locations, collecting them at an agreed time and then reporting to the military headquarters that weapons had been stolen.[4] In response to the PLA's actions, Mao Zedong issued a directive to "arm the leftists" (meaning here the "leftist" rebel factions).[5] According to this directive, Jiang Qing ordered the rebel faction to "attack with words, defend with force." In this context, the confrontation between the conservatives and the rebel factions intensified, and after July, large-scale armed battles took place all over the country. The sad state of conflict between the two during this period can be seen in the following examples.

- On May 6, 1967, a conflict between conservatives and rebels occurred in a factory in Chengdu, Sichuan Province. According to the records of the rebel group, the conservative group, the Industrial Army, had between 1000 and 2000 firearms. The battle between the two factions lasted more than six hours, with about 300 rounds fired and more than 200 casualties.[6] In addition to automatic rifles and machine guns, hand grenades and poison gas were used. Armed fighting broke out again the following day, with the Industrial Army mobilizing thousands of peasants to fight and attack the rebels from the city's north, west, and south. As of the previous day, automatic rifles, machine guns, and hand grenades were used. The fighting between the two factions lasted more than a few hours, leaving more than 100 dead and 3000 wounded.[7]
- From May 13 to 15, 1967, an armed struggle between conservatives and rebels occurred in Yibin, Sichuan Province. According to information recorded by rebel groups, more than 300 people died, and more than 100 were injured in the armed struggle, which lasted for three days.[8]

- Between July and September 1967, Guangzhou was the scene of frequent arming fighting between conservative groups, mainly the Demon Beheading Platform (*Dizong*). A rebel group called the Flag Faction (*Qipai*), with incidents including 7–23, 8–11, 8–13, 8–18, 8–20, 8–31, 9–11, 9–22, and other occurrences of the armed struggle taking place.[9] According to rebel groups, more than 160 people were killed or injured in the 8–20 incident alone.[10]
- Starting in April 1967, conservative groups in Wuhan received support from the army. They expanded the organizations they had rebuilt earlier in the year, uniting with similar groups in May to form a million-strong army, the Million Heroes. The rebel groups that had split during the seizure of power in January 1967 also joined again in June to create the Wuhan Revolutionary Rebellion Headquarters to counter the suppression of the army and the attacks of conservative groups. The armed struggle, which had already begun in late April, intensified in June. The rebel faction recorded 174 firefights from 29 April to the end of June, among which 133 occurred between 4 and 26 June. During the two months from April to June, when fighting was frequent, about 70,000 people were involved in armed conflict, resulting in 158 deaths and 1060 serious injuries.[11] In response, the central leadership issued several orders banning armed fighting but failed to curb the deterioration into violence.[12] In July, the Central Cultural Revolution Group sent Wang Li and others to Wuhan to resolve the situation. The visit, however, only intensified the confrontation, and the army's long-standing discontent with the Central Cultural Revolution Group erupted at once. The PLA soldiers of the 8201 and 8199 units, tasked with "supporting the left" in Wuhan, joined forces with the Million Heroes to detain Wang Li and others, unleashing the Wuhan Incident, which was a blatant confrontation with the Central Cultural Revolution Group.[13]

The fourth period was the period of reorganization of class ranks in the summer of 1968 when a nationwide reorganization campaign was launched in May. During this period, under the pretext of reorganizing, the conservative organizations, after obtaining the support of the Maoist propaganda team of the PLA, inflicted collective violence on the rebel organizations. Wang Nianyi points out that the campaign to reorganize the class ranks was related to a change in Mao's understanding of the target of struggle during the Cultural Revolution. In other words, in the summer of 1967, Mao adjusted the goal of the Cultural Revolution from overthrowing "capitalist roaders within the party" to purging the mass organizations of bad elements. The Three-In-One Regime, conceived by Mao in 1967, brought about armed fighting between factions in many areas, and the regime's establishment was delayed. In response, Mao believed this was due to mixing bad elements within the mass organizations.[14] Thus, in 1968, to remove these obstacles, a campaign to reorganize the class ranks of the mass organizations was carried out while the regime reconstruction was proceeding. However, the central government was ambiguous about defining what kind of people should be removed as bad elements. Coupled with the fact that regime reconstruction in most places meant that the old pre-Cultural Revolution bureaucrats were back in power, it also meant that the beneficiary groups were given a new chance. Thus, the restoration of party organizations in most places expanded the conservatives' power and weakened the rebels' power with smaller party membership. Moreover, the Maoist PLA propaganda team, the key bearer of the 1968 re-establishment of power and restoration of law and order, supported conservative organizations on most occasions. Thus, the campaign to reorganize the class ranks limited anarchist competition between factions and evolved in some areas into an act of retaliation by conservative organizations to inflict collective violence on rebel organizations in the name of exposing bad elements.

During this period, many former members of the rebel faction identified as "class dissidents" were tortured and imprisoned.[15] According to Wang Nianyi's research, millions of people were persecuted in the campaign to reorganize class.[16] Most of those persecuted were members of black five groups placed outside the category of the people and former rebel factions. In some areas, there was even renewed armed fighting between conservative and rebel groups due to rebel group resistance. For example, between May and June 1968, several armed fights took place in Guangzhou between the Flag Faction, a rebel group, and the General Faction, a conservative group supported by the PLA and the provincial and municipal revolutionary committees.[17]

1.2 Intra-Factional Collective Violence

Collective violence within factions refers to armed clashes (mostly) within rebel groups. This type of collective violence occurred primarily in 1967 when a split emerged within the rebel groups around the seizure of power early that year. The confrontation between the two factions escalated into an armed struggle. For example, in January 1967, a battle broke out within the Guangzhou rebel groups over the party's seizure of power and political institutions. When the Red Flag and other factions announced their seizure of power, anti-seizure of power actions were aroused in other factions excluded from the seizure. In addition, violent clashes occurred within the factions participating in the seizure because of the unequal power distribution.[18]

It should be noted that in April 1967, with the resurgence of conservative groups in some areas, the conflict between conservative and rebel groups became more important than the conflict within the rebel groups. In this case, the rebellion reached a renewed coalition against the conservative groups supported by the army. The armed struggle in the summer of 1967 was characterized by collective violence between conservative and rebel groups in areas where the conservative groups were strong and by collective violence within the rebel groups for power in areas where the conservative groups were weak.

For example, in February 1967, the Chongqing rebel group split into the 8–15 Faction and the Anti-fundamentalist Faction (*Fan daodi pai*) due to differences in attitude (support or opposition) toward the municipal Revolutionary Rebellion Joint Committee. From the summer of 1967 to the summer of 1968, the two factions launched 31 armed struggles, including two with automatic rifles, machine guns, hand grenades, anti-aircraft guns, warships, combat vehicles, etc. The battle in late August 1967 is said to have involved the firing of some 10,000 rounds of anti-aircraft firearms overnight. During that year, countless people were wounded, and some 645 died due to the fighting.[19]

The period from July to September 1967 was a period of armed struggle between factions throughout the country. The collective inter-factional violence of this period was highly chaotic and complex. Thus, it can be assumed that there were areas where both types of collective violence, inter-factional and intra-factional, occurred simultaneously or alternately. The armed struggles of this period were eventually brought under control in late 1967 by strengthening the authority of the PLA, the fall of some members of the Central Cultural Revolution Group, and the weakening of political power held by mass organizations.

1.3 Collective Violence Between People's and Non-People's Groups

Collective violence between people and non-peoples groups differs from the two-way violence between and within factions in that it is one-way violence of the people's groups inflicted on non-people's groups. It is characterized by the absence of the right of the abused to resist and defend themselves. This type of collective violence can be subdivided into two main types.

The red five Red Guards carried out the first type of violence against the black five, the black seven, the "monsters and freaks," and their families in the summer of 1966. In late May, the red five Red Guards, who were mainly cadres' children, organized the "breaking of the Four Olds," raiding homes and using "shower" and "jet" methods of torture to carry out the "red terror." In only a month, from mid-August to mid-September 1966, the red terror spread throughout the country as similar groups developed. The extent of the violence was huge. For example, in the 40 days from mid-August to late September 1966, some 1700 people were beaten to death in Beijing, 336,000 families were raided, and 850,000 people identified as black five and their families were forcibly sent to the countryside.[20]

In addition, on August 27, 1966, when the Red Guards' rebellion storm swept across China, mass abuses occurred in Daxing County, south of Beijing. From 27 August to 1 September, a total of 325 people from 22 households ranging from 88 years old to infants were abused and killed throughout the county, with the death count reaching 20.[21]

Second, after October 1967, a new form of violence emerged in which conservative organizations throughout the country persecuted local people of black five class origin and their families. This type of violence against black five was similar to that inflicted by the Red Guards in 1966. However, the persecution committed by the conservative organizations differed from the former in that they persecuted people and families of black five origin as if they were behind the rebel organizations. For example, from October 1967 to October 1968, there were frequent incidents in Guangxi where conservative groups brutally murdered five black and their families. The first incident occurred in Dongshan Commune, Quanzhou County, where 76 people, including one-year-old babies, were killed in two days. In Bingyang County, also in Guangxi, 3681 people were killed in 11 days between July 26 and August 6, 1968, including 176 whole families.[22]

In addition, from August 13 to October 17, 1967, a 66-day-long mass killing of the black five and their families took place in the Daoxian area of Hunan Province. This incident affected ten districts, 36 people's communes, 468 production brigades, 1590 production squads, and 2778 families, of whom 4193 were murdered and 326 were forced to commit suicide.[23]

1.4 Collective Violence Perpetrated by Factions Against State Agents

Two main types of collective violence perpetrated by factions against state agents exist. The first is the violence committed by the red five Red Guards against schoolteachers, i.e., "reactionary academic authorities" and school agents, at the beginning of the Cultural Revolution.[24] After the *People's Daily* published the editorial, *Let Go of the Masses and Thoroughly Defeat the Counter-Revolutionary Gangsters*, violence by students against party and government cadres and faculty members who were considered "black gangs" began to occur in schools. On 18 June, students at Peking University set up the "ghost fighting platform" and the "demon beheading platform," where more than 60 "black gangs," such as Lu Ping, wore tall hats and were attacked.[25] This action spread widely, and school violence spread to universities and high schools everywhere. Some students of black five class origin were also targeted. This type of violence reached a climax in late July and mid-August 1966 when the group withdrew. According to Wang Nianyi's research, tens of thousands of university, high school, and primary school teachers were persecuted that summer of 1966.[26]

For about two years after October 1966, the direction of the movement shifted from "sweeping away monsters and freaks" to overthrowing "capitalist roaders within the party." As such, the object of violence also changed dramatically. During this period, rebel organizations persecuted the "capitalist roaders within the party" with violence, i.e., heads of party and government organs and Work Units at all levels. Though the recipients of violence changed, how it was carried out was the same as that of the red five Red Guards at the beginning of the Cultural Revolution, inheriting the red terror. This is the second type of collective violence perpetrated by factions against state agents. During this time, the rebel movement persecuted three-quarters of the senior cadres (vice ministers and above in central and local party and government organs). More than 1200 senior cadres were persecuted in the Expose the Traitors Campaign, mainly organized by the student rebels of Nankai University in Tianjin.[27] Not only were senior cadres persecuted in this campaign, but ordinary cadres were also affected. According to records, more than 1400 cadres were persecuted in the Expose the Traitors Campaign in Shenze County, Hebei Province.[28]

2 Causes of Collective Violence

As mentioned earlier, this book views the collective violence of the Cultural Revolution as a result of the interaction between the state and society. Therefore, based on the analytical framework discussed in Chapter One of this book, we analyze the causes of collective violence in four aspects: (1) the violent struggle within the state; (2) the competition between factions for interests; (3) the competition within factions for interests; and (4) the interaction between the state and society, using specific examples from the Cultural Revolution.

2.1 *Power Struggles and Collective Violence Within the State*

The modern state provides order in society through its monopoly on the apparatus of violence. Still, in China during the Cultural Revolution, internal strife within the state weakened its ability to control violence. Moreover, as the following analysis will show, the power struggles within the leadership led the government or individual state agents to encourage violence. This resulted in the state, which was supposed to provide social order, becoming the root cause of the social disorder.

As discussed in earlier chapters of this book, the Cultural Revolution began with political struggles within the party. The Cultural Revolution was a social movement due to the interaction between the socialization of political struggles within the party and the politicization of competing interests among the masses. Mao became dissatisfied with the bureaucracy and the guidelines of the first-line leaders of the Central Committee, such as Liu Shaoqi, who ran the system. For this reason, Mao began to emphasize class struggle and defined class antagonism as between "capitalist roaders within the party" and the worker and peasant masses. This definition gave a new meaning to the Theory of Class Struggle. Mao's aim in launching the Cultural Revolution was to mobilize the masses for class struggle, to remove the forces of Liu Shaoqi, who valued formalization and bureaucratization, to transform the administration, and to purify the ideology. Here, the Theory of Class Struggle was both a tool for Mao's political mobilization of the masses and an ideological goal that the Cultural Revolution sought to achieve.

The Theory of Class Struggle is an integral part of the ideology of socialist China. However, it is essential to note here that the ideology of class struggle has covered more than just the same content at different times. Two theories of class struggle defined by Mao existed in China in the 1950s. One defined class antagonism as between the exploiting classes, such as the landlords and capitalists of old China, and the exploited worker and peasant masses. The theory was based on the fact that the old exploiting class was trying to use new means to achieve restoration, so class struggle had to be waged. This theory was widely propagated and emphasized in various political movements in the early 1950s, such as the land reform movement, the "purge," and the "three-anti and five anti campaigns." The other was to extend the class struggle to intellectuals who retained the values and habits of the old ruling class. This theory emerged during the anti-rightist movement in the latter half of the 1950s. In this context, in the 1960s, Mao revived the idea of class struggle and gave it a new significance by extending the target of class struggle from the old exploiting class, the intellectuals, to a section of party bureaucrats. In other words, the black five class and their families, intellectuals, and state party cadres became the targets of collective violence, which was already ideologically legitimized on the eve of the Cultural Revolution. Moreover, since Mao's Theory of Class Struggle contained three different elements, the state agents who held leadership in the Cultural Revolution and the general masses who participated in the Cultural Revolution were free to define their rights based on different purposes and thus selectively used the Theory of Class Struggle to justify their actions.

On the other hand, Xu Ben points out that Liu Shaoqi did not have the power to define ideology as Mao Zedong did. Still, he did have the power to interpret the Theory of Class Struggle, turning the Cultural Revolution into another anti-rightist campaign.[29] In July 1966, when Liu Shaoqi was leading the Cultural Revolution, he instructed that "the main task of the Cultural Revolution in secondary schools was to review the teaching staff."[30] In this context, school cadres and "reactionary academic authorities," as well as staff members who were considered to be "monsters and freaks," were the first victims of collective violence.

Moreover, at the level of the participants in the movement, the cadres' children who were threatened by the middle-class children in the competition for higher education at the beginning of the Cultural Revolution and who gradually lost their privileged status were the first to rise in rebellion in response to the call for class struggle. They took up the Theory of Class Struggle put forward by Mao Zedong in the 1950s and developed it into the bloodline theory during the Cultural Revolution, trying to exclude those who came from outside the red five class. Although their actions deviated from Mao's Cultural Revolution concept of mobilizing the masses to transform the bureaucracy, they provided Mao with a way to mobilize the masses at the beginning of the Cultural Revolution and thus received strong support from Mao. The red five Red Guards fought against "reactionary academic authorities," "monsters and freaks," and the "exploiting classes." The revolutionary actions and the resulting red terror violence spread throughout the country.

In early July, violence in many Beijing high schools was brought under control. Controlling violence is the state's proper action, but the Working Team's actions deviated from Mao's concept of the Cultural Revolution. That is, the state was the only legitimate institution to control violence in society, but one of Mao's aims in launching the Cultural Revolution was to use the power of society to transform the bureaucracy. To achieve this, Mao actively called on the masses to revolt against the central government, thus encouraging social unrest. For Mao, social unrest was "achieving great governance through great chaos." Therefore, Liu Shaoqi's action of sending the Working Team was criticized by Mao as "suppressing the student movement by resisting the Cultural Revolution."[31] In other words, due to the political struggle of the central leadership, the state let go of the control of collective violence between factions and, by August, even began to reward collective violence.

As mentioned, in late August 1966, before the abuses in Daxing County, Xie Fuzhi, alternate member of the Political Bureau and Minister of Public Security of the State Council, said at an expanded meeting of the Beijing Public Security Bureau, "I disapprove of the masses killing people, but the masses hate the bad guys to the bone, and we can't dissuade them, so don't force them. The civil police should side with the Red Guards and provide them with information."[32] Xie Fuzhi's speech was also conveyed to the public security organizations in Daxing County, which provided information about the black five in their jurisdiction to the Red Guards, thus causing mass abuse.[33] During the same period, in late August, the public security organs and the PLA issued the "Regulations on the Strict Prohibition of the Use of Police to Suppress the Revolutionary Student Movement" and the "Regulations on the Absolute Prohibition of the Use of Armed Forces to Suppress the Revolutionary Student Movement."[34] In addition, the leaders of the Central Government emphasized the legitimacy of violence in their statements on various occasions. For example, on July 28, 1966, Jiang Qing told the crowd, "We don't advocate beating people up, but beating people up is no big deal" and "bad people deserve to be beaten up by good people; when good people beaten up by bad people, good people are honored; when good people beaten up by good people, it was a misunderstanding and did not fight each other again."[35] Then, Chen Yi also defended the violence of the red five Red Guards: "It's no big deal for a mass movement to kill a few landlords."[36] In this way, collective violence was legitimized by the encouragement of the state and its agents.

From August to the end of September 1966, the bloodline theory of the red five Red Guards centered on the cadres' children was used by some local party and government cadres to maintain their political status, which hindered the progress of the Cultural Revolution. Therefore, in October 1966, Mao Zedong corrected the course and aimed the campaign at "those in power who follow the capitalist road" within the party. This revision of the Cultural Revolution gave the children of the middle class, who had hitherto been dissatisfied with class policies and excluded from the Red Guards at the beginning of the Cultural Revolution, the opportunity to express their discontent and to participate in politics legally. As a result, this group began to form rebel groups, using the class struggle theory of "red degeneration" put forward by Mao Zedong in the 1960s to participate in political movements while also pursuing collective interests. Their actions were used to the fullest extent by Mao in late 1966 as a tool to purge the party of the Liu–Deng clique. This reorientation of the Cultural Revolution struggle brought about a change in the formation of political opportunities and a change in the distribution of and competition for new forms of citizenship. The results intensified confrontations and triggered violent clashes within conservative and rebel organizations. For example, a study of factionalization and conflict at Tsinghua University shows that the armed struggle between the conservative Red Guards in Beijing and the 3rd HQ rebel group in late 1966 was closely related to the adjustment of state policy and the support of the Central Cultural Revolution Group for the 3rd HQ.

Likewise, the armed struggles between the various factions of the workers' rebel groups were inextricably linked to the intensification of factional rivalries and the acquiescence of the state and its agents in rewarding armed struggle. Specifically, the construction of the Three-In-One Regime intensified the competition between factions over power and authority. On the other hand, the relationship between Zhou Enlai and other State Council cadres and the Central Cultural Revolution Group, between PLA cadres and the Central Cultural Revolution Group, between PLA cadres and local party and government cadres, and between PLA cadres and the Central leadership was complicated and contradictory. These contradictory relations influenced and complicated the inter-factional rivalry. It is worth noting that, as the analysis of the examples in the previous chapters shows, the mass organizations were not passive participants. Although mass organizations were used as instruments in the power struggle, they used political opportunities in the power struggle for their benefit. The Cultural Revolution's chaos resulted from an intricate entanglement of "control and counter-control" and "use and counter-use."

For example, the Wuhan incident in July 1967 reflected both the competition for power and interests between conservative and rebel groups around the Three-In-One Regime and the confrontation between the Central Cultural Revolution Group and the Wuhan Military Region. While the conservative and rebel organizations were each supported by the Wuhan Military Region and the Central Cultural Revolution Group and were manipulated by them in their political struggle, both groups also tried to use the political struggle between the Wuhan Military Region and the Central Cultural Revolution Group to destroy rival organizations.[37] the previous chapter's study of the Workers' General HQ shows that the *Jiefang Daily* Incident and the Kangping Road Incident between the Scarlet Guards and the Workers' General HQ of Labor in Shanghai were good examples. This series of armed conflicts between the Shanghai Scarlet Guards and the Workers' General HQ was influenced by the change in the policy of the central leadership and the inconsistency of internal political views, as well as the loss of legitimacy of the party and government organs in Shanghai to control society and the decline of their ability to manage disorderly social movements.

Like the red terror of the summer of 1966, the inter-factional armed fighting that took place throughout the country between the spring and autumn of 1967 was directly encouraged by state agents. When Jiang

Qing received representatives of the Anhui rebellion in September 1967, she said, "I do not advocate armed fighting, but whoever wants to fight me, I will defend myself and fight back."[38] Similar statements were found in Zhang Chunqiao's remarks when he received the rebel group in Shandong.[39] Thus, in the summer of 1967, forged Mao's top instructions about giving guns to the left spread throughout the country, giving further legitimacy and courage to rebel groups in their confrontation with the conservative groups.

As mentioned earlier, collective violence is linked to the state. On the one hand, when collective violence occurs, the state arguably does take control measures. However, these efforts to control were highly ineffective due to the low degree of institutionalization and interaction between the state and society. Even when it can temporarily prevent immediate violence, its measures created new problems and triggered new acts of collective violence. For example, in August 1966, when the red five Red Guards committed violence against black gangs, the black five, and their families, the Central Committee of the CCP ordered the military, police, and other security forces not to suppress the Red Guards. As such, Zhou Enlai ordered the leaders of the Red Guards to form Red Guard pickets to control violence, which could be seen as a demand for self-discipline within the Red Guards. In the context of the rebellious mobilization orders of the time that "the masses should not be suppressed" and "the masses should liberate themselves," this method can be seen as an emergency method adopted by Zhou Enlai under abnormal circumstances. In September 1966, the Xicheng Branch Command of the Capital Red Guard Pickets (hereinafter referred to as "Xicheng Pickets") issued general orders prohibiting armed struggle.[40] From the point of view of social control methods, Red Guard pickets such as the Xicheng Pickets, which had been given the power to control violence by Zhou Enlai, used the method of using violence to counter violence and therefore were unable to gain the support and cooperation of other Red Guard organizations.[41] More importantly, because almost all members of the Red Guard pickets were propagandists of the bloodline theory, they met stiff resistance from the rising rebel Red Guard groups, and the attempt to prevent violence by the Xicheng Pickets was seen as a monopoly on the exercise of violence by privileged bureaucratic-class children, as was the bloodline theory. This intensified the confrontation between the conservative Red Guards and the rebel Red Guards and even caused the conflict of interests between the two to develop into armed combat. As the support of the Central Cultural Revolution Group for student organizations shifted to rebel student organizations such as the 3rd HQ, the Xicheng Pickets, which cadres' children dominated, were accused by the Central Cultural Revolution Group of being "gendarmes of the bourgeois reactionary line."[42]

In late 1966, not only did armed fighting occur between the conservative Red Guards and the rebel Red Guards, but the students' southward movement also brought about armed fights between conservative workers' organizations and rebel workers' organizations at the local level. Under these circumstances, Mao ordered the PLA to intervene in the Cultural Revolution, attempting to use the PLA to control the collective violence. However, as the analysis of the factional struggle at Tsinghua University and the rise and fall of the Workers' General HQ in Shanghai shows, the low degree of institutionalization of the state and the full-scale intervention of the PLA led to the simultaneous eruption of various conflicts within and among the state and various factions of society. These conflicts were exacerbated by the 1967 campaign to "expose a small group of reactionary elements within the army," which led to collective violence among mass organizations.

In the absence of the rule of law, with state institutions wholly paralyzed and the police, which was supposed to maintain social order as a state apparatus, caught up in strife, the PLA was left to clean up the chaos. Thus, after September 1967, Mao again gave the PLA the authority to restore social order and control factional conflicts. As a result, the Cultural Revolution broke the iron rule of the Chinese Revolution that "the party commands the gun" as the military intervened in politics and grew in power. As Kazuko Mori points out, "The Mao Zedong-Lin Biao system, in which military power prevailed, aroused the military's ambition to impose a dictatorship, which eventually led to the eruption and latter suppression of the coup d'état of the Lin Biao faction (The Lin Biao Affair of September 1971)."[43]

2.2 *Inter-Factional Competition for Interests and Collective Violence*

The collective violence that occurred during the Cultural Revolution was closely linked not only to the state but also to inter-factional rivalries. In short, collective violence resulted from intensified competition between factions for citizenship.

Generally speaking, if the element of the state discussed is removed, it can be argued that factional rivalry is difficult to generate violence under two conditions. First, in societies where the private sphere is the exit point, factional competition in the public sphere can hardly be transformed into violence. This is because participants can retreat to the private sphere if they lose the competition or if there is an inconsistency between the conduct of the movement and the participants' original goals. Second, factional rivalries governed by institutions and rules are less likely to generate violence.

The factional competition during the Cultural Revolution occurred without these conditions. First, it took place without the private sphere and exit options, so inter-factional competition over power and interests could quickly become radicalized. For the masses, all the citizenship needed is monopolized by the state and distributed hierarchically to various groups. The state monopoly of citizenship and the hierarchical distribution of citizenship stimulates factional ambitions to monopolize political resources, giving an exclusive character to the competition between factional interests. This is because the public goods, such as political and economical, and cultural rights created by competition, are shared by all members of the collective and even the faction. In other words, once a group or faction has gained a high political status, all members enjoy its benefits. Conversely, once a group or faction is placed at a disadvantage in the power or benefit distribution structure, as with the treatment of the black five class, all members must suffer the consequences. In addition, the scarcity of resources, i.e., the extreme lack of opportunities for upward mobility, such as giving promotions and obtaining a party-youth league membership, created a desire on the part of all groups and factions to keep opportunities private from other groups and factions. Exclusionary competition for interests in a public sphere that lacks avenues of exit can easily lead to violent conflict. This is one reason why inter- and intra-factional violence is so frequent in the context of the establishment of the Three-In-One Regime.

Second, inter-factional competition for interests is not governed by institutions and rules and is conducted in anarchy. Inter-factional competing interests in such a poorly institutionalized situation have the following three effects on collective behavior, making them highly susceptible to collective violence.

First, the low ability of the factions to govern themselves. Most of the factions during the Cultural Revolution were not formed spontaneously but under the political mobilization of the state. For example, the red five Red Guards emerged from the political mobilization in the spring and summer of 1966, and the large number of rebel Red Guards that occurred in October of that year was in response to the central government's call to criticize "capitalist roaders within the party." Moreover, as the Shanghai example shows, the division and organization of the conservative and rebel factions within the workers was the result of the mobilization of the masses by the local party organizations at the beginning of the Cultural Revolution to control the direction of the movement and to protect themselves, and the assembly and organization of people excluded by this mobilization. Although the factions organized during the Cultural Revolution pursued their groups' interests, they fundamentally differed from the institutionalized autonomous organizations that had developed over time in civil society, and their autonomy was extremely low.

Secondly, there was no synergy, cooperation, and dispute resolution mechanism between factions, only non-institutional competition. The opposition between conservatives and rebels was formed before the factions were created. In other words, one faction was formed to oppose the other faction and was already in the camp of opposition to the other faction when the organization was formed. For example, an analysis of the rise and fall of the Old Red Guards shows that the red five Red Guards, centered on the cadres' children, had already been competing with the intellectual children and other middle-class people over interests such as further education and membership in the party before they were formed. Once organized, they criticized the non-red five classes and their children. On the contrary, the Red Guards, a rebel group with a large concentration of middle-class children, was organized in opposition to the bloodline of the cadres' children and had the characteristics of a Red Guard opposed to the red five even before it was founded. The confrontation between the two turned violent in late 1966. Analysis of Shanghai shows the same result: the Scarlet Guards, a conservative workers' organization, was formed to confront the Workers' General HQ, a workers' rebel group. Soon after the two organizations were organized, there were armed fights, such as the *Jiefang Daily* Incident.

During the Cultural Revolution, two major rival organizations, the conservatives, and the rebels, were within almost all schools and factories, and other Work Units. To overcome the opposing camp, these rival organizations often sought the support of external organizations, establishing alliances with organizations outside the Work Unit. However, in establishing these coalitions, actions are not based solely on political opinion, and the greatest goal of most factions is to use external factions with solid political power against the opposing organizations within the Work Unit. In some cases, there are examples of cooperative relationships between organizations with radically different political views. For instance, at the beginning of the Cultural Revolution, workers at the Shanghai Diesel Engine Factory split into the "pro-Zhu faction" and the "anti-Zhu faction," around whether they supported or opposed a particular cadre surnamed Zhu. The anti-Zhu faction formed the Eastern Red Headquarters on September 15, 1966; a week later, on 22 September, the pro-Zhu faction formed the Shanghai Revolutionary Rebellion Joint Command. The Eastern Red Headquarters had most of the party and league members, activists, and veteran workers, while the Shanghai Revolutionary Rebel Command had a majority of general workers. Considering the widespread phenomenon of the Cultural Revolution, the former should have joined the camp of conservative organizations, and the latter should have joined the camp of rebel organizations. Still, the actual situation was the exact opposite. The Eastern Red Headquarters was subordinate to the Workers' General HQ, a municipal workers' rebel organization, and was active as the leading force in the Anting Incident, the *Jiefang Daily* Incident, and the Kangping Road Incident. On the other hand, the Shanghai Revolutionary Rebel Command was subordinate to the Scarlet Guards, a joint organization of workers' rebels in the city.[44] According to Li Xun's research, the formation of this situation had nothing to do with the political opinions of the two sides. Still, it was the result of a very fortuitous situation. In other words, at the time of the formation of the two organizations, both sides wanted to unite with the powerful Workers' General HQ. As the Eastern Red Headquarters had contacted the Workers' General HQ first, the Shanghai Revolutionary Rebel Command had no choice but to join the Scarlet Guards, the opponent of the Workers' General HQ.[45]

Third, as mentioned, the factions during the Cultural Revolution had yet to form autonomously. Still, due to political mobilization, they needed more stability compared to the autonomous organizations of civil society in terms of their goals and the competition mechanisms between them. Specifically, the ultimate goal of the factions during the Cultural Revolution was to pursue political and economic interests. The claims they sought were not determined by themselves but were defined by the state. This decided that the sectarian interests pursued by the factions were subject to the political goals set by the Cultural Revolution and that they needed to adjust their behavior in response to changes in political objectives. Many temporary worker rebel groups emerged when the trend of economism was spreading. These factions explicitly shone a light on the demands of economic interests. Though appearing to have collective interest goals independent from the state, they were derived from the state's political objective, which was to criticize Liu Shaoqi. In other words, the spread of economism began when social groups discontented with economic interests took advantage of the opportunity to attack Liu Shaoqi to pursue their interests. If Liu Shaoqi's pre-Cultural Revolution line were criticized as a bourgeois reactionary line, then those who had suffered injustice under his policies (such as hiring temporary workers) would have been justified in demanding vindication in the campaign to criticize Liu Shaoqi. In other words, what appears to be a rebellion of temporary workers with independent economic goals is a group using the political goals of the Cultural Revolution to pursue its economic interests. Of course, what must be seen here is that their demand for economic benefits was then consciously used as a tool for political struggle within the state. Or rather, the opportunity for these groups to organize themselves to pursue economic interests was dictated by the creation of political opportunity provided by the need for political struggle. Leaders of the Central Cultural Revolution Group and Zhou Enlai made repeated statements around October 1966 encouraging the rebellion movement of half-working (peasant) students and temporary workers. Local party cadres tried to meet their economic demands as much as possible to save themselves. In this way, the encouragement of state agents contributed to the spread of the trend of economism, and organizations pursuing economic interests popped up throughout the country in a brief period.

In short, these economic interest-seeking organizations result from political mobilization by the state. The economic interests they pursue are not collective goals independent of the state's political goals, so the survival of factions and the legitimacy of their goals depend entirely on the state's political objectives. Moreover, it is much more difficult for factions pursuing political interests to establish their own factional goals than organizations seeking economic interests. During the Cultural Revolution, the state's political goals were constantly changing in response to the internal power struggle and the advancement of the Cultural Revolution.

This unstable external political environment affected factions, making it extremely difficult to establish factional goals, institutionalize groups within factions, and create mechanisms for institutionalizing factional competition.

2.3 Competing Intra-Factional Interests and Collective Violence

In examining intra-factional competing interests, it is necessary to analyze them in two ways. The first is intra-factional fragmentation, that is, the fragmentation of a faction or coalition into several factions and how the competing interests of these fragmented factions affect collective violence. The second is how the relationship between leaders and members within a faction affects collective violence.

Competing Interests Within Factions

The logic of competition for interests within a faction is the same as competition between factions. When an organization or a coalition splits into several factions, competition among the factions takes on the same nature as the competition among factions. Intra-factional competition for interests during the Cultural Revolution can be divided into two main types: competition between factions within a collaborative organization and between factions formed by the division of an organization.

The relationship between factional rivalry and collective violence within the first type of joint organization is described first. As discussed earlier, during the Cultural Revolution, there were usually two rival organizations within schools and factories, and these organizations often sought support from the outside to overcome rivals. They would establish alliances with organizations outside the Work Unit. However, there needed to be institutionalized cooperation between these joint organizations. Thus, even if the various factions form temporary alliances based on immediate common interests and goals, once the interests and goals change, the basis for the alliance's survival is lost, and the possibility of disintegration and fragmentation of the alliance increases. Among these divided organizations, collective violence resulting from disorderly competition is likely to occur for the same reasons as the rivalry between conservative and rebel organizations.

For example, the 3rd HQ, a university rebel group in Beijing, was formed by rebellious students who had been suppressed during the period when the Work Team held the leadership of the Cultural Revolution in their schools to restore their reputation and to seek the right to participate in the Cultural Revolution. After October 1966, the 3rd HQ received the support of the Central Cultural Revolution Group. As its power grew and its goals were achieved, it lost its internal unity and split into the *Tian* Faction and the *Di* Faction. The core organization of the *Tian* Faction was the New Beijing University Commune, headed by Nie Yuanzi, the *Jinggangshan* Corps of Tsinghua University, headed by Kuai Dafu; and the Red Flag Combat Team of the Beihang Institute, headed by Han Aijing. On the other hand, the core organizations of the *Di* Faction were the *Jinggangshan* Red Guards of Beijing Normal University and the Eastern Red Commune of the Beijing Geological Institute. With the collapse of the conservative student organizations at the end of 1966, the confrontation between the *Tian* Faction and the *Di* Faction became the most prominent conflict within the Red Guards of the Beijing universities. The *Tian* Faction and the *Di* Faction were large unions of factions that lacked institutionalized collaboration. As a result, there were conflicts between their respective internal organizations. For example, the *Tian* Faction had a competition between Nie Yuanzi and Kuai Dafu to gain control of the Red Guards in Beijing, which led to constant conflict. The situation was much the same for the conflict between the internal organizations of the *Di* Faction.[46] The analysis of factionalization and collective violence at Tsinghua University in the preceding section shows that the violent clashes during the Zhongnanhai sit-in in the summer of 1967 were entirely a conflict between rebel groups triggered by competing loyalties. The conflict was not triggered by differences in political views or class differences in origin but by competition between organizations within the rebellion over which side was more revolutionary.

The following section describes the violence that results from the competing interests of derivative organizations within the second type of faction. In civil society, when a part of the membership breaks away to form a new organization because the organization does not meet the interests and demands of its members, the breakaway organization often establishes new organizational goals, programs, and methods of achieving them based on institutions and rules to pursue collective interests. In contrast, the most important enemy for organizations separated from their parent organizations during the Cultural Revolution was the parent organization. In other words, the purpose of separation was to counter the parent organization. In some cases, the aim was to destroy the parent organization, even to the extent of creating a coalition with the former rival organization. For, in the words of the time, "the enemy of our enemy is our friend."

An example of factionalism at Tsinghua University shows that the rebel group, the *Jinggangshan* Corps, had internal differences of opinion around the construction of the Three-In-One Regime and eventually split into two ground, the 4–14 Faction and the Regiment Faction. The 4–14 Faction joined the *Di* Faction, which opposed the *Jinggangshan* Corps, to confront the Regiment Faction.[47] The same happened in the summer of 1967 during the arrest of Xu Shiyou campaign. The 4–14 Faction also intended to participate in the arrest of Xu Shiyou campaign but was criticized by the Regiment Faction as being outwardly opposed but inwardly supportive of Xu Shiyou. The Regiment Faction and its allied organizations attacked them. As a result, the 4–14 Faction eventually chose the path of joint struggle with the Fengbo Faction, which supported Xu Shiyou.[48]

It is important to note that almost all violence arising from intra-factional rivalries during the Cultural Revolution occurred within rebel groups. To the author's knowledge, there are few such instances within conservative organizations. This is important in understanding factional behavior during the Cultural Revolution. Since it is not the focus of this book, I will not analyze it in depth but will offer the following two points for reference as to why.

First, in terms of the composition of the organizations, conservative organizations generally have a majority of progressives, such as party members and activists. In contrast, rebel organizations have a majority of ordinary people who were placed outside the vested interest groups before the Cultural Revolution. Thus, in terms of the organizational mechanism of intra-organizational and inter-organizational alliances, the former, like the latter, needed a more rigorous organizational framework. Still, because of the preponderance of party-youth league members and activists, it can be argued that the organizational mechanism within the original bureaucracy and the organizational ability and mobilizing the power of faction leaders still exerted some form of influence on conservative organizations. In contrast, the internal stability of the rebel groups was much more challenging to maintain than that of the conservative groups since most members were mobilized and organized before the Cultural Revolution and had not received training in the ability to organize and mobilize themselves. An example of this can be found in

Wang Shaoguang's study of the Wuhan conservative alliance, the Million Heroes, and the rebel alliance, Wuhan Rebel General Headquarters.[49]

Second, regarding the homogeneity of their organizational goals, the conservatives joined the movement to protect the bureaucracy being destroyed. Of course, this aim was closely linked to their own political and economic interests since they were part of the vested interests of the bureaucracy. Wang Shaoguang points out that preserving what they already had during the Cultural Revolution was the primary goal for members of the conservative faction, so it was easier to maintain internal homogeneity.[50] In contrast, the common goal of rebellion members was improving the status quo through participation in the Cultural Revolution. However, the pursuit of interests under this common purpose was very complex. For some faction leaders, such as Wang Hongwen, participation in the Cultural Revolution was about becoming an official, but for most participants, it was about self-preservation. While participation in the Cultural Revolution might not necessarily be profitable, non-participation might turn themselves into a target of criticism. In the words of a famous saying during the Cultural Revolution, "No revolution is not far from a counter-revolution." Of course, some participants were followers of the Cultural Revolution philosophy who joined the movement because they resonated with Mao's anti-bureaucratic philosophy of the Cultural Revolution. In this way, there were different goals within the rebellion. It took much work to maintain long-term homogeneity of action as the organization continued to fragment and reorganize in response to changes in the external political environment and the internal division of interests. Even if the organization can maintain relative unity of purpose within the organization, allies with common aspirations are potential competitors because only a few people will ultimately hold power. With the collapse of the conservatives as a common rival group, inter-factional rivalry within the rebellion became necessary. This explains the frequent armed fights within the rebellion during the power grab period. In addition, the competition for interests between the leaders of the different factions and the mobilization of members within their factions also intensified inter-factional conflicts. This will be discussed in more detail in the next section.

Interaction Between Faction Leaders and Members

In examining collective violence arising from competing interests, it is essential to discuss the interactions between faction leaders and members since most collective violence during the Cultural Revolution resulted from interactions between faction leaders and members.

Violence is a means of showing loyalty to the leader for faction members. To participate in the Cultural Revolution, an individual must join a faction. Joining a faction was the only way to participate in the movement. In other words, individual competition for citizenship had to be achieved through the collective behavior of factions. Gong Xiaoxia points out that for one faction to prevail over an opposing faction, it needed motivated and combative members. Further, motivated and combative members tend to be trusted and respected within the faction for promotion opportunities. As a result, some militant members actively used violence to prove their revolutionary nature and gain promotion within the faction.[51] For example, during the red terror in August 1966, the "black gangs and black five" classes were targeted. For this reason, people within the Red Guards competed to use violence to demonstrate their revolutionary nature, using the number of beatings to measure their revolutionary performance.[52] At the time, Liu Shaoqi's fourteen-year-old daughter bragged about beating three people to death.[53] On the other hand, those excluded from the Red Guards, to prove that they too were qualified to join the revolution, were more fervent than Red Guard members in their violence against black gangs and people of lower political status than themselves.[54] In this context, the number of people beaten to death in Beijing in late August reached hundreds per day.[55]

On the other hand, there are many factions within the same camp under various names, all competing with each other. As mentioned earlier, factions are not institutionalized autonomous organizations. Therefore, the mobilization of leaders plays a vital role in the survival of factions. In other words, factions are sustained by the mobilization of their leaders. Therefore, conflicts between factions are, in a sense, also conflicts of interest between faction leaders. To win over the leaders of other factions competing for conflicting interests, the faction's leaders mobilize the members of their faction to join the competition, resulting in active inter-factional conflict. In other words, inter-factional conflicts can, in a sense, result from faction leaders mobilizing the masses to express loyalty to the state agents they support or to pursue personal political interests. It is the conflict of interests between leaders that leads to inter-factional conflicts.

For example, Nie Yuanzi, Kuai Dafu, Han Aijing, Tan Houlan, and Wang Dabin, the leaders of the Five Students' Organization in Beijing, clashed with each other over the right to command the Red Guards and the exclusive support of the Central Cultural Revolution Group. Their conflict of interest brought about disputes between the *Tian* and the *Di* Factions. Conflicts among their internal organizations led to armed struggles between some of them.[56]

The analysis of the factional division at Tsinghua University also shows that the Hundred Days War between the Regiment Faction and the 4–14 Faction resulted from the leaders of the two factions fighting for control of the university. It was, in fact, a conscious attempt by Kuai Dafu to make a big fuss and attract the intervention of the central government to use the power of the central leadership to destroy the rival 4–14 Faction. In other words, violence became a tool used by the faction leaders to pursue their political interests. In this case, the violence was not only the result of the radicalization of competing interests but the violence itself was also used as a method for competition. When used as such, it is likely to be more violent than that produced by radicalizing competing interests.

2.4 State–Society Interactions and Collective Violence

The preceding analyses of the impact of intra-state power struggles and inter-factional and intra-factional competing interests on collective violence have been presented. These analyses have addressed the effect of the interaction between the state and society on collective violence. In addition to what has already been discussed, a critical issue needs further exploration: the impact of power struggles at the central level, inter-factional and intra-factional competing interests, and policy communication between state agents at the decision-making level and factions on the collective violence.

The state is a collection of hierarchical political institutions and agents of the state. During the Cultural Revolution, the agents of the state at the policy decision-making level were divided into several groups due to differences in policy, power struggles, and differences in their views on nation-building and the means and methods of dealing with problems. These groups competed with each other, as were the factions of society and the members of the factions themselves. The competition between them thus influences the actions of the factions through state policy and rhetoric. This influence can be summarized in the following two categories.

First, state agents consciously induced factional behavior through rhetoric to achieve their goals of power struggles and other political ends.

For example, as noted earlier, in the summer of 1966, Mao Zedong, to mobilize a social revolt, ordered the army and police in the name of the CCP to refrain from suppressing mass movements, a move that served to encourage violence. Moreover, in the summer of 1967, Jiang Qing and others called the rebel groups to "attack with words, defend with force" to confront the anti-Central Cultural Revolution Group forces in the military. As a result, the rebel groups were given the legitimate right to exercise violence, thus intensifying the armed struggle.

Second, for state agents, there may not have been an explicit aim to encourage violence. Still, adjustments in state policy allowed factions to pursue their interests, indirectly intensifying competition between factions. For example, in April 1967, Mao weakened the authority of the army to prevent the PLA from suppressing the rebellion. While this did not mean he favored rebel groups attacking the PLA, it weakened the army's authority, which at the time controlled violence and maintained social order, resulting in a resurgence of collective violence and social disorder. Immediately after that, the rebels loosened from the control of the PLA and launched a counterattack against the army and army-backed groups.

The same is true of the rhetoric of state agents. State agents may not have explicitly intended to incite violence, but their rhetoric had an indirect inciting effect on many occasions. For example, when Mao Zedong first received the Red Guards at Tiananmen on August 18, 1966, he replied, after asking the name of Song Binbin, who had given her a Red Guard armband, "*yao wu ma*." The Chinese word pronounced "bin" is commonly used in the phrase "*wenzhi binbin*," meaning elegant and courteous. As a result, the media reported Mao's reply, and it spread widely throughout the country. Mao's message may have been to encourage the spirit of rebellion rather than physical violence, but it encouraged the Red Guards to be violent.[57] Between the 18 August meeting and early September, the red five Red Guards-led red terror reached its zenith peak.[58]

Factions had two tendencies in understanding state policy and the rhetoric of state agents. First, they received clear instructions from the state and its agents to promote violence.[59] For example, to criticize military cadres such as Xu Shiyou of the Nanjing Military Region, Han Xianchu of the Fuzhou Military Region, and Chen Xilian of the Shenyang Military Region, Kuai Dafu of Tsinghua University directed the Regiment Faction in Nanjing, Fuzhou, and Shenyang to provoke violent conflict between local factions. Wang Hongwen and Geng Jinzhang also received instructions from Zhang Chunqiao to mobilize workers' rebel groups to suppress the rebel Red Guards who had planned the "January 28th bombarding of Zhang Chunqiao." Still, they were unable to act due to the aborted plan.[60]

Second, violence that results from misunderstanding state policies and the words and actions of state agents or the conscious misinterpretation of state policies and the words and actions of state agents in the pursuit of a group or even individual interests. The mobilized participants in these instances were sometimes not necessarily passive participants. Analysis of the Old Red Guard cases, the *Jinggangshan* Corps at Tsinghua University, and the Workers' General HQ in Shanghai show that mobilized participants may take advantage of the state's opportunities to pursue interests spontaneously. Participants in the Cultural Revolution often interpreted state policies and the statements and actions of state agents based on their interests. For example, Mao's Theory of Class Struggle was interpreted differently depending on the faction's interests. For the red five Red Guards, the class struggle was directed against the old intellectuals and the black five. But for the rebel Red Guards, the class struggle was directed toward the bureaucratic privileged class. Both used Mao's Theory of Class Struggle to pursue the interests of their groups.[61] In addition, during the seizure of power in Shanghai, Zhao Quanguo, the leader of the rebel Red Guards, falsely claimed to have received "four instructions" from Zhou Enlai to seize power from the Shanghai party and government organs.[62] There were many such incidents during the factional struggles of the Cultural Revolution, where misunderstandings and misinterpretations of state policies and the statements and actions of state agents intensified the competition between factions and gave rise to violence.

3 Heretical Thinking and Dropouts

3.1 *Heretical Thinking*

Heterodoxy is another type of radicalization of factional competition. Unlike radical actions such as collective violence, hereticalism is a departure from the orthodox ideology of the Cultural Revolution. It can be seen as the result of the radicalization of factions in the process of competing ideologies of the Cultural Revolution. During the three years of the Cultural Revolution from 1966 to 1969, more than ten types of social ideologies were judged as heretical and criticized by the Central Committee of the CCP. For example, at the beginning of the Cultural Revolution, the Red Guards, mainly cadres' children, put forward the birth bloodline theory, which advocated the idea of privilege. Yu Luoke raised the class origin theory to counteract the birth bloodline theory, which advocates equal rights. At the end of 1966, the conservative Red Guards in Beijing''s universities, who were distrustful of the Central Cultural Revolution Group and Lin Biao, and other authoritative figures of the Cultural Revolution, advocated "doubting everything," the Skeptical Trend. In 1967, around the formation of the Three-In-One Regime, the 4–3 Faction Red Guards in Beijing's high schools demanded a redistribution of property and power and put forward the 4–3 Trend which advocated the theory of changes in class relations, while the 4–14 Faction in Tsinghua University advocated for the idea of unchanged class relations, which was called the "4–14 trend." In the summer of 1967, the May 16 Corps of the Capital, in opposition to Zhou Enlai's eclecticism, put forward the "May 16 Trend." From the end of 1967 to the summer of 1968, the Shengwulian, a group in Changsha, Hunan Province, and the Wuhan-based "Bei, Jue, Yang" group, who were both distrustful of the Three-In-One Regime, proposing the "Shengwulian Trend" and "Bei, Jue, Yang Trend," to pursue a democratic system like the Paris Commune. In 1968, the student rebels, excluded from the Three-In-One Regime organized to oppose Zhang Chunqiao and the Workers' General HQ monopoly on party and government power in Shanghai, proposing the "Shanghai New Trend," which called for a second revolution in Shanghai.[63] Additionally 1974, during the post-Cultural Revolution period, the "Li Yizhe Trend" of Guangzhou, which emphasized democracy and the rule of law, was also widely popularized throughout the country.[64]

The political contexts that contributed to the emergence of heretical ideas differed from one period to another, as did the content of the heretical ideas. However, these heresies generally developed in the same way as collective violence competing for sectarian interests. Only these competing interests occurred on the ideological front.

The heretical thinking during the Cultural Revolution can be broadly divided into two types: that which developed during factional rivalry and that which originated during the conflict between factions and the state.

Factional Competition and Heretical Thinking

Regarding the heretical trends developed during the factional competition, the confrontation mentioned above between the birth bloodline theory and the class origin theory, the 4–3 ideology, and the 4–14 ideology all belong to this category. This heresy mainly reflects the competition between factions, which can be said to have arisen to justify their actions. Specifically, the birth bloodline theory and class origin theory were developed during the first sub-movement of the Cultural Revolution. During this time, the Old Red Guards, mainly cadres' children, and the rebel Red Guards, particularly intellectual and other middle-class children, competed for interests and created them to justify their actions in the ideological sphere. As discussed earlier, before the Cultural Revolution, there was already competition between the cadres' and intellectuals' children for higher education opportunities. During the Cultural Revolution, the cadres' children led the movement. Applying the Theory of Class Struggle put forward by Mao Zedong in the 1950s, they made those outside the red five class the target of their criticism and tried to monopolize the right to participate in the Cultural Revolution, thus putting forward the couplet "a good son is a hero; a counterrevolutionary son is a bastard." The aim was to justify their actions and to institutionalize their privileged political, economic, cultural, and educational status through the birth bloodline theory. The birth bloodline theory was the ultimate expression of the Communist Party's class line and could be said to be no different from the class line policy. Therefore, even if Mao Zedong and other central leaders disapproved of it at the beginning of the Cultural Revolution, they did not take explicit measures to prevent it. However, from September to October 1966, the birth bloodline theory was used by some central and local cadres as a political tool for self-defense, and it became clear that the action of the cadres' children to turn the criticism of outside the party became an obstacle to the advancement of the Cultural Revolution toward its stated goal. In this context, with the reorientation of the Cultural Revolution at the Central Work Conference in mid-October, the birth bloodline theory was judged to be anti-Marxist and anti-Maoist.[65]

On the other hand, amid the birth bloodline theory debate, the class origin theory of Yu Luoke, who advocated equality of political power and educational opportunities, emerged and soon became very influential among young people without privileged status, serving as a theoretical tool in pursuing their interests. The main idea of the birth theory was that all young people are equal. According to Yu Luoke, due to the class line policy implemented during the seventeen years of the state's founding, a new privileged class was formed in China, followed by a new oppressed class. As the class division became fixed and unshakable, class oppression was created. Therefore, using the Marxist doctrine, he criticized the "emphasis on performance" argument in the class policy of the CCP, pointing out that if class origin and class composition determine the attitude toward political participation, the "emphasis on political performance" and "emphasis on class" in the party's policy would have lost its meaning and become "one-sided in terms of class composition." On this basis, he advocated: "All youth are equal before the performance. It is not determined in the mother's womb who is the backbone. We do not recognize any right that is not attained through individual effort."[66]

The theory of birth, which proclaims equality, can be said to be a declaration of human rights in the era of class origin. However, it is essential to note that although he emphasized the equality of rights, he did not oppose class struggle. In *Class Origin Theory* and a series of related articles published afterward, he stated that the principal contradiction during the Cultural Revolution was between the revolutionary masses and "capitalist roaders within the party." In other words, Mao's theory of "red degeneration" was used as a theoretical tool to justify the theory of equality. In this sense, he was using Mao's class theory of the 1960s to criticize the class theory of the 1950s.[67] The *Class Origin Theory* was officially published after October 1966, just after the CCP Central Committee judged the birth bloodline theory as promoting the wrong view of "lopsided class composition."[68] In addition, the article quoted Mao Zedong and Jiang Qing extensively. Considering these circumstances, *Class Origin Theory* reflects the interests of young people who do not enjoy privileged benefits. Despite the political constraints, it cannot be denied that *Class Origin Theory* motivates young people to show their loyalty to the state and its agents.

The 4–3 and 4–14 ideologies were also the product of competition between factional interests, as with the bloodline and class origin theories. The rivalry between factions intensified in 1967 around establishing the Three-in-One Regine. The rebellion split into stable and radical factions in many areas. The 4–3 camp reflected the radicals' ideology, while the 4–14 camp reflected the ideology of a more prudent group. The former pointed out that the essence of the Cultural Revolution was a struggle over the redistribution of property and power and that the power structure of the Communist Party itself fostered corruption. Therefore, they opposed the end of the Cultural Revolution and called for the redistribution of property and power to bring about social revolution and to crush the privileged class.[69] In contrast, the latter demanded that the Cultural Revolution enter a phase of closure and consolidation, arguing that since class relations in China had remained unchanged during the seventeen years before the Cultural Revolution, radical rebels couldn't achieve power.[70]

Although the 4–14 Faction, which advocated the 4–14 trend, was a breakaway from the rebel group, it absorbed many conservative members during the decline of the conservative group who supported the pre-Cultural Revolution bureaucracy and the birth bloodline theory. Thus, shades of the birth bloodline theory could be seen in the 4–14 trend. In this sense, the controversy between the 4–3 Trend and the 4–14 trend was merely a continuation of the class origin theory and the birth bloodline theory.

Conflict Between Factions and the State at the Ideological Level

The second type of heretical thinking developed as factions clashed with state agents at the conceptual level of the Cultural Revolution. Among the heretical trends of the three years of the Cultural Revolution listed at the beginning of this section, the Skeptical Trend, the May 16 Trend, the Shanghai New Trend, the Shengwulian, and the Bei, Jue, Yang Trend belong to this category. This type of heretical thinking is the same as the first type in that it reflects the interest competition between factions and demonstrates the rationality of their behavior. However, compared with the first type, it is more significant because it heavily demonstrates the conflict between factions, state ideology, and agents, especially Mao Zedong's thought on the Cultural Revolution. It expresses the faction's thinking about the Cultural Revolution. Here, the Shengwulian Trend, which significantly impacted the whole country, is a good illustration.

The Shengwulian (formally known as the "Great United Committee of Proletarian Revolutionaries of the Hunan Provincial Assembly") was formed on October 11, 1967 to oppose the Three-In-One Regime. The prominent leader of the faction was Yang Xiguang, a high school student at the time.[71] The faction was formed by more than 20 mass organizations in Hunan Province that had been excluded from the Three-In-One Regime.

The main arguments of the Shengwulian can be summarized as follows. The Three-In-One Regime, dominated by the old bureaucrats, had degenerated into a tool for the bureaucratic privileged class, which had long been defeated, to regain power and was incompatible with the formation of the regime in the style of the Paris Commune through the introduction of full elections, as stipulated in the Sixteen Points. To build a genuinely democratic government like the Paris Commune, the existing state power and the bureaucratic privileged class must be ultimately defeated, and power must be seized by arming the people and rebuilding the party and the army. Therefore, the Cultural Revolution should not be ended at this stage but should be continued.[72]

Song Yongyi points out that the heretical trend was, in fact, a logical extension and development of Mao's democratic promises when he launched the Cultural Revolution.[73] As discussed in previous chapters, the Cultural Revolution was both a political struggle within the central leadership and an experiment by Mao to find a socialist form of democracy in the state. The transformation of the hierarchical and bureaucratic administrative system by the power of the people and the elimination of the cadres' sense of privilege was one of the aims of the Cultural Revolution. Thus, at the beginning of the Cultural Revolution, Mao put forward very radical ideas with democratic overtones. However, because the Cultural Revolution was Mao's search for a new form of state, the goals often changed and were adjusted as the political situation changed. After 1967, especially after the nationwide upheaval of the summer of 1967, it can be said that radical, destructive ideas had declined in Mao's idea of the Cultural Revolution, and the idea of seeking stability and construction became mainstream. However, for the participants, who were enamored with Mao's democratic ideas, who had risen to criticize the bureaucratic privileged class, and who had learned to pursue their interests during the movement, ending the Cultural Revolution by establishing the old bureaucrat-dominated Three-In-One Regime was an unacceptable outcome for them. It was unacceptable not only in terms of political ideas but also in terms of practical interests. Restoring the existing bureaucracy meant a return to the old structure of the distribution of citizenship. For the rebels, this would mean losing all the benefits they had gained during the Cultural Revolution. Moreover, from experience, there was a high risk of reprisals from the cadres who had regained power. Therefore, they clung to the ideas of Mao Zedong at the beginning of the Cultural Revolution and strongly opposed its ending.

In this sense, heretical thinking developed due to the conflict between factions and the state at the ideological level. Similar to the organization formed by factions under the political mobilization of the state, the ideas of factions are influenced by and derive from the views of the state's political mobilization. When the state's political goals coincide with the goals of factional interests, there is an ideological consensus between the faction and the state. However, when the political goals of the state and the interests of the factions diverge, the factions begin to reject the state in its entirety. In advancing the concept of the state based on the interests of the factions, a "heretical trend" emerges that escapes orthodoxy.

3.2 The Emergence of Dropouts

Those who did not join a faction during the Cultural Revolution were called the "dropouts." In previous studies of collective behavior during the Cultural Revolution, the modes of action of this group have been almost entirely ignored. This book argues that the emergence of dropouts, like collective violence and heresy, resulted from the intensification of factional competition. In other words, the dropout falls within the category of "exit" discussed in Chapter 2 and is a manifestation of radicalization. The result of the radicalization of competing interests between factions has led one part to try to resolve the conflict with violence and the other part to try to use ideological means to argue for the justification of actions and to push the movement in a direction that is in its interest. In contrast to these two modes of action, the dropout has taken the opposite course of action of withdrawing from the competition. The proliferation of dropouts made them visible during the Cultural Revolution and played a significant role in the state's control of the mass movement. Because of the substantial withdrawal of members within the mass organizations, only a few radical members of the movement remained. The decline of the mass movement made it possible for the state to control the extreme behavior of the minority by sending the PLA Maoist and the workers' propaganda teams, even before the system of social control had been restored. The same methods would have likely been entirely unworkable in late 1966 and before the summer of 1967 when the movement was at its peak. During the Cultural Revolution period, dropouts can be classified into three types: those who were unqualified to join a faction, those who gave up competition, and those who withdrew.

The first dropout category refers to those not eligible to join the faction and consists mainly of people from the black five class and their families. During the Cultural Revolution, people from this class were defined as one of the targets of criticism and struggle. They belonged to a group that was deprived of political rights to participate. In other words, these people only had the right to be fought but not the right to fight others during the Cultural Revolution. At the policy level, children of the black five had the right to participate in politics. Still, neither conservative nor rebel organizations were willing to accept them into their organizations. Since the end of 1966, as the Cultural Revolution was underway, some children of the black five class joined the Red Guards, but only a few. According to a survey by Stanley Rosen in Guangzhou, of the fifty classes surveyed, seventeen percent of the children of rich peasants had joined the faction, and only seven percent of the children of bad elements had joined.[74]

The second type of dropout comprises people eligible to join the movement but do not care about political movements or who, after weighing the pros and cons, feel that it is better to refrain from joining a faction for various reasons. For example, compared to the student and worker groups, the intellectuals were not considered active participants in the movement.[75] One reason for this was that many intellectuals were persecuted as "reactionary academic authorities" during the early years of the Cultural Revolution. It also has to do with the experience of the anti-rightist movement and the political status of intellectuals who were constantly forced to undergo ideological reform during the seventeen years before the Cultural Revolution. This experience taught intellectuals that the safest way to protect themselves was not to speak or act. It can be assumed that this understanding has also affected their children. According to Stanley Rosen's survey, sixty-seven percent of the children of senior intellectuals, such as researchers and professors, joined factions in the fifty classes surveyed, much lower than the ninety percent of senior military cadres' children and the ninety-five percent of senior party and government cadres' children.[76]

The third type of dropout refers mainly to those who leave a faction because of a change in the direction of the movement. This includes two types: active and passive dropouts. The former refers to people who choose to leave a political movement of their own accord, either because of the faction's radicalization, disintegration, or the loss of identification with the faction's or the political philosophy of the state. The latter, in contrast, are those who have chosen to leave after the disintegration of the organization and the rejection of the opposing organization. It can be assumed that faction leaders and radical participants predominate among the passive dropouts.

Of these three types, the unqualified dropouts and the renouncers of competition for power, two of which had existed since the initial stages of the Cultural Revolution, increased or decreased not much in number until the end of the Cultural Revolution. In contrast, the third type of dropouts emerged during the Cultural Revolution due to factional competition and changes in the relationship between the state and factions. Like the conservative and rebel groups, the third type of dropout is itinerant, with dramatic fluctuation in numbers. As this significantly impacted the rise and fall of the Cultural Revolution, it is briefly explained below.

During the three sub-movements of the Cultural Revolution, as the movement declined, the dropouts emerged due to changes in the direction of the movement, and former dropouts became participants in the movement. The children of cadres were the first to rise up in rebellion during the Cultural Revolution, and they were also the first to leave the movement. In October 1966, Mao Zedong and the Central Cultural Revolution Group withdrew their support for the Red Guards of cadres' children in secondary schools (the Old Red Guards) and the conservative Red Guards in universities, which led to their collapse. With the change in the direction of support from the central leadership, an essential political resource, most of the Old Red Guards and conservative Red Guard members shifted to rebel groups. At the same time, some members of the Old Red Guards and most of the leaders of the conservative Red Guards chose to leave the movement.[77] Some of those who left formed the *Liandong* in late 1966 to rejoin the movement, but their actions eventually ended when the core members were arrested. Although the arrested *Liandong* members were released in April 1967, by this time, they had lost their right to participate in politics, as the rebel groups immediately criticized their actions as attempts to restore their bloodline, and most of them had to remain as dropouts.[78]

On the other hand, an analysis of the examples of factionalization at Tsinghua University shows that, in contrast to the original participants who were transformed into dropouts after October 1966, those former dropouts who had previously been excluded from the Red Guards because they were not from the red five began to join the rebel group and start to participate in the movement. However, as the movement developed, the leaders of the factions, especially those of the rebel factions, mobilized and manipulated faction members to fight for power, resulting in an escalation of conflict between factions. Against this backdrop, the ranks of dropouts expanded overwhelmingly after the summer of 1967, as armed fighting became widespread throughout the country and participants in the Cultural Revolution withdrew in large numbers. Dropouts also emerged within workers' organizations for the same reason. However, the collapse of the workers' conservative organizations occurred mainly in early 1967, so the transformation of workers' conservative members into dropouts occurred much later than in the case of the Old Red Guards and conservative Red Guard organizations among the students. Despite these differences, the emergence of a large number of dropouts among workers, both in the conservative and rebel camps, took place after the summer of 1967, as did the student movement.[79]

As discussed in Chapter 2, exit here is not an exit from the public sphere but rather a shift from the more politically risky areas of the public sphere to the less risky ones, which also implies an exit from the patron–client relationship provided by a state agent. Since the private sphere did not exist in China during the Cultural Revolution, the public sphere can be divided into several sub-domains, with areas of greater and lesser danger depending on the political risk. State policy and state agents determine the degree of political risk at the state policy decision-making level. The state decided the legitimacy of factions in the Cultural Revolution. When a faction had the explicit support of the state and even of the authorities of the Cultural Revolution period, such as Mao Zedong and the Central Cultural Revolution Group, the faction's behavior was deemed revolutionary. In such a situation, joining a faction is a safe option for the participants, although they may not be able to gain as much as the faction leaders. However, the political costs and risks of confronting a faction supported by political authority are high for the opposing faction. For example, the actions of Kuai Dafu and others against the Work Team and the red five Red Guards during the early years of the Cultural Revolution were politically risky. On the other hand, when the support of political authority, an essential political resource, shifts from one faction to another, that is, when authority withdraws its support from one faction and gives it to another or when the political authority itself falls in the political struggle, the political risk changes, and previously safe areas may suddenly become extremely dangerous. For example, when the bloodline theory was criticized, the actions of *Liandong* members were accompanied by a high level of political risk. Conversely, the political stakes for those rebel groups that oppose the bloodline theory begin to diminish sharply. In this sense, in the summer of 1967, when the state had explicitly ordered the cessation of armed conflict to stabilize the social order, withdrawing from factional conflict was the safest option to carry out the state's instructions. But even being an at-large faction was not the same as withdrawing from the public sphere, and the required citizenship could still only be obtained by the allocation of the state. That is, there is no private space where their rights can be guaranteed to accommodate the movement's dropouts. Therefore, even if they withdraw from factional conflicts, they cannot use legal means independent of the party to defend their rights. Therefore, in the subsequent series of political movements, such as the hunt for May 16 elements movement and the cleansing of the classes, they, like the participants in the radical violence and the advocates of the heretical trend, once they had withdrawn, could not use legal means to defend their rights. Once they were defined as the target of the purge, they were left to their own devices and had no way out. Thus, the exit of dropouts was not an exit from the public sphere. In this sense, it can also be an argument from the opposite side that in China, before Reform and Opening-Up, there did not exist any private sphere where the political and economic rights of citizens were guaranteed by legal means. This was true even of the then State President Liu Shaoqi, who, when disagreements with Mao Zedong over his philosophy of governance eventually escalated into a power struggle, did not have the option of resigning or the option of using legal means to protect himself, and was ultimately buried in the Cultural Revolution.

4 The Decline of Factional Conflict

The above analyses of the impact of intra-state conflict, inter- and intra-factional competing interests, and state–society interactions on factional conflict have been presented. These analyses show that factional conflict and collective violence are closely linked to inter- and intra-faction competition for interests and interaction between factions and the state. The radicalization of factional rivalries had produced collective violence, heretical thinking, and many dropouts. Likewise, the decline of factional conflict has been linked to both factions and the state. The impact of factional competition on the drop of factional conflict, as discussed in the analysis above, is briefly summarized as the creation of collective violence and the derailment of the ideological dimension that allowed the state to suppress mass movements. In addition, the increase in dropouts has created favorable preconditions for the state to harness the behavior of radical minorities. This section focuses on the role of the state in controlling factional conflict.

As the analysis in the previous chapters shows, the three years from 1966 to 1969 were marked by factional conflicts. The sub-movements produced different types of collective violence and heretical ideas of varying content. These factional conflicts, particularly the collective violence, were linked to the state. In other words, the state and its agents tolerated and sometimes directly encouraged collective violence to achieve their specific political goals. But it is important to note here that the creation of collective violence was not the goal of the Cultural Revolution. More specifically, the Cultural Revolution represented an experimental phase where Mao Zedong explored various forms of democratic socialism. Throughout the process, Mao Zedong did not have specific preconceived plans but rather continuously sought methods to express his ideas or correspond to reality through the practice of the movement. In this process, various types of collective violence emerged due to the entanglement of complex factors, including political struggles within the state and competition among factions, exerting a significant impact on the political, economic, and social order of that time. Furthermore, collective violence also challenged the state's control over factions.

Even during the Cultural Revolution, the Central Government would immediately control it whenever there was collective violence. But at that time, because of the low level of institutionalization of both the state and society, the masses who rose in rebellion under political mobilization were not so easy to control. As shown in the previous section on the analysis of collective violence, the state took measures to stop it each time it occurred, but rarely to any effect. As one type of violent behavior was controlled, new forms of violence would emerge. The ultimate containment of collective violence occurred toward the end of 1968.

Why, then, was the state able to finally control the collective violence at this point? Specifically, the ultimate control of the collective violence was related to issuing the "3 July Announcement." On 3 July, to deal with the armed struggle in Guangxi, the CCP Central Committee, the State Council, the Central Military Commission, and the Central Cultural

Revolution Group jointly issued the 3 July Announcement on the immediate cessation of armed struggle. After that, the armed struggle gradually stopped.[80] Earlier, the CCP Central Committee issued several notices to stop armed fighting in 1967. For example, in June, it issued the 6–6 Circular banning armed fighting; on 13 July, it issued the Proclamation on the prohibition of provoking the peasants to enter the city for armed struggle; on 5 September of the same year, it issued the Circular on Prohibiting the Seizure of Weapons, Equipment, and Various Military Supplies of the People's Liberation Army.[81] But all these notices and orders did nothing to stop the armed conflict. Why was the collective violence brought under control after the issuance of the July 3rd Announcement? The author believes that there are three main reasons for this.

First is the increase in the number of dropouts, a social movement whose decline has created favorable preconditions for the state to control the radical behavior of a minority.

Second, the state's policy toward the Cultural Revolution had changed. From late 1967 onwards, the state had positioned the Cultural Revolution as having entered its final phase. During this stage, the reconstruction of state power, the stabilization of social order, and the opposition to sectarianism became the most important policy objectives of the state. While power struggles and policy differences among state agents persisted at the decision-making level, it can be argued that key state agents such as Mao Zedong, Zhou Enlai, and Lin Biao unified their understanding of stabilizing social order and opposing violent behavior. Against the backdrop of controlling violent behavior becoming the paramount policy objective of the state, the situation where state agents tolerated or even rewarded collective violence for achieving certain political goals, as seen in the early stages of the Cultural Revolution, no longer existed.

Third, in addition to issuing notices ordering the cessation of collective violence, the state took more robust measures to stop sectarian conflicts than at the beginning of the Cultural Revolution. The state's methods of controlling factional disputes were repression, reorganization, and the Send to the Countryside Movement. Beginning in 1968, as part of restoring order, a nationwide campaign was launched to criticize sectarianism and anarchism. During the campaign, many factions were disbanded under the name of reorganization.[82] Additionally, in the summer of 1968, a campaign was launched to reorganize the class ranks. To promote this campaign, PLA Maoist propaganda teams were sent to units where factional struggles were fierce. Under the leadership of the PLA Maoist propaganda team, some leaders and radical members of rebel organizations were suppressed as bad elements inciting sectarianism and factional struggle. During this period, most organizations advocating heretical thinking, such as the *Shengwulian*, were banned as counter-revolutionary organizations, and many of their leaders were arrested. In China, all aspects of workers' lives were constrained within their units, and their actions were restricted accordingly. With the re-establishment of power in the work units, the restoration of the function of the party, and the class reorganization campaign, collective violence among the various workers' rebel groups was brought under control. The situation of the students was slightly different from that of the workers. Since the students were not as restricted by their work units as the workers were, they were more challenging to control regarding speech and behavior than the workers' groups. Two special measures were taken. The permanent dispatch of Maoist propaganda teams brought the whole country, from universities to secondary schools, under the management of Maoist propaganda teams, mainly soldiers and workers, with the result that the Red Guard movement, which was based in schools, was wholly deprived of its base of activity. In addition to the purpose of controlling the Red Guards, the practice of sending youth to the countryside had the dual role of sending young people who had received school education to the countryside to help modernize agriculture and alleviate the burden of employment in the cities caused by the slow growth of industry and the economic downturn.[83] In 1968, some 60,000 to 70,000 young people left for the countryside after finishing their studies. Under these measures, the Red Guard movement quickly declined, and order was restored in the cities. By early 1969, the social movement, which had lasted for three years under solid state control, had declined entirely.[84] Subsequently, the Cultural Revolution entered a period dominated by organized political mobilization by party and government organs. During this period, political struggles within the state and competing factional interests continued, but no large-scale social unrest occurred as the state tightened its control over society.

Notes

1. As analyzed in Chapters 6 and 7 of this book, the "3rd HQ" intervened in the attack on the Central Cultural Revolution Group by

the conservative Red Guards to express their loyalty to the Central Cultural Revolution Group and defeat the rival organizations in a revolutionary competition. On the contrary, in the "Kangping Road" incident, the "Workers' General HQ" intervened in the Scarlet Guards' attack on the Municipal Committee to prevent the fruits of the mass organizations' criticism of the "capitalist road in the Party" from being monopolized by the Scarlet Guards. Therefore, Workers' General HQ used the opportunity to defeat the opposing faction of the Scarlet Guards.

2. During this period, armed fighting was fought between the military and rebel groups in some areas. This was the case, for example, in the Shihezi incident in the Xinjiang Uyghur Autonomous Region (Wang Nianyi, op. cit., p. 202). This incident started with an attack on the PLA by a rebel group, and the PLA then opened fire on the rebel group. The shooting of mass organizations by the PLA during the Cultural Revolution was undoubtedly closely related to the internal conflict within the state. Still, this book views it primarily as state violence exercised by the state to maintain social order, not as collective violence, and therefore does not discuss such violence.
3. Wang Shaoguang, op. cit., pp. 101–108.
4. Xu Youyu, *The Freedom of Speech* (*Ziyou de yanshuo)* (Changchun Press, 1999), p. 52.
5. Issue No. 65 (August 26, 1967) of *The Dynamic Newspaper (Dongtaibao)*, edited by the Beijing Correspondence Institute's Eastern Red Commune (Beijing hanshou xueyuan dongfanghong gongshe), carried the article "Mao's determination: give guns to the Yanbian rebels" ("Mao Zedong de juexin: gei yanbian zaofanpai faqiang"). The main points of this story are as follows. According to Comrade Wang Li's speech to the Xinhua News Agency on August 15, the military divisions in the Yanbian area of Jilin Province have rallied the local "capitalist party power-holders" to support the conservative organizations and suppress the rebel groups. The conservative organizations burned down even the post and telecommunications building. After this incident, Mao Zedong issued more than 1000 firearms to the rebel groups. This was the first instance in the country of giving guns to the rebels.
6. "Chengdu Urgent Telegram" ("Chengdu jidian"), The Propaganda Group of the Red Rebel Headquarters of the Ministry

of Electricity and Water (Shuidianbu dianyansuo hongse zaofan zongbu xuanchuanzhu), *Dynamic Information* (*Dongtai ziliao*), no. 85 (1967, May 9).

7. "After the May 6 Incident in Chengdu" (Chengdu 5–6 shijian yihou), op. cit., *Dynamic Information*, no. 85.
8. "Yibin expedited telegram" ("Chengdu jiaji dianbao"), op. cit., *Dynamic Information*, no. 93 (1967, May).
9. Wen Yushan, "The Rise and Fall of the Red Flag in Guangzhou" ("Guangzhou hongqipai de xingwang"), edited by Liu Guokai, *The History That Could Not Be Sealed* (*Fengsha buliao de lishi*), published by the Editorial Department of the Reevaluation of the Cultural Revolution Series (Chongxin pingjia wenhuadageming congshu bianjibu), 1996, p. 93.
10. Wang Xiang, "Experiences and Lessons of the Great Proletarian Cultural Revolution" ("Wuchanjieji wenhuadageming de jingyan he jiaoxun"), edited by Liu Guokai, op. cit., p. 222.
11. Wang Shaoguang, op. cit. pp. 116–129.
12. Ibid., p. 127.
13. Ibid., pp. 131, 146.
14. Wang Nianyi, *The Age of Great Unrest*, op. cit., pp. 298, 299.
15. Wang Shaoguang, op. cit., p. 206; Wen Yushan, op. cit., p. 161.
16. Wang Nianyi, *The Age of Great Unrest*, op. cit., p. 299.
17. Wen Yushan, op. cit., pp. 169–172.
18. Ibid., p. 50.
19. Chen Xiaowen, "Sketch of the Red Guard Cemetery in Chongqing", ("Chongqing hongweibing mudi sumiao"), edited by Liu Qingfeng, op. cit., pp. 163, 173–175.
20. Zhou Ming, *History Meditates Here* (*Lishi zai zheli chensi*), vol. 5 (Huaxia Press, 1986), p. 73.
21. Wang Nianyi, op. cit., pp. 69–70.
22. *The Chronology of Events of the Cultural Revolution in Guangxi* (*Guangxi wenhuadageming dashi nianbiao*), edited by The Editorial Team of the Guangxi Cultural Revolution Chronology (Guangxi wenhuadageming dashi nianbiao bianxiezu), op. cit., pp. 53, 111.
23. Wen Yu, *China's Left Scourge*, op. cit., p. 444.
24. On the violence inflicted by students on teachers, see Wang Youqin, "1966: The Revolution of Students Beating Teachers" ("1966:

xuesheng da laoshi de gemin"), collected in Liu Qingfeng, ed. op. cit., pp. 17–36.

25. Ibid., p. 18.
26. Wang Nianyi, op. cit., p. 86.
27. The former collection of materials, "*Cultural Revolution*" *research materials* (part 1), p. 353.
28. *Hebei Daily News*, 14 December 1978.
29. Xu Ben, "On Fear and Violence in the Political Culture of the Cultural Revolution" ("Lun wenge zhengzhi wenhua zhong de kongju he baoli"), edited by Luo Jinyi and others, op. cit., p. 63.
30. Yan Jiaqi, Gao Gao, op. cit., p. 47.
31. Wang Nianyi, op. cit., p. 48.
32. Ibid., p. 69.
33. Ibid.
34. *"Cultural Revolution" Research Materials* (part 1), op. cit., pp. 90–91.
35. Wang Nianyi, op. cit., p. 74.
36. "Comrade Chen Yi's Speech of August 30" ("Chen Yi tongzhi 8 yue 30 ri jianghua), *Compilation of Materials on the Great Proletarian Cultural Revolution*, op. cit., p. 147.
37. For a study of the Wuhan incident, see, Wang Shaoguang, *Failure of Charisma: The Cultural Revolution in Wuhan*, op. cit.
38. "Important Speech by Comrade Jiang Qing ("Jiang Qing tongzhi zhongyao jianghua"), (September 5, 1967)", compiled by Xinhua Commune (Xinhua gongshe), The Red Flag Combat Team (Hongqi zhandoudui), p. 1.
39. When Zhang Chunqiao received representatives of the Shandong Province rebel organization, he gave the following instructions: "We do not approve of armed struggle, but if the other side (referring to the conservative organizations, author's note) starts it first, we will respond with armed struggle." ("Remarks by the Central Leaders in Receiving the Shandong Delegation ("Zhongyang lingdaoren jiejian Shandong daibiaotuan de jianghua" (April 27, 1967)" Reprinted by The Propaganda Group of Beihang's Red Flag (Beihang hongqi xuanchuanzu).
40. The Capital Red Guard Pickets Xicheng Branch Command (Shoudu hongweibing jiuchadui Xicheng fendui zhihuibu), "Circular Order No. 6" ("Di 6 hao tongling") (September 9, 1966), "Circular Order No. 9" ("Di 9 hao tongling") (September 14,

1966), and "Circular Order No. 10 ("Di 10 hao tongling") (September 14, 1966), printed by the Capital Revolutionary Yong Hong War School (Shoudu yonghong zhanxiao, formerly Beijing Seventy-seventh High School), September 1966.

41. Jiang Pei, *The Red Guard Storm* (*Hongweibing kuangbiao*) (Henan People's Publishing House, 1994), p. 77.
42. "Speech by Comrades Jiang Qing, Chen Boda, Zhou Enlai, and Kang Sheng at the Pledging Conference to Criticize the Bourgeois Reactionary Line in Beijing High School on December 16" (Jiang Qing, Chen Boda, Zhou Enlai, Kang Sheng tongzhi 12 yue 16 ri zai Beijing zhongxue pipan zichan jieji fandong luxian shishi dahuishang de jianghua), *Compilation of Materials on the Great Proletarian Cultural Revolution*, op. cit., p. 569.
43. Kazuko Mori, *Modern Chinese Politics*, op. cit., p. 188.
44. Li Xun, *The Great Collapse—A History of the Rise and Fall of the Shanghai Workers' Rebellion*, op. cit., pp. 371–372.
45. Ibid., p. 372.
46. Jiang Pei, op. cit., p. 129.
47. William Hinton, op. cit., p.169.
48. Ibid., p. 168.
49. Wang Shaoguang, op. cit., pp. 117–121 (Chinese version).
50. Ibid., p. 119.
51. Gong Xiaoxia, "The Origin and Development of Mass Violence in Culture Revolution", *Hong Kong Journal of Social Sciences*, no. 7 (1996), p. 108.
52. Mitsuyuli Kagami, *The Chinese Revolution as a Paradox*, op. cit., p. 22.
53. Wang Youqin, op. cit., 1966: "The Revolution in which Students Beat Their Teachers", p. 31.
54. Zhong Weiguang, op. cit., p. 13.
55. Wang Youqin, "Between Hitting a Teacher and Hitting a Classmate" ("Da laoshi he da tongxue zhijian"), in Liu Qingfeng, ed., op. cit., p. 47.
56. Jiang Pei, op. cit., p. 129.
57. Wang Youqin, op. cit., "1966: The Revolution of Students Beating Teachers", p. 29.
58. Wang Youqin, op. cit., "Between Hitting a Teacher and Hitting a Classmate," p. 47.
59. William Hinton, op. cit., pp. 162–168.

60. Li Xun, op. cit., *The Great Collapse—A History of the Rise and Fall of the Shanghai Workers' Rebellion*, p. 349.
61. Lijun Yang, "The Social Origins of the Red Guard Movement", op. cit., pp. 209–221.
62. Li Xun, *The Great Collapse—A History of the Rise and Fall of the Shanghai Workers' Rebellion*, op. cit., p. 327.
63. This book defines the three years from May 1966 (the publication of the "May 16 Notice") to April 69 (the convening of the 9th National Congress of the CCP) as the "Cultural Revolution" period, and the period from April 69 to October 76 (the death of Mao Zedong and the collapse of the Gang of Four) as the "Post Cultural Revolution" period.
64. For an explanation of the specific content of heretical thinking during the Cultural Revolution, see Song Yongyi and Sun Dajin, *The Cultural Revolution and Its Heretical Thinking*, op. cit.
65. Chen Boda, "The Two Lines in the Great Proletarian Cultural Revolution" ("Wuchanjieji wenhuadageming zhong de liangtiao luxian") (Speech at the Central Working Conference on October 16, 1966) (1966nian 10 yue 16 ri zai Zhongyang gongzuo huiyi shang de jianghua), *"Cultural Revolution" Research Materials* (part 1), op. cit., pp. 139–140.
66. The Beijing Research Group on Family Issues (Beijing jiating wenti yanjiu xiaozu), "On the Origin" ("Chushen lun"), the inaugural issue of the *Journal of Middle School Cultural Revolution* (*Zhongxue wenge bao*), The Capital Middle School Revolutionary Command (Shoudu zhongxuesheng geming zaofan silingbu), January 18, 1967.
67. The Beijing Research Group on Family Issues, "What Does the Liangdong's riots tell us?" ("Liandong de saoluan shuoming le shenme?"), the third issue of the *Journal of Middle School Cultural Revolution*, op. cit.
68. Chen Boda, "The Two Lines in the Great Proletarian Cultural Revolution (Speech at the Central Working Conference, October 16, 1966)", op. cit., p.140.
69. The Eastern Red Resistance Electric School (Dongfanghong kangda dianxiao), *April 3rd War Journal* (*4-3zhanba*), no. 1 (1967, June 11).
70. Song Yongyi and Sun Dajin, op. cit., pp. 365–371.
71. Ibid., p. 267.

72. The main ideas of "*Shengwulian*" refer to the following articles. Yang Xiguang, "Where is China Going?" (January 6, 1968) (Zhongguo xiang hechu qu?); "Report on the Inspection of the Intellectual Youth Movement in Changsha" (November 1967) (Changsha zhishi qingnian yundong kaocha baogao); "Proposal for the Establishment of a Maoist Group" (October 1967) (Guanyu jianli maozedong zhuyi xiaozu de jianyi); Zhang Yugang, "Our Programme" (March 1968) (Women de gangling); and "Decision of the *Shengwulian* on Certain Issues in the Current Hunan Proletarian Cultural Revolution" (December 1967) (Shengwulian guanyu muqian hunan wuchanjieji wenhuadagemingzhong de ruogan wenti de jueding), in Song Yongyi and Sun Dajin, op. cit., pp. 274–333.
73. Song Yongyi and Sun Dajin, op. cit., p. 28.
74. Stanley Rosen, op. cit., 1982, pp. 148–149.
75. For the actions of intellectuals during the Cultural Revolution, see the following studies. Yin Hongbiao, "Intellectuals and the Mass Rebellion" (Zhishifenzi yu qunzhong zaofan yundong, Luo Jinyi et al., eds, *Beyond the Catastrophe*, op. cit., pp. 310–335.
76. Stanley Rosen, op. cit., 1982, pp. 148–149.
77. Mi Hedu, *The Generation of Red Guards*, op. cit., p. 234.
78. Ibid., p. 233.
79. Wang Shaoguang, op. cit., pp. 163–164 (Chinese version).
80. "Notice of the CCP Central Committee, the State Council, the Central Military Commission, and the Central Cultural Revolution Group" (Zhonggongzhongyang, guowuyuan, Zhongyangjunwei, Zhongyangwenge bugao) (July 3, 1968), *"Cultural Revolution" Research Materials* (part 2), op. cit., p. 138.
81. *"Cultural Revolution" Research Materials* (part 1), op. cit., pp. 492, 500, 561.
82. Wang Shaoguang, op. cit., p. 204.
83. Uwe Richter, op. cit., p. 107.
84. Ibid., p. 106.

CHAPTER 10

Collective Violence and Social Order

In the previous chapter, the author analyzed the phenomenon of violence in collective behavior during the Cultural Revolution from the perspective of the relationship between the state and society. Types of collective violence, its causes, and consequences were discussed. The analysis in the preceding chapters focused on exploring the macro-level institutional elements that contributed to collective violence during the Cultural Revolution. Building upon the previous discussion, this chapter addresses the following important questions. First, what are the similarities and differences between collective violence during the Cultural Revolution and the collective violence that arose during the various political movements that preceded it (e.g., the violence inflicted on landlords during land reform, etc.)? Second, as the previous chapters have shown, four types of collective violence during the Cultural Revolution manifested differently at different times. What, then, contributed to the typological shift in collective violence? Third, despite the widespread collective violence during the Cultural Revolution, which caused far more chaos than expected, why did the social order not collapse completely? What was it that maintained social order in the Cultural Revolution? Was there a unique method of maintaining order in the Cultural Revolution? What exactly is the correlation between collective violence and social order?

Y. Lijun, *Contest for Citizenship and Collective Violence during China's Cultural Revolution*, IPP Studies in the Frontiers of China's Public Policy, https://doi.org/10.1007/978-981-97-0906-9_10

To answer this set of questions, this chapter first discusses the methods of maintaining social order before and during the Cultural Revolution and, on this basis, further analyzes the correlation between these methods of maintaining social order and violence.

1 "Movement Order" and Collective Violence

This book contends that the collective violence during the Cultural Revolution resulted from a clash between the pre-existing social order and the idealized social order that Mao Zedong sought to establish through the Cultural Revolution. As discussed in previous chapters, Mao Zedong had multiple reasons for initiating the Cultural Revolution: to dismantle the old bureaucratic system and construct a new social order. In this sense, the Cultural Revolution can be seen as an experimental process for Mao Zedong to establish a new social order. This book defines the social order maintained during the Cultural Revolution in its state of constant movement as the "movement order." Given that large-scale and sustained social movements represent an abnormal state, the social order that persists under such abnormal conditions is also an abnormal social order. Typically, social order is understood as the relationships between various social roles, the methods employed to sustain these relationships, and the legitimacy from which these methods derive widespread acceptance. Therefore, understanding the "movement order" during the Cultural Revolution entails investigating whether and how the relationships between roles changed under the state of constant movement, how the methods of sustaining these relationships differed from before, and whether there were changes in the sources of legitimacy upon which these methods relied. We can only comprehend the reasons behind maintaining the "movement order" during the Cultural Revolution by deconstructing these three questions. On the other hand, the "movement order" here embodies the unique, abnormal "social order" that existed during the Cultural Revolution and served as a bridge between the pre-existing order and the idealized social order under construction. Its actual manifestation lies in the tension between these two orders. Therefore, to grasp the actual state of the "movement order" and the factional behavior that occurred within this state, it is necessary to explore the methods of maintaining the pre-existing social order and the idealized new social order that Mao Zedong attempted to establish before the Cultural Revolution, as well as the interactions among the key actors and their impact on the direction of the movement during the process of transforming the existing social order into the idealized new social order. In other words, to understand the collective violence during the Cultural Revolution, it is essential to analyze the roles and interactions among various system components that sought to maintain or change the existing order in response to the transformation of the social order.

Providing social order is the most crucial function of modern states, and China is no exception. In pre-Cultural Revolution China, social order consisted of four components: political leaders at the top of the political power pyramid, state institutions in the middle (including party and government officials), the general population as the social foundation, and the class enemies targeted for rectification. Within this structure, the integrated state institutions of the party, government, and military played an extremely important role in safeguarding social order. State policies, which embodied the guiding principles of the political leaders, were transmitted from the central level to the grassroots through various levels of party and government institutions. The implementation of policies, along with social feedback, was then reported back to the central level through these institutions. From the establishment of the People's Republic of China until the Cultural Revolution, this period witnessed the gradual establishment of state institutions, during which these institutions played a dominant role in strengthening the political foundation, achieving social integration, pursuing economic development, and creating social order. However, as the state institutions gradually became more established, issues such as administrative rigidity, institutional expansion, and the privileging of party and government officials began to emerge. In particular, the recovery period of economic construction after the Great Leap Forward in the early 1960s exacerbated these problems. Administrative rigidity and institutional expansion within bureaucratic institutions were common problems faced by modern states, especially in socialist countries under a planned economy where the state monopolized all economic activities. Consequently, bureaucratic institutions became more administratively rigid and expanded, while party and government officials enjoyed a privileged status. However, it is important to note that the pre-Cultural Revolution bureaucratic institutions completely deviated from Mao Zedong's conception of the relationship between the state and society in his founding ideology.

So, what kind of state–society relationship did Mao Zedong originally hope to build through the Cultural Revolution? In simple terms,

Mao's idea was to establish a social order centered around "equality" and "justice," which, of course, had class implications. Furthermore, bureaucratic institutions were unimportant in his ideal conception of the state–society relationship. As Satoshi Amako states, "For Mao Zedong, the state was at most a means and a tool." The society he sought to build was a "people's commune-style political power that integrated workers, peasants, merchants, intellectuals, and soldiers, as well as combining the township (government) and the commune (cooperative)."[1] Constructing such a society was Mao Zedong's consistent policy since the founding of the nation. The decentralization of power from the central to local levels in the late 1950s was also aimed at correcting the excessive expansion of bureaucratic institutions and the privileging of party and government officials, thus serving as a political experiment to establish Mao Zedong's ideal state–society relationship. During the Cultural Revolution, Mao Zedong instructed the reform of bureaucratic institutions to achieve socialization and streamlining. For example, the *Resolution on the Great Proletarian Cultural Revolution* adopted at the 11th Plenum of the 8th Central Committee of the CCP in August 1966 (known as the "16 Points") stated, "Cultural Revolution groups, revolutionary committees, and representative assemblies are the best new organizational forms for the masses to educate themselves under the leadership of the Communist Party. They are the best bridges connecting our Party with the masses and the power organs of the proletarian cultural revolution."[2] Mao Zedong regarded the new type of organization that combined power institutions and mass organizations, cadres, and the masses, as a communal type of power institution.[3] In this sense, the Three-In-One Regime of the Cultural Revolution (a regime combining revolutionary cadres, soldiers, and masses) was developed from this idea.

In short, Mao believed that the fundamental aspect of transforming the bureaucracy lay in bridging it with the masses. He gave the power to change the bureaucracy to those outside the power structure. In this case, as the state institutions that originally safeguarded and maintained social order were set up as the objects of transformation in the movement, the result was a blurring of the functions of the state institutions and confusion in policy implementation. In addition, the mass mobilization approach, which left out the party and government institutions, caused the relationship between the supreme leader Mao Zedong, the state institutions at all levels from central to the local level, and members of society in general not only to break the original top-down linear state structure but also to complicate and put the relationship between the three in a constantly changing and unstable dynamic.[4]

Specifically, as shown in the Fig 1, the relationship between the three changed from a top-down vertical leadership relationship before the Cultural Revolution to a triangular relationship, as shown on the right. In this diagram, the members of society as the ruled and the state organs as the ruling body were under Mao's leadership. In other words, Mao abandoned his original practice of relying on the bureaucracy as he advanced his national goals and mobilized the masses, establishing a direct authority relationship with them. Although the original vertical system of authority still existed and functioned, its power had been significantly weakened. And as the Cultural Revolution progressed, the interests of several elements, including party and government agencies, the military, and mass organizations, waxed and waned and merged, causing the authority of the vertical system to change. This change also affected the relationship between the various actors within the movement order. Specifically, on the one hand, Mao gave the members of society the authority to remove the "capitalist factions in power" from the state apparatus. On the other hand, to restrain the measures of the factions and prevent the mass movement from deviating from its intended direction, he again strengthened the authority of the party, government, and PLA organs at all levels. This meant that the authority to construct social order during the Cultural Revolution was fluidly distributed between two groups, the mass organizations and state agencies, including the party, government, and PLA. At the same time, competition around political resources, among and between mass organizations, party, and government organs, was also affected by these changes in authority, sometimes intensifying and sometimes temporarily pausing. The movement order never collapsed during the Cultural Revolution partly due to the unquestioned supreme authority of Mao Zedong, the legitimacy of power, and the ideology he led and controlled. It was also because the authority to lead the movement changed and constantly shifted between the party, government, military, and mass organizations. This prevented any part of the social order from obtaining long-term legitimate dominance. Of course, the adjustment of authority also gave rise to new struggles for obtaining political resources, leading to new acts of violence. In this sense, it can be said that the movement order produced violence, and in turn, violence maintained the movement order (Fig. 1).

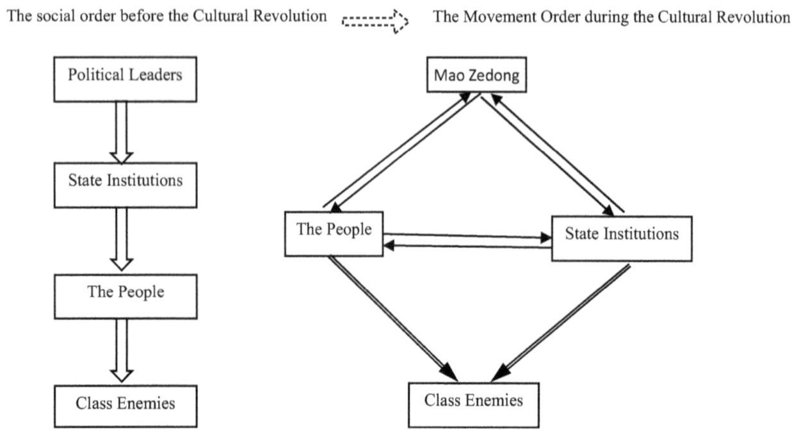

Fig. 1 Diagram of the changing roles of actors (*Note* The three groups from the People's level and above seen in the diagram were given the right to participate in the Cultural Revolution, while the Class Enemy below them had only the right to be criticized and was not allowed to "rebel." This book is limited to discussing classes other than class enemies)

In summary, the interactions among Mao Zedong, state institutions, and social members determined the form of the movement order during the Cultural Revolution. With the changing stages of the Cultural Revolution, the interaction dynamics among the key players involved in the movement also continuously evolved without a lasting balance. As a result, the movement order underwent constant changes based on the relationships among the three parties. Moreover, this change in the movement order also dictated how the main actors of the movement behaved and interacted with each other.

The interaction among the top leader, state institutions, and social members determined the nature of the movement order during the Cultural Revolution. Therefore, in analyzing the relationship between collective violence and social order in the Cultural Revolution, it is necessary to emphasize the interaction between the state and society, as this book has consistently highlighted. At least three aspects need to be considered. First, it is crucial to approach the state institutions as the actors. This entails studying the conflicts among various levels of state institutions, the response of state institutions to the top leader and the state policies representing their will, and the interaction between state institutions and social factions. In short, it is essential to analyze how the integrative nature of state institutions changed during the Cultural Revolution, how such changes affected the power stakeholders and social factions, and at the same time, what countervailing effects they had. Second, the perspective on social members as a key actor. Here, examining the conflicts among factions, the interaction between factions and the top leader, and the interaction between factions and various levels of state institutions is essential. Third, it is crucial to consider the top leader as an active agent. This entails analyzing the interaction between the top leader, Mao Zedong, social members, and state institutions.

As mentioned in the preface of this book, when examining collective behavior during the Cultural Revolution, the elements of "time" and "space" are of great importance. The "space" element refers, on the one hand, to the institutional arrangements of the state and society that remained in place, which determined the behavior of each participant during the Cultural Revolution. On the other hand, it refers to the geographical scope in which the Cultural Revolution unfolded and the administrative departments within the governmental system. The relationships between party, government, and military in different regions, the interactions among key leaders at the local level, as well as their relationships with higher-level leaders, and their attitudes toward the Cultural Revolution, all contributed to regional disparities within the Cultural Revolution.

The element of "time" refers to the dynamic process of the movement. Let's examine the interactions among the three roles mentioned earlier, revolving around the competition for power and resources, and place these interactions on a timeline. The three years of the Cultural Revolution can be divided into three stages, coinciding with the three sub-movement stages mentioned in previous chapters.

2 Social Order Transformation and Collective Violence

2.1 Breaking the Old Order and Structural Violence

As specifically analyzed in previous chapters, the earliest instances of collective violence during the Cultural Revolution were the acts carried out by the people in the name of "revolution," unilaterally imposed upon non-revolutionary segments of society. This type of violence emerged almost concurrently with the Cultural Revolution and reached its zenith approximately two weeks from mid-August to early September 1966. From early June to the end of July 1966, the Red Guards, composed of the red five classes, launched campaigns within schools to criticize the party, government officials, and teachers and students who did not belong to the five red. In August of the same year, especially after Mao Zedong met with the Red Guards in Tiananmen Square on August 18, this type of violence expanded from schools to society at large, targeting party and government officials, members of non-Communist parties, former capitalists, former bank employees, landlords, former Kuomintang military personnel, and others who were labeled as "counterrevolutionary black elements" becoming the objects of violence.

Why did the early acts of violence during the Cultural Revolution take the form of the people targeting non-revolutionary sectors? This book argues that this relates to the structurally violent elements inherent in the social order before the Cultural Revolution.[5] In his book *Structural Violence and Peace*, Johan Galtung categorizes violence into two types: direct violence, which targets the physical body, and structural violence, which operates through political, social, and economic systems within society. Galtung identifies four forms of structural violence: exploitation, penetration, division, and marginalization. Exploitation refers to the unequal division of labor that benefits one party; penetration involves the ruling party occupying a central position by controlling the thoughts of the ruled; division refers to the fragmentation and isolation of the ruled, dividing and conquering them; and marginalization places the ruled in a marginalized position.

Applying Galtung's concept to the analysis of collective violence during the Cultural Revolution, it can be observed that the acts of violence discussed in previous chapters fall within the category of direct violence defined by Galtung. However, it is also important to recognize that behind this direct violence lies the legitimacy embedded in the social structure and the ideology of governance. For example, the people exerted direct violence on non-revolutionary sectors due to the perceived legitimacy and legality derived from the concept of proletarian dictatorship and the Theory of Class Struggle. It can be said that direct violence, especially in the early stages of the Cultural Revolution, is closely related to structural violence. Therefore, the discussion also incorporates the pre-Cultural Revolution structural violence to better explain the relationship between collective violence (direct violence) and social order. In other words, the social order in China before the Cultural Revolution, the system that ensured stability, and the means of maintaining this system all contained elements associated with structural violence. This structural violence had already been internalized within the political structure of Mao Zedong's era in China.

The structural violence within the political system of Mao Zedong's era in China primarily manifested in the ideological focus on class struggle, the mass movement-style form of criticism and struggle, and the reward and punishment system based on political criteria for the distribution of citizenship. As discussed below, the political culture and behavioral characteristics shaped by the pre-Cultural Revolution structural violence influenced direct violence, especially during the early stages of the Cultural Revolution.

Like the series of political movements after the founding of the PRC, such as agrarian reform, the Cultural Revolution was initially carried out in the form of mass political propaganda with the participation of the masses. In other words, the participation of certain segments of the population who responded to political mobilization and engaged in revolutionary actions against the targets to be eliminated had been a recurring pattern since the founding of the Communist Party of China. In this sense, the origins of the collective violence of the summer of 1966 are highly relevant to structural violence. However, beyond commonalities, it is essential to note that the development and expansion of violence during the Cultural Revolution differs from the direct violence inflicted on the targets of struggle in previous political movements. The violence of agrarian reform and other political mass movements before the Cultural Revolution resulted from the excesses of political movements carried out under the leadership of party organizations and administrative departments. Although the violence inflicted on the targets of struggle in political movements was often used as a "ritual" to reinforce authority and promote policy, the activities were always conducted under party and government leadership in which the party had absolute control. By contrast, the collective violence of the Cultural Revolution was not merely "movement overkill"; it was the result of a mass movement carried out in the absence of party and government authorities leading the political movement or carried out with inconsistent instructions.

In the early summer of 1966, Mao Zedong launched an open political mobilization of society and forbade the party and government agencies from suppressing the mass movement. At the level of implementation, this not only caused confrontations between the party-government agencies and mass organizations such as the Red Guards and severely damaged the existing system of maintaining social order. Specifically, the Red Guards were given the authority to overthrow "those in power within the party who were following the capitalist road," causing the party and government, which were responsible for implementing and supervising state policies, for losing their leadership of the Cultural Revolution movement in their areas of jurisdiction and became paralyzed for a time. Moreover, the police and military, originally the coercive apparatus tasked with maintaining social order, were also instructed to cooperate with the rebellious actions of the Red Guards, leading to changes in their operational functions. As a result, this political movement to transform the bureaucracy with mass organizations was carried out without constraint mechanisms, unlike movements during the pre-Cultural Revolution. Under these circumstances, two significant problems arose: achieving policy communication and maintaining social order.

In the early years of the Cultural Revolution, in 1966, the function of conveying instructions from the top leaders was performed by the media, such as *Xinhua News Agency, People's Daily*, and *Red Flag*. It was through these three media that the top leaders' ideas and policies on the Cultural Revolution and the country's governance were generally disseminated to the masses. However, delegating functions from party and governmental institutions to the media gave rise to two major issues related to collective violence.

First, the media had enormous power in manipulating public opinion on the one hand and extreme weakness in controlling the mass movement on the other. From the summer to the end of 1966, to curb the Red Guards' collective violence and, at the same time, redirect the targets of the struggle of the Cultural Revolution, which had turned to criticize the black five and teachers by the Red Guards, back to criticism of Liu Shaoqi and Deng Xiaoping, the Central Cultural Revolution Group and senior leaders such as Zhou Enlai frequently met with mass organizations in an attempt to control their actions. But subsequent events have shown that it was impossible to control mass organizations without relying on the party and government institutions and solely on the extraordinary charisma of a few political figures. Moreover, as the public speeches and the actions of Jiang Qing and other members of the Center Cultural Revolution Team in the autumn of 1966 show, the political orientation of political figures very often exacerbated the confrontation among mass organizations.

Second, although the media informed society of the aims and direction of the Cultural Revolution, they did not openly reveal the underlying elements, such as intra-party struggles, that lay behind the grand cause of the revolution. They also did not disclose the detailed rules of the movement to the public. As a result, difficulties arose in communication between the top leadership and the mass organizations. After learning about them through the media, factions interpreted Mao Zedong's intentions based on their own understanding. This sometimes led to misunderstandings of Mao's intentions and, at times, intentional distortions for personal or even group interests. For example, Mao Zedong's understanding of the breaking of the four olds movement was very different from that of the red five Red Guards. Where the Red Guards understood the breaking of the four olds at the literal level, carrying it out in action, Mao understood it from more of an ideological level.[6] In addition, how those in charge of the media understood and reacted to Mao's words and actions could also significantly impact the mass movement.

As shown through the actual course of the movement, Mao Zedong attempted to utilize the rebellious actions of the Red Guards to overthrow Liu Shaoqi and rectify his political line. However, even Mao Zedong found it difficult to control the actions of the Red Guards solely based on his own power and extraordinary charisma, without relying on party and government institutions. Therefore, in October 1966, to redirect the focus of the mass movement from targeting individuals outside the party to those within the party, he allied with the rebellious factions within universities, thereby weakening the stubborn "birth bloodline theory" and the red five Red Guards who fought against people outside the revolutionary ranks. The growth of the rebel factions organization was closely intertwined with Mao Zedong's support. However, similar to the red five Red Guards, this organization soon deviated from the set direction of the movement, engaging in power struggles and factional conflicts, ultimately leading to more intense collective violence known as "armed clashes" or "struggles" (*wu dou*).

2.2 *Building a New Order and Collective Violence*

The collective violence in the summer of 1966 was mainly led by the red five Red Guards of middle and senior high schools. However, the collective violence from November 1966 to the end of 1967 involved the greater part of Chinese society. This period's collective violence included confrontations between conservatives and rebels and within rebel groups. These two types of collective violence sometimes occurred simultaneously, sometimes consecutively, and took on different forms in different regions. The high incidence of armed struggle was from July to September 1967.

In 1967, due to the active promotion by the rebel factions, the so-called "capitalist reactionary line" represented by Liu Shaoqi and Deng Xiaoping was eliminated, and the Cultural Revolution began rebuilding the Three-In-One Regime. However, significant changes occurred within the political structure during the six months of advancing the Cultural Revolution's agenda. From the beginning, the mass organizations were in a state of mutual opposition, and the development of the Cultural Revolution further intensified the conflicts between these organizations and led to divisions within the state institutions.

The reasons for these divisions were complex, with various factors contributing to them. However, the divisions changed the state institutions dominated by a single party. Specifically, factions emerged within various levels of party and government institutions, engaging in political competition as seen in a multi-party system. It is important to note that this competition was not based on a legal framework. Instead, under the leadership of distinct political figures, each faction established and expanded their own information networks and organizational structures. Their objective was to mobilize mass organizations and gain influence through competition.

As a result, multiple and intricate interactive relationships emerged among Mao Zedong, various mass organizations, and party, government, and military organizations at different levels.

In early 1967, two major rival factions were formed within schools, factories, and other work units. The struggle for power and profit between the factions was further intensified around the formation of the Three-In-One Regime and focused on how to get their representatives into the new organization. To achieve this, factions had to gain the support of the local "left-supporting" PLA and establish cooperation with party and political cadres who might be eligible to join the Three-In-One Regime. In obtaining these political resources, the factions struggled intensely with each other.

Within a work unit, one faction often sought support from external organizations to overcome opposing factions. They formed alliances with factions at the provincial and municipal levels while cooperating with factions from other units. This network of relationships enabled them to obtain external support, and victories achieved by allied factions benefited their faction. However, if a faction's behavior within the network violated the prevailing ideology at that time, it resulted in the loss of political resources for all factions in the network. They might even have faced questioning of their ideology's legitimacy by opposing factions. Given the complex variables influencing factional fate, achieving long-term success became exceedingly difficult. Factional struggles persisted amidst this diverse and uncertain environment.

The struggles within the party and government institutions were just as complex as those in factional conflicts. As mentioned earlier, the state institutions, including party organizations, comprised administrative bodies and state representatives. The Chinese political system's personnel appointment system created strong interdependencies among party and government officials at various levels. The political fate of a party cadre not only had a high likelihood of impacting the officials in higher and lower departments but could also extend to officials in other departments at the same level within the party and government institutions.

Before the Cultural Revolution, there was a high degree of integration within the party and government institutions. However, during the Cultural Revolution, the integration within these institutions was weakened due to the impact of the "struggle against the capitalist roaders within the party" campaign. From the early summer to the end of 1966, party and government officials at all levels responded to the rebellious actions of mass organizations in various ways, including indulgence, resistance, and manipulation.

Before October 1966, the Cultural Revolution was confined to the cultural and educational sectors, and the Work Teams in factories responsible for production tasks still held the leadership of the Cultural Revolution. Criticism of Liu Shaoqi and Deng Xiaoping had not been made public then. Although party and government organizations had lost their leading role in the Cultural Revolution at many provincial and municipal levels, they still maintained a certain degree of integration before the end of 1966. However, in 1967, as the movement developed, the internal divisions within party and government organizations also became apparent.

The power struggles within the party and government institutions not only influenced the power struggles in other relevant departments but also significantly influenced the factional struggles among the masses. It is undeniable that the factional struggles among the masses, in turn, affected the internal party struggles. In such a situation, party and government officials frequently manipulated mass organizations to gain an advantage in power struggles. Conversely, mass organizations actively intervened in the power struggles of party and government officials to obtain political resources and support officials affiliated with their factions.

Without a properly functioning coordination mechanism within party organizations and bureaucratic institutions, the mediation of power struggles within party and government organizations at all levels could only rely on the paramount leader, Mao Zedong.

To understand the armed struggle during the 1967 factional struggle, it is necessary to address the role played by the PLA in it. In January of that year, Mao ordered the PLA to intervene in the Cultural Revolution to smooth the formation of the Three-In-One Regime. However, it should be noted that the PLA in this period did not take coercive actions to control factional rivalries but instead mediated disputes. In other words, the PLA was not given absolute power to prevent social disorder. Not only that, the authority of the PLA to control the social order was constantly changing due to the influence of the power struggle between the Central Cultural Revolution Group and the top echelons of the CCP, including Lin Biao.

The uncertainty of the PLA's function weakened the army's ability as a coercive device to guarantee social order. At the same time, it was difficult to exercise effective control over factional struggles. In most cases, the PLA's involvement even complicated the rivalry's structure by intervening in disputes. This problem was manifested in the relationship between the PLA and the party and government and the relationship with mass organizations.

In brief, although the PLA existed independently of the bureaucracy, it had close ties with local party and government organizations, such as the People's armed forces department, regarding working interactions and human relations. There was cooperation between PLA and local cadres, but there were also differences of opinion on work and even conflicts of character and personalities. Therefore, the PLA's involvement in the Cultural Revolution often resulted in their entanglement in power struggles among local party and government officials. Consequently, the internal power struggles within provincial and regional-level institutions affected the power struggles within administrative levels such as cities and counties. Likewise, reverse influences were inevitable.

At the level of relations with the mass organizations, the PLA could not be entirely impartial to opposing mass organizations, often supporting one directly or indirectly and suppressing the other. As a result, disputes between factions were exacerbated. The Wuhan Incident, for example, was an extreme case in which the PLA openly joined hands with one of the factions to confront the other and the political forces behind it. In summary, when opposition occurred within mass organizations and state institutions and became embroiled in widespread power struggles, the PLA's intervention only served to exacerbate various contradictions within the state, between state and factions, and among factions themselves. Thus, in 1967, as the flames of the Cultural Revolution spread to the military with the campaign to "root out a handful of reactionary elements within the army," various conflicts were manifested in the form of armed struggle between factions. This was precisely the reason why the collective violent behavior among mass organizations became even more intense during the summer of 1967.

Amid the complex web of struggles mentioned above, by the summer of 1967, not even the paramount leader Mao Zedong could effectively suppress the chaotic situation solely through his extraordinary charisma. Also, loyalty to Mao became part of the factional struggle. Whenever Mao issued instructions during the Cultural Revolution, a new round of competition between political figures and mass organizations revolved around access to political resources. In this sense, at the beginning of the launching of the Cultural Revolution, Mao was the main body that mobilized the masses. Still, as the Cultural Revolution developed, Mao became the most crucial political resource, changing from the subject to the object of the movement. For Mao, it can be argued that 1967, when his transcendent charisma reached its peak, was also the year when his will was least able to be carried out.

Mass organizations could use one of Mao's words against another, and the little red book, *Quotations from Chairman Mao Zedong,* became both a weapon and a talisman. The "invincible" Maoist ideology used as a tool by the mass organizations had been far beyond Mao's control. Furthermore, Mao Zedong himself was overshadowed by his quotations and the cult of Mao movement. It is impossible in any system for a leader's will and policies to be effectively implemented at the grassroots level without a highly organized bureaucratic system.

Mao desperately tried to maintain order in the movement by distributing political resources and power among the members of society, the party apparatus, and the PLA in a fluid manner. However, this method only temporarily curbed the chaos. It soon triggered a new round of struggles over political resources and power, giving impetus to factional struggles while bringing about new forms of violence. Eventually, in the summer of 1967, the whole of China was plunged into chaos.

2.3 *Restoration of Order and State Violence*

In May 1968, a nationwide campaign was launched to "clean up and reorganize the ranks of the class." Two kinds of violence took place during this period. One was the collective violence committed by conservative organizations against rebel groups and the black five in the name of reorganizing class ranks with the support of the PLA's Mao Zedong Thought Propaganda Team and the local militias. Another type of violence was the state violence inflicted on factions by the PLA and the local armed forces to end factional struggles. Because this book focuses on inter-factional violence, the discussion of state violence will not be expanded here.

Both types of violence were linked to the policy adjustments to end the pandemonium in 1967. From that autumn onwards, the party's functional and power restoration became the state's primary objective. At the same time, mass organizations were instructed to implement a coalition within schools and factories. In the highly chaotic political environment of the time, it took work to implement these two policies. Thus, the PLA was given absolute authority as the executor of state policy. In the summer of 1968, with the establishment of revolutionary committees throughout the country, the functions of the party and government organizations were largely restored at all levels. During that autumn, under the leadership of the PLA, the armed struggle between factions, which had lasted for more than a year, was mostly under control. However, the PLA continued to use violent means, directly or indirectly, to curb further conflict.

By restoring the functions of party and government institutions and suppressing factional struggles, the collective acts of violence during the Cultural Revolution were finally brought under control by the end of 1968, and social order gradually stabilized. It should be noted that the restored social order was not the integrated state–society social order that Mao had aspired to but a return to the social order of the pre-Cultural Revolution. The new Three-In-One Regime, and the revolutionary committee, were nominally composed of revolutionary cadres, soldiers, and masses, but, in reality, the cadres and soldiers were appointed by higher authorities, and the revolutionary masses were elected from mass organizations with a patron–client relationship with high-ranking individuals. The revolutionary masses who joined the regime had only a symbolic presence, with few substantive functions. From the perspective of the political system's pyramid structure, the social order of political leaders → state institutions → people, → class enemies remained largely unchanged. If there was anything different from the pre-Cultural Revolution era, it was the symbolic inclusion of the masses in political power and increased military influence within state institutions, as mentioned earlier. In the author's opinion, this is the price to be paid for total reliance on the PLA to restore social order. Of course, the increased power of the military within the political system also sowed the seeds for the high-level political struggles of the 1970s.

3 Concluding Remarks

This chapter primarily examines the relationship between social order and collective violence during the Cultural Revolution, attempting to answer why the large-scale collective violence during this period did not result in a complete collapse of social order. The previous discussions show that social order and collective violence in the Cultural Revolution are mutually causative. The relationship between the two can be summarized into the following three points.

First, structural violence already existed in the pre-Cultural Revolution social order. The earliest acts of collective violence during the Cultural Revolution were related to this structural violence. Mao Zedong carried out mass mobilization to break the existing social order and establish a new one. As a result, the structural violence embedded in the existing social order surfaced and expanded as the state institutions that maintained the social order were weakened. Although Mao Zedong curbed this type of collective violence by depriving the Red Guards, who belonged to the Red categories, of their right to rebel and instead supported the rebel factions, this solution simultaneously gave rise to a new form of collective violence.

Second, during Mao's breaking of the existing social order and establishing a new one, conflicts occurred between the forces representing the old and new orders. Moreover, conflicts also occurred within the forces representing the new order itself, giving rise to the *wu dou*, or armed struggle, as a new form of collective violence. Moreover, the movement order, maintained by the fluid distribution of political resources and power among members of society, the party, government institutions, and the PLA, was in danger of collapsing due to the conflicts within and among themselves.

Third, as a result of the setback in establishing a new social order, the full intervention of the PLA, representing the violent institutions of the state, was necessary to restore stability to the social order. As a result of this intervention and the restoration of the functions of the party apparatus, the social order returned to its previous state. The structural violence inherent in the pre-Cultural Revolution social order resurfaced in this process. In some areas, there were even acts of collective abuse of the black five and other social groups that suffered from political discrimination, such as landowners and their children, by factions supported by authorities.

Despite the great chaos during the Cultural Revolution, the author believes three critical factors prevented the complete collapse of social order. First, the unquestionable ideology of Mao Zedong and class struggle, as well as the absolute authority of Mao himself, played a decisive role. Although mass organizations had the right to interpret Mao's ideology and thought, the decision on the philosophy of the Cultural Revolution and related policies was always in Mao's hands. Any mass organization deemed "incorrectly oriented" lost its political legitimacy and was disarmed in the movement. On the other hand, the absolute unquestionability of Mao's ideology and the ideology of the Cultural Revolution triggered competition among the mass organizations, which also resulted in deepening violence.

Second, the role played by the army in the ultimate control of social order must be addressed. This had much to do with the army's discipline, and the fervor of mass rebellion had less effect on the PLA, comparatively speaking. Although the military intervened in the Cultural Revolution, they intervened as a solid unit. There were rarely revolts in the military among subordinates and superiors. Though a few confrontations occurred between different units during the Cultural Revolution, they were isolated incidents and did not cause significant unrest.

Third, the existence and maintenance of a movement order in which no single organization had exclusive access to political resources also contributed to the survival of social order. Still, the fluid distribution of political rights and resources gave impetus to inter-factional rivalries, leading to repeated and escalating acts of collective violence.

In summary, violence is inherent within the social order of China. Whenever there is a reform of the social order, violent conflicts arise within state institutions, between different social groups, and ultimately between state institutions and society. The violence within state institutions and the violence among social members interact, exploit, and influence each other. Looking back at history, as demonstrated by China's modernization process since the *Xinhai* Revolution, the formation of a new social order is often accompanied by violence. Considering the relationship between the creation of a new social order and violence, how to achieve institutionalized social order remains a lingering challenge facing contemporary China.

Notes

1. Satoshi Amako, *China, Changing of a Socialist Power* (Tokyo University Press, 1992), pp. 78–79.
2. The People's Liberation Army National Defense University, Party History, Party Construction and Political Work Teaching and Research Department, ed., "*Cultural Revolution*" *Research Materials* (Part 1) (People's Liberation Army National Defense University Press, 1998), p. 75.
3. Satoshi Amako, The Fetal Movement of a Giant Dragon—Mao Zedong vs. Deng Xiaoping, Kodansha, 2004, pp. 196–198.
4. In order to better explain the relationship between collective violence and social order in the Cultural Revolution, this paper restricts the members of society to the category of the people who had the right to participate in the Cultural Revolution, because the social group outside the category of the people, called "class enemies," did not have the right to participate in the movement, much less the right to commit violence.
5. For a study of structural violence, see, John Galtung, translated by Sakio Takayanagi, etc., *Structural Violence and Peace* (Chuo University Press, 1991).

6. For a related discussion, see Yin Hongbiao, "The Culture and Politics of the Red Guards' 'Breaking the Four Olds'"("Hongweibing 'posijiu' de wenhua yu zhengzhi"), in Shi Gang, ed., *Institutions and Culture in Modern China* (*Xiandai zhongguo de zhidu yu wenhua*) (Hong Kong Social Science Press Ltd., 2004), pp. 51–97.

CHAPTER 11

The Impact of the Cultural Revolution and the Transformation of Social Movements

The collective behavior of the Cultural Revolution has had a significant impact on all aspects of China, and it can be said that it has continued to this day. Concerning national policies during the Reform and Opening-Up period and the social movements that arose under the new political conditions, the impact of the Cultural Revolution was particularly great. Therefore, to study the Cultural Revolution, it is necessary to investigate the historical events themselves, their effect on subsequent social development, and the lessons we should learn from them. In other words, studying the historical legacy is as essential as studying history. This chapter will outline the book's main arguments and briefly discuss the impact of the Cultural Revolution on subsequent nation-building and social movements.

This chapter consists of four sections. The first section will provide a comprehensive summary of the main arguments regarding the factional divisions and collective acts of violence during the Cultural Revolution, as discussed in this book. The second section will analyze how that factionalization and collective violence have affected the state and society relations in the country during the Reform and Opening-Up period. The third section will examine how social movements have changed structurally during the Reform and Opening-Up period, compared to the Cultural

Y. Lijun, *Contest for Citizenship and Collective Violence during China's Cultural Revolution*, IPP Studies in the Frontiers of China's Public Policy, https://doi.org/10.1007/978-981-97-0906-9_11

Revolution. Finally, the fourth section briefly overviews the relationship between social movements and nation-building.

1 The Main Causes of Collective Behavior During the Cultural Revolution

Exploring the causes of factionalization and collective violence during the Cultural Revolution is the central theme of this book. A brief overview of what has been discussed in the previous chapters is provided below.

Among the leading causes of factionalization and collective violence during the Cultural Revolution, the institutional spatial configuration of the state and society and the system of distribution of citizenship implemented in this environment can be ranked as the most critical factors. This institutional spatial configuration, in which the public sphere comprehensively squeezed out the private sphere, was essential for Cultural Revolution-period social movements. As the analysis in Chapters 3 and 4 shows, in the mid-1950s, after the founding of the PRC, the state introduced a socialist system. It gradually implemented radical public ownership reforms in various areas. This institutional change re-created the country not only in terms of its economic system but also in terms of state and society relations in the new China. As the public ownership reform progressed, the old private enterprises and civil organizations were gradually dismantled and replaced by a new system of people's communes and work units. These combined production with living and political functions. This system carried social welfare and assistance functions and enabled the state to control life "from the cradle to the grave." With the Communist Party permeating Chinese society and establishing the mass organizations under its leadership (trade unions, the Communist Youth League, the Women's Federation, the Young Pioneers, etc.), state power reached all corners of society. As a result, the private sphere gradually diminished and vanished while the public sphere expanded and became highly politicized. By reorganizing state–society relations, the state monopolized all political, economic, and cultural resources—citizenship—and the power to allocate them. Thus, with the establishment of various systems and policies, the state created a hierarchical system of citizenship distribution through such institutions and policies as the system of class origin, the Household Registration System, the labor and personnel system, the administrative hierarchy system of party and government cadres, the employment system of enterprises, enterprise ownership management system, and institutions of the selection of activists. This method of state monopoly of citizenship and political distribution created a hierarchical social structure and determined the conditions in which the population competed for access to citizenship—competing for interests through political means.

The collective behavior of the Cultural Revolution analyzed in this book is a competition for citizenship within the institutional configuration of the state–society discussed in this book. The analysis in this book shows that, on the surface, the collective behavior of the Cultural Revolution was characterized by irrationality and violence. Still, behind these was a competition for citizenship in which the participants, as individuals, took advantage of the political opportunity structure provided by the political movement and rationally pursued their interests in a closed space where there was no possibility of exit.

Let's consider the precondition that there is no institutional space configuration from which withdrawal is possible. The following situation becomes clear when we understand collective behavior during the Cultural Revolution. Only a few options were available to both faction leaders and participants. In an institutional spatial configuration with no private sphere, and the state monopolized all citizenship, participants were left with only one path to pursue their interests to the maximum in the politicized public sphere.

Competition over citizenship did not lead to social unrest before the Cultural Revolution, but why did it lead to unstoppable chaos during the Cultural Revolution? The author believes that the low degree of institutionalization of social governance in China at that time also led to factionalization and collective violence. The degree of institutionalization can be examined at three levels: the state, society, and the interaction between the state and society. In Maoist China, the degree of institutionalization at all three levels was meager. At the state level, this was mainly reflected in the lack of a legal system, the monolithic leadership of the party, the centralization of power, and the extraordinary personal charisma of Mao Zedong. When there was a power struggle at the top of the central leadership, the absence of an institutional and legal check and balance mechanism led to political chaos at the national level. Consequently, it weakened the state's primary function to provide societal stability and order. Moreover, due to the low level of institutionalization of the state, the top leaders were also unable to defend their interests and political positions in political struggles based on laws and institutions. Hence, a section of the population turned to society for support and manipulated the masses with the intent to win in political struggles. However, factions were not just passive participants. They maximized the use of opportunities provided by the state and its agents to shape state policies in favor of their interests. As a result of this interaction between the state and society, the Cultural Revolution as a movement developed its internal dynamics, leading to a shift from one "sub-movement" to another.

In addition, the low-level institutionalization of society also hindered the capacity for autonomy and the development of autonomous organizations. In other words, as the private sphere shrank after the founding of the PRC, society lost its autonomy and was placed in a totalitarian system of political mobilization. Although there were managed mass organizations such as work units, the Communist Youth League, and trade unions, they were all organized by the state and controlled by state policy and agents. On the surface, these organizations seem to reflect a highly organized society. Still, they were instruments of state domination and mobilization of individuals for political movements and did not play any role in pursuing society's rights and interests. It was precisely because members of society could not use managed mass organizations as tools for pursuing citizenship that factions emerged during the Cultural Revolution.

The composition of factions is closely related to the hierarchical social structure before the Cultural Revolution. In other words, a hierarchical social structure made it easier for people to assemble quickly according to political or economic perceptions. Moreover, in the hierarchical social structure before the Cultural Revolution, political and economic resources were distributed to different groups in a decentralized manner. Workers and peasants were regarded as the masters of the country and had a high political status but a low economic status. As a reformed class, intellectuals had a low political but high economic status. The party and government cadres, especially those above the middle level, who hailed themselves as servants of the people, had all kinds of privileges that the people could not enjoy. This decentralized way of distributing resources made members of society more or less discontent and drove their motivation to rebel. It also greatly facilitated the reorganization of factions during the Cultural Revolution. This is because the scattering of resources for allocation provides people with multiple collective understandings, and this understanding brings more choices. People could gather based on common political or economic identities or goals. Moreover, when the goals of the Cultural Revolution changed, and a faction failed to meet the members' demands or their value alignment with the faction shifted, individuals could easily adjust their affiliations, join other factions, or form new ones.

During the Cultural Revolution, a large number of factions emerged, but all of them had very little autonomy. There are four reasons for this. First, the state determined the survival of factions, which did not allow development into autonomous organizations. Second, the state played a decisive role in determining the collective perception of the participants in the factions. In other words, the goals of the state and the changing power structure among central leaders during the Cultural Revolution affected the stability of factions and hindered their autonomy. Third, the competing interests of faction leaders impeded the ability of factions to develop autonomy and create inter-factional ties. Fourth, the development of faction autonomy itself requires a considerable amount of time. Even without the previous three reasons, it would have been challenging for factions during the Cultural Revolution to achieve autonomy quickly.

For these reasons, the capacity for autonomy within factions during the Cultural Revolution was extremely low, and there needed to be an institutionalized mechanism of coordination and control between factions. As a result, when the state's ability to control society was weakened, and society could not self-regulate and resolve conflicts of interest, the competition among the masses was left in a state of anarchy, eventually leading to radicalization and violent conflict.

In addition, factional relations during the Cultural Revolution, whether between factions of the same camp or factions of opposing camps, were essentially non-institutionalized competition. Even if some factions established temporary alliances based on immediate interests, the alliances would turn to non-institutional competition once external conditions changed. In addition, for some factions, from the time of separation from the parent faction, their objectives would take on the character of a rivalry with the parent faction. This non-institutionalized competition between factions eventually led to the radicalization of factional behavior, and collective violence emerged.

Moreover, the Cultural Revolution period was characterized by a shallow degree of institutionalization of the state and society and their interaction. State control over society was not legal but political and can be summarized in two points. First, the state-controlled society by emphasizing class struggle as an ideology; second, it controlled society through an exclusive and hierarchical approach to the distribution of citizenship. The Theory of Class Struggle changes its connotations accordingly with the change in political goals and the necessary adjustment of political struggle. The method of citizenship distribution is also subject to a great deal of instability, influenced by state goals, policies, the intentions of state agents at the policymaking and policy implementation level, and the power struggle between state agents. Thus, in the spatial configuration of the state and society, the state has complete control over society through its ideology and organization but a shallow degree of institutionalization in terms of its methods of governing.

Viewed from a societal perspective, it becomes evident that a dearth of institutional and legal channels existed for political participation. As a result, the engagement of the masses in politics took the form of utilizing opportunities for political mobilization to protect their rights and pursue their interests. Consequently, political movements became the stage where individuals competed to attain citizenship.

This dimension in the non-institutionalization of state–society interaction can be understood using the example of mass mobilized forms of political participation. In Mao-era China, politically mobilized participation was the only way the masses could participate in politics. In such situations, the ideology of class struggle was an essential means of political mobilization. The emphasis on the doctrine of class struggle aided the task of pursuing national goals and strengthening social unity. Still, it was often used as an instrument of a power struggle when political differences arose between leaders at the top level of central leadership. Also, the state's overemphasis on class struggle and its expansion injected the masses with a sense of hierarchy and class antagonism.

Moreover, during political movements that have long emphasized class struggle, the people learned to use the "righteousness" of class struggle to pursue their interests.

Why is it necessary for the state to adopt a mass mobilized form of political participation? The main reasons for this can be considered in four ways. First, the CCP began as a revolutionary party. After the founding of the PRC, it still invoked its politically mobilized methods of nation-building efforts from the revolutionary period. Second, nation-building after the state's founding occurred in a scarcity of social resources, so it was necessary to carry out political mobilization to draw resources from society. Third, because of the low degree of institutionalization of the state, when power struggles or policy differences occurred between the top leaders, Mao and Liu Shaoqi, there was no institutionalized solution to the problem. As such, Mao tried to solve the problem using mass mobilized political participation. Fourth, Due to the low institutionalization of state and society interactions, society did not have normal channels of political participation, nor did the state have mechanisms to curb the abuse of power. Against this background, political mobilization during the Cultural Revolution was used by Mao as a means of reflecting public opinion. In this sense, it can be said that political mobilization and institutionalization hindered each other's progress. That is to say, on the one hand, the use of political mobilization by the state resulted in slowing the development of the institutionalization of the state, society, and the interaction between them; on the other hand, it was the low degree of institutionalization in these three areas that made political mobilization necessary. At the same time, the use of political mobilization further resulted in the stagnation of institutionalization. Such a vicious cycle continued until the Reform and Opening-Up period.

Political mobilization has significantly impacted the occurrence, development, and decline of social movements. While political mobilization has created opportunities for the emergence of social movements, it has also constrained their goals, methods of resistance, and survival. Empirical research presented in this book demonstrates that the rise of radical or conservative factions in student and worker organizations is closely related to the state's political mobilization. Moreover, the competition between factions during the Cultural Revolution has directly linked to the state's policy objectives and the political mobilization to achieve them. Therefore, it is clear that political mobilization has played a dual role in shaping the landscape of social movements during this period.

For these reasons, factional rivalries and collective violence were inevitable during the Cultural Revolution. This, coupled with the pervasive Chinese spiritual construct of converting loyalty into profit, created by the institutional environment at the time, shed an irrational light on the rational act of competing for citizenship, further contributing to social unrest. The upheavals of the Cultural Revolution also significantly impacted policy adjustments during the Reform and Opening-Up period, which began in the latter half of the 1970s. State policies were substantially adjusted to address the institutional factors that led to collective violence discussed earlier to prevent the recurrence of collective violence similar to that of the Cultural Revolution. This will be examined further and in more detail in the next section.

2 The Impact of the Cultural Revolution and the Transformation of State–Society Relations During the Reform and Opening-Up Period

In 1969, with the closing of the Ninth National Congress of the CCP, the Cultural Revolution, as a social movement, ended. Subsequently, the state strengthened its control over society. As a result, no large-scale social unrest occurred, although political struggles within the state remained. After 1969, the central leadership began a "rectification and institutional reconstruction" campaign as instructed by Mao Zedong. However, due to the ongoing political struggles within the central leadership and the prevailing influence of the Cultural Revolution's emphasis on class struggle, the overall structure of China's political system remained largely unchanged during the post-Cultural Revolution period, from 1969 until the downfall of the Gang of Four in 1976 and the era of Hua Guofeng's leadership. Only after 1978, when China entered the Reform and Opening-Up period, did the government start rebuilding its core political system. During the Reform and Opening-Up period, the central leadership undertook a massive state-rebuilding effort to modernize the country. Economic development and social stability were the most important goals of state reconstruction. To achieve these two goals, the central leadership, with Deng Xiaoping at its core, learned from the lessons of the Cultural Revolution and embarked on state-rebuilding in the following areas.

2.1 Restructuring of the Spatial Configuration of State–Society Institutions

First, during the process of national reconstruction, the revitalization and development of the private sector were permitted by the state. In other words, state ownership had been gradually transformed from a single form of ownership, namely ownership by the whole people and collective ownership, to multiple ownership systems, such as individual ownership, foreign-capital enterprises, and Sino-foreign joint ventures. The people's commune system, characterized by collective management, was finally dismantled in the agricultural sector. The Household Contract Agricultural Production Responsibility System was introduced in the countryside, with family management at its core. The system allows farming households to manage agricultural production on their initiative while the farmland remains in the ownership of the rural collective. In the industrial sector, non-state enterprises such as township enterprises, individual enterprises, private enterprises, and foreign-capital enterprises were loosened from state control and restrictions and shifted to the flexible business system of the market economy. Through relentless business efforts, the private sector has rapidly developed and become the engine of China's economic growth. Simultaneously, state-owned enterprises underwent reforms starting with profit retention and progressing toward "profit-tax separation," corporatization, and stockholding. These reforms significantly enhanced the vitality of state-owned enterprises.[1] Today, although state-owned enterprises still dominate the Chinese economy, the resulting system of multiple ownership coexistence provides the economic system that guarantees the private sphere. The increase in non-state enterprises brought about changes in the allocation of institutional space for the state and society. People began to derive economic benefits from the private sphere, which is external to the state-controlled public sphere. The enhancement of the private sector made it impossible to maintain the institutional spatial configuration of the state and society that had been in place.

In addition, the existence and development of the private sector have also brought about changes in the distribution of citizenship and the competitive status of its acquisition. The state monopoly of citizenship and its distribution is no longer possible, and members of society can acquire citizenship not only in public but also in the private sphere. At the same time, the competition for access to citizenship has diversified and pluralized, and the competition that existed during the Cultural Revolution period, which revolved solely around political resources, no longer exists.

2.2 *Advancement of Institution-Building*

As the country was rebuilt in the Reform and Opening-Up period, there was a significant increase in institutionalization compared to the Maoist era, as illustrated in the following three areas.

Institutionalization at the National Level

Institutionalization at the national level is manifested in the following three main areas.

First, institutional formation of government bureaucracy. During the Cultural Revolution, Mao Zedong opposed the hierarchical and bureaucratic nature of the bureaucratic system. He attempted to establish a new state system that integrated the government and the people through political mobilization. However, during the era of Reform and Opening-Up under the leadership of Deng Xiaoping, the central leadership shifted the focus of political reform toward institutionalizing and standardizing bureaucratic institutions. The party, government, and military reconstruction was primarily aimed at institutionalization and standardization.

Second, the promotion of the rule of law. The lack of a legal framework was a crucial factor contributing to the emergence of collective violence during the Cultural Revolution, both from the perspective of the state–society relationship and the interaction between the state and society. As a result, as discussed earlier, the masses in the Cultural Revolution harshly criticized the targeted groups, including bureaucrats, in a brutal manner and engaged in extremely violent and cruel power struggles among different factions.

Recognizing this, during the era of Reform and Opening-Up, emphasis was placed on the rule of law as a governing method and establishing a legal society as one of the goals of national reconstruction. Although the construction of a legal state is an ongoing process to this day, the fact that it has been established as a national goal represents a significant advancement compared to the Mao Zedong era in China.

Third, the transformation of the party function through institution-building. The transformation of the party's functions and the institution-building manifested itself in many ways; the two main ones relevant to this book are as follows. The first is the institutional reconstruction of power succession within the party. In Mao-era China, the indication of "revolutionary" was ranked first and took precedence over knowledge and ability in evaluating party cadres. When disagreements within the party arose over policy, they were resolved mainly by power struggles. During the Reform and Opening-Up period, Deng Xiaoping vigorously promoted cadres' rejuvenation, intellectualization, and professionalization while introducing a retirement system. This resulted in institutionalizing the party's criteria for hiring and retiring cadres. But such political reforms did not eliminate the power struggle within the party, especially among the top central leadership. Of course, power struggles exist in any society, but whether or not the political system has a mechanism for institutionalizing them determines how they manifest themselves and their outcome. The political reform advocated by Deng Xiaoping in the 1980s led to a reduction in power struggles within the party compared to the Mao era. However, it did not eliminate them. In addition, because the system of power succession within the party, especially at the top, had yet to be completed, power struggles among top leaders occurred repeatedly. Periods of power succession are times of frequent power struggles within the party. Nevertheless, compared with the power struggle between Mao Zedong and Liu Shaoqi during the Cultural Revolution, it can be seen that power struggles during the Reform and Opening-Up period attached more importance to social stability. In other words, while power struggles during the Cultural Revolution were conducted by mobilizing the masses and socializing political struggle, those during the Reform and Opening-Up period were confined to the party, particularly those within the central leadership. This approach prevents the socialization of the party's struggle from negatively impacting social stability.

Another point is the transformation of the party's functions. While during the Reform and Opening-Up period, the state abandoned the ideology of class struggle, it also abolished the policies associated with it. It switched to a policy of class harmonization. In the 1980s, the government took measures to encourage the existence and development of private enterprises. In the 1990s, after Deng Xiaoping's Southern Tour, the development of a capitalistic economy was fully boosted. In the twenty-first century, the CCP not only legalized the private economy in the constitution but also allowed the entrepreneurs of private enterprises to join and participate in the CCP. In other words, the CCP transformed from a revolutionary party with the goal of the proletariat eliminating the bourgeoisie to a ruling party that accepts the capitalist system and the bourgeoisie and represents pluralistic interests.

Institutionalization of the Social Dimension

The institutionalization of the social dimension during the Reform and Opening-Up period has been marked in other ways besides developing the private sphere. In particular, the state has gradually allowed the existence and development of social organizations with autonomous characteristics. However, it is important to note that these social organizations were limited to non-political ones, and their establishment and activities required the approval, management, and supervision of the relevant government departments.[2] Expanding freedom of association has led to a dramatic increase in social organizations and their operational autonomy.[3]

The expansion of the private sphere and autonomous social organizations have changed state–society relations.[4] However, it is essential to note here that the relationship between the state and society in China, and the nature of autonomous social organizations, cannot, at this stage, be equated with social organizations in European and American democracies. This is because the state remains involved in society and firmly controls it. In contrast to Maoist China, the state's control of social organizations after Reform and Opening-Up has shifted from the previous complete control to how it is controlled through the leaders of social organizations. The leaders of social organizations are party and government cadres and staff of government agencies, and many social organizations are registered as subordinate organizations of government departments.[5] In this sense, most social organizations in China during this period do not possess the same level of autonomy as those in Western democratic countries. Many organizations are essentially semi-governmental social organizations operating under the control and influence of the government.

The Strengthening of Institutionalization of State–Society Interactions

As mentioned earlier, although the state was still involved in society during the Reform and Opening-Up period, there has been an increase in institutionalization in the interaction between the state and society compared to the Maoist period, with more significant changes in political participation.

The government gradually allowed more political involvement at the grassroots level, establishing a mechanism for political participation that differed from that during the Cultural Revolution. It demonstrated its importance to public opinion by introducing a system of direct, competitive, and preparatory elections of people's representatives at the county level and below. In particular, Since the mid-1990s, a system of competitive, direct elections was introduced to select village committees, and the will of the voters has gradually been reflected in the election results.[6] Of course, whether grassroots electoral political participation can be extended to society still requires the democratization of institutions and channels. Still, it is a big step forward compared to the mass mobilized political participation in the Maoist era.

In addition to the electoral system reforms mentioned above, the government introduced democratic practices, incorporating public opinion in selecting and appointing cadres and personnel. The Pre-Appointment Public Notice System for Leading Officials was recently introduced at the local level as one example. In other words, the power of appointment and dismissal is ultimately in the hands of the party committee, but the public also has the right to voice their sentiment. To reflect public opinion in political decisions, the government even openly published drafts of the constitution and created a channel for the people to participate in politics through various means, including public consultation.[7]

Judging from the efforts to reflect public opinion in political decision-making and the expansion of society's right to participate in politics, it can be said that the interaction between the state and society is moving toward institutionalization and rationalization compared to the era of Mao Zedong.

Ending Methods of Political Mobilization

During the Reform and Opening-Up, the state completely abolished the mass mobilization approach to political participation. Political and social stability was established as a policy objective of the state. The party sometimes engaged in political mobilization, but the purpose was to promote policy implementation in a manner quite different from the Maoist era.

In addition, from the late 1970s onwards, the policy of family class origins was ended, and the political classification of red five and black five no longer existed. With the abolition of class policy, the government removed the doctrine of class struggle from the state's ideology.

Furthermore, with the abolition of class policies, the concept of class struggle was removed from the state ideology by the government. Starting from the late 1970s, policies that divided people into different classes were abolished, and the political classification methods of the "Red Five Categories" and "Black Five Categories" ceased to exist. Of course, opposition to the current regime is still not allowed, and acts against the government were still considered "counter-revolutionary crimes." Still, the criteria for judging them were not as arbitrary as they had been during the Cultural Revolution. In the 1990s, the politically charged concept of "counterrevolutionary crime" was changed to the legal term "crimes against national security," a change legal scholars have highly praised as having removed the political aspect from the law.[8]

As a result of these changes, the institutional environment of the Cultural Revolution, which spawned social movements, no longer exists. In other words, social movements like those during the Cultural Revolution are no longer possible. This does not mean that social movements have disappeared from China, but rather the changes in the institutional environment have become a breeding ground for new forms of social movements.

3 Transformation of Social Movements in the Reform and Opening-Up Period

3.1 The Institutional Context in Which Social Movements Emerge

The state reconstruction of the Reform and Opening-Up period has, on the one hand, invigorated the internal development dynamics of the state and society and brought about rapid economic growth. On the other hand, it has also given rise to many unforeseen social problems and created an institutional environment in which new social movements are nurtured. The social problems arising from state reconstruction are manifested in many ways, and the four main social problems relevant to this book are as follows.

First, the gap in income distribution has widened. The economic reforms of the Reform and Opening-Up period destroyed the egalitarian system of income distribution and introduced a market economy-based system. This resulted in economic development, but wealth distribution and relative poverty reduction were highly unequal. In other words, the economic reforms and their development have yet to narrow the gap between social groups, regions, and urban and rural areas. A widening tendency has been seen when compared to the Mao Zedong era.

The disparity in income distribution during the Reform and Opening-Up period can be classified into four main types of disparity within rural areas, urban areas, between urban and rural areas, and between regions. According to the survey results, these four types of inequalities have been widening.[9] In particular, the widening income gap has created a sense of dissatisfaction and relative deprivation, creating an incentive for social protest movements to emerge.

The second social problem is that of bureaucratic corruption. As mentioned earlier, the Reform and Opening-Up is a government-led reconstruction of the country. The government vigorously promoted economic reform and created the miracle of continuous economic growth. However, due to the government's excessive involvement in the economic sphere, the corruption of party bureaucrats has caused severe social problems to this day.[10] Especially into the 1990s, there has been a tendency for the number of offenses committed by party and government bureaucrats to increase yearly, most of which are related to corruption.

The ruling party is not unaware of the seriousness of the problem. The government has launched several anti-corruption campaigns but have failed. The corruption of party bureaucrats has significantly impacted society, and the public has come to distrust the government and doubt whether it can govern the country fairly and legally. In other words, the widespread and routine corruption of party bureaucrats shook the legitimacy of the Communist regime. This distrust was then extended to social justice, the fairness of the existing system, etc. The deterioration of societal security and increased crime in the 1990s cannot be entirely unrelated to losing confidence in the government and the system. The perception that the law is dictated by power is widespread, and distrust of law enforcement has led some to act openly in violation of the law to maintain their understanding of social justice and fairness.[11]

Third, the retreat of ideology has produced a crisis of faith. During the Reform and Opening-Up period, the government consciously weakened its ideology to develop the market economy. Although this created an environment conducive to economic development, national identity declined. It led to various problems, including declining integration within the Communist Party, bureaucratic corruption, and a crisis of faith spread throughout society. The main reason for these problems was that moral and spiritual ethics, the new spirituality necessary to create a better future, and the communist ideology had provided the image of the future beyond the country's economic power. After the collapse of the communist faith, there was no alternative value system. Religious revival is strongly linked to this crisis of faith. Although the government did not explicitly prevent the growth of religion, the confrontation between the government and religious organizations arose when the society-unifying function of the Communist Party in ideology and organization was threatened. The rise of the Falun Gong in the late 1990s occurred in this context.

Fourth, the increase in non-state enterprises has given rise to labor disputes. During the Reform and Opening-Up, the government implemented preferential policies for foreign-invested enterprises to obtain capital from abroad. The increase in foreign-invested enterprises has contributed to economic prosperity and solved employment problems for some. However, labor disputes in foreign-capital enterprises have gradually attracted attention. The government enacted the Labor Law to protect workers' rights and interests. However, the law's implementation was hindered by "localism," and the rights and interests of many workers in foreign-capital enterprises were not fully protected. That is to say, local governments had to compete with each other to attract foreign capital to promote the development of the local economy. They grew a negative attitude toward implementing the Labor Law. Against this background, social protest movements arising from labor disputes in non-state enterprises frequently occurred.

3.2 Social Movements in the Reform and Opening-Up Period

As mentioned earlier, social problems brought about by the country's reconstruction since Reform and Opening have generated new social discontent. The institutional reforms carried out simultaneously created an institutional environment for social movements. Against this background, social movements occurred throughout the country, especially in the 1990s. Among them, the protest movement of the Falun Gong received worldwide attention, but this was only the tip of the iceberg of social movements in China. The recent frequency of social movements can be seen in the 1998 survey report of the State Planning Commission of China.[12] The revival of social movements since the Reform and Opening-Up has also aroused the interest of many scholars, and social movements in China have become a prominent focus in the social science research field.[13]

Some social movements during the Reform and Opening-Up period were aligned with national interests and had no clear antagonistic relationship with the government. For example, the women's rights movement had ideological legitimacy from the outset. Because the Chinese government advocated gender equality, the government took a supportive and leading role in the women's rights movement. As a result, the women's rights movement achieved excellent results in pursuing equal rights during the Reform and Opening-Up period. In addition, the environmental movements that have taken place throughout China since the 1990s also have an aspect that complements the state's policy goals and therefore has the support and cooperation of the government. Conversely, other aspects of social movements had an antagonistic relationship with the government. Peasant/worker and religious movements that were frequent at the end of the last century were highly alarming to the government and are described in detail below.

Demonstrations by peasants and workers include non-violent protests, such as marches and petitions, and violent protests, such as attacks on government institutions and bureaucrats, blocking traffic and causing explosions.[14] Among the non-violent demonstrations, some involve "policy-based resistance," which uses laws, policies, and other official communications to defend their rights against the activities of local governments and corporations.[15] Many researchers have pointed to overburdened livelihoods as the root cause of collective peasant and worker protests.[16]

The causes of overburdened livelihoods come from various sources. In addition to the general factors such as income distribution disparity and bureaucratic corruption discussed earlier, the tax-sharing reform and the reform of state-owned enterprise system was carried out in the 1990s also increased the burden of peasants' and workers' livelihoods and became the main reason for the peasant and worker protests at the end of the last century. According to the results of Wang Shaoguang et al.'s survey,[17] from the mid-1990s onwards, peasant-worker collective protests showed an increasing trend. Among them, protest against unreasonable levies by local governments and marches by laid-off state-run enterprise workers accounted for a large proportion.

Economic reforms have brought about dissatisfaction among peasants and workers, and the corruption of bureaucrats and corporate executives has exacerbated their economic difficulties and further reinforced their hardship. When contrasting their situation with the corruption seen in bureaucrats and enterprise managers, peasants and workers are apt to become frustrated with society and develop a motive for protest action. Of course, to combine the motivation for dissatisfaction with specific protest actions, there must be inducing factors, organization, mobilization, and the institutional environment that determines the endurance of the social movement. In other words, the proliferation of collective protest movements during the Reform and Opening-Up period was also related to changes in the institutional spatial configuration of the state and society. Simply put, as the total dependence of the individual on the state dissolved, the cost of participation in collective protest movements fell. At the same time, the government's avoidance of overt repression of protests by dissatisfied peasants and laid-off workers also reduced the risk of participation, resulting in a proliferation of collective protest actions.

The emergence of religious movements is linked to the reorientation of government policy toward religion. Religious revival at the end of the last century was related not only to the decline of the Communist Party's ideology and the faith needs of the masses but also to the government's adjustment of religious policy. During the Reform and Opening-Up period, the government stopped the total control of religion of the Maoist era and adopted an authoritarian style of governance. That is to say, it tolerated the existence of religious organizations and put religion under the supervision of the government by controlling the heads of religious organizations. Once a religious organization was out of government control, and the teachings of its organization became antagonistic to the Communist Party ideology, the organization was no longer allowed to continue to exist. In particular, the government altogether banned religious organizations that were highly organized and political, such as the Falun Gong.

In China, many religious organizations are apolitical. However, the pursuit of religious legitimacy and independence is inherently political. Moreover, even if the group does not openly pursue legitimacy and independence, the organization's expansion can threaten the government's ability to integrate into society. Religion is not just a belief but can also be an essential means and tool for social integration. Thus, at this stage, the government must deal with religious organizations, not only the Falun Gong. Under such circumstances, how the relationship between the government and religion is handled can be considered a critical issue for the communist regime.

3.3 Features of Social Movements in the Reform and Opening-Up Period

Compared to the politically mobilized social movements of the Mao era, the social movements of the Reform and Opening-Up period have the following three characteristics. The first is the spontaneity of social movements. While the government in Mao's era launched social movements through top-down political mobilization, the Reform and Opening-Up period social movements had nothing to do with government political mobilization, and spontaneity was their main feature.

The second feature is the regional and decentralized nature of social movements. Since the social movements of the Mao era took place under the state's political mobilization, the nationwide scope was a characteristic of the movements of that era. In contrast, most social movements during the Reform and Opening-Up period were not nationwide and were characterized by their regional and decentralized nature. For example, almost all of the protest movements of laid-off workers were confined to a single work unit. There were very few horizontal joint actions between different enterprises and industries in different fields.[18] Two main reasons can be identified for this feature. One is that different groups pursue different interests, so taking joint action is challenging. The other point is that the state suppresses national social movements; when they do occur, it takes a relentlessly repressive approach.

The third feature is the diversity of protest targets. During the Maoist period, spontaneous social movements could not exist within the institutional spatial configuration of a monolithic state–society; the social movements of the time took advantage of the political opportunities provided by the state's political mobilization. Consequently, protest movements could also not be divorced from the critique targets prescribed by the political movements of the state. In contrast, the social movements of the Reform and Opening-Up period, because they were spontaneous, had a variety of targets, ranging from the central government to those targeting local governments, government bureaucrats, factory managers, and the heads of foreign-funded enterprises.

4 Concluding Remarks

As already discussed in the literature of the history of the study of social movements, social movements are a constant phenomenon in society. No country is immune to social movements regardless of their political system. Social movements exert political pressure on governments, sometimes leading to social unrest, but not all of their effects on governments are negative. As the analysis in this book shows, the creation of social movements is strongly linked to political and economic institutional elements and reflects structural problems. In this sense, social movements are a driving force that leads to social development and change. In the case of China, during the Cultural Revolution, the state and its agents confronted social problems by launching political movements, which resulted in social movements. In the Reform and Opening-Up period, the central leadership not only halted the method of political mobilization to maintain social stability but also took measures to prohibit social movements by actively promoting political and economic reforms to eradicate the institutional factors that gave rise to social movements. As a result of these reforms, some of the institutional elements that brought about social movements were controlled. However, in the course of reform, new institutional factors gave rise to new social movements. Thus, the emergence of new social movements again became a driving force for institutional change in the ruling party. In this way, social movements contributed to the development of society in confrontation with the government.

Notes

1. Katsuji Nakakane, *China's Economic Development* (Yuhikaku Publishing, 1999), p. 243.
2. Regulations of the People's Republic of China on the Management of Associations (Zhonghuarenmin gongheguo shetuan guanli tiaoli).
3. Tang Liang, "State-Social Relations in China during the Reform Era", *Asian Studies*, vol. 46 (April 2000), p. 22 pp.
4. For a study of changes in state and social relations, see Yang Lijun and Shan Wei, eds., *Governing Society in Contemporary China* (London and Singapore: World Scientific Publishing, 2016); Shan Wei and Yang Lijun, eds., *Changing State-Society Relations in Contemporary China* (London and Singapore: World Scientific Publishing, 2016).
5. Sun Bingyao, et al., *Reform and Opening Up and the Organization of Associations in China* (*Gaige kaifang yu Zhongguo de shetuan zuzhi*) (China Development Press, (Zhongguo fazhan chubanshe), 1993), p. 243.
6. Tang Liang, op. cit., p. 23.
7. Ibid.
8. Ibid.
9. More than eighty percent of rural farmers have incomes below the national average of farmers' incomes. See, Shaoguang Wang,

"The Social and Political Implications of China's WTO Membership", *Journal of Contemporary China*, vol. 9, no. 25 (2000), pp. 379–80. According to data published by the *World Bank* in 1997, the rural Gini coefficient rose from 24.2 in 1981 to 33.3 in 1995. See, The World Bank, *China 2020: Sharing Rising Incomes* (Washington, D.C.: The World Bank, 1997), p. 17. Increasing rural unemployment has had a dramatic impact on the income gap within the peasantry. Although the rapid development of township enterprises in the 1980s provided employment opportunities for many farmers, this situation did not last long. Since the mid-1990s, employment in rural areas has fallen sharply due to the difficulties of operating township enterprises. According to the National Bureau of Statistics, the employment rate of peasants fell by 4.8% in 1997 compared with the previous year, and by a further 18.7% in 1998. See, China Statistical Yearbook: 1999, edited by the National Bureau of Statistics, China Statistics Press, 1999, p. 388.

In the cities, the income gap between urban residents increased sharply into the 1990s, with some 30 million urban residents earning only one-third of the national average annually. See, Wang, 2000, op. cit., p. 385. According to the World Bank, the urban Gini coefficient rose from 17.6 in 1981 to 27.5 in 1995. See, The World Bank, *China 2020: Sharing Rising Incomes*, op. cit., p. 16. In 1990, the difference between the highest and lowest levels of average household income among urban residents was 4.2 times, and by 1998 it had widened further to 9.6 times. See, Ru Xin et al., "1998–1999: Analysis and Estimation of Employment, Income and Information Industry in China", in Ru Xin et al., eds., *1999: Analysis and Measurement of China's Social Situation* (Social Science Literature Press, 2000), p. 34.

In addition, the income gap between urban and rural areas continues to exist. On the eve of the Reform and Opening-Up policy, the average income in urban areas was 2.6 times higher than that in rural areas. See, Wang, 2000, op. cit., p. 386. The gap between the two decreased during the Reform and Opening-Up period due to the economic reform in rural areas, but after the economic reform in cities in 1984, the income gap between the two showed a widening trend. The income gap between urban and rural areas exists not only in China but also in other countries. The problem, however, is that the income gap between urban and rural areas in China is significantly higher than the international average. In most countries, the average income of rural residents is equal to sixty-six percent of the average income of urban residents. In China, however, the average income of peasants fell from fifty-nine percent of the average income of urban residents in 1983 to forty percent in 1995. See, The World Bank, op. cit., p. 17. According to 1999 statistics, the annual per capita income of peasants was 2162 *yuan*, compared to 7662 *yuan* for urban residents. See, China Statistical Yearbook: 1999, China Statistics Press, pp. 133, 161.

With regard to geographical disparities, according to the World Bank, the situation is slightly more positive than the three types of disparities within regions mentioned above. For example, in 1992 the average income in coastal areas was fifty percent higher than inland areas, a much smaller gap than the two times difference in average income between urban and rural areas in the same period. See, The World Bank, op. cit., p. 22. But Wang Shaoguang and Hu Angang point out that the income gap between regions is also widening and the situation is getting more and more serious, especially the income gap between coastal and inland regions is accelerating after the 1990s. see, Wang Shaoguang and Hu Angang, *The Political Economy of Uneven Development: The Case of China* (Armonk, NY.: M. E Sharpe, 1999), pp. 56–57. For example, urban per capita incomes in coastal areas, except Guangxi, have exceeded the national average, with Shanghai being 4.5 times higher than the national average. On the other hand, with the exception of Heilongjiang, none of the central and inland regions of China has a per capita income above the national average. In particular, the per capita income in Guizhou province is only 38% of the national average. See, Wang and Hu, 1999, pp. 56–57.

10. The Organization Department of the CCP(Zhonggong zhongyang zuzhibu bianxiezu),*China Survey Report 2000–2001: A Study of the Internal Contradictions of the People in the New Situation* (*2000–2001 Zhongguo diaocha baogao: xinxingshixia de renmin neibu maodun yanjiu*), (The Central Compilation and Translation Press (Zhongyang bianyi chubanshe), 2001), p. 86.
11. "Killing 14 people in a row just to get rid of corrupt officials?—The Case of Hu Wenhai in Yuci city, Shanxi Province (liansha shisiren zhiwei chu tanguan? Shanxisheng yucishi huwenhai an shimo), see, http://news.sohu.com/26/85/news147518526.shtml.

12. Wang Chunguang, "1997–1998: A Survey of China's Social Stability" ("1997–1998: Zhongguo shehui wending zhuangkuang de diaocha"), in Ru Xin et al., eds., *1998: Analysis and Measurement of China's Social Situation* (Social Science Literature Press, 1998 (*1998nian: zhongguo shehui xingshi fenxi yu yuce*, zhongguo shehui wenxian chubanshe, 1998), pp. 121–132.
13. For example, Yanqi Tong and Shaohua Lei, *Social Protest in Contemporary China, 2003–2010: Transitional pains and Regime Legitimacy* (London: Routledge, 2014).
14. Wang Chunguang, op. cit., p. 129.
15. On peasants' "Policy-Based-Resistance", see, Lianjiang Li and Kevin J. O'Brien, "Villagers and Popular Resistance in Contemporary China", *Modern China*, vol. 22, no. 1 (January 1996), pp. 28–61.
16. For a study of peasant protest movements, see Li and O'Brien, op. cit., 1996; For a study of labor protests, see Feng Chen, "Subsistence Crisis, Managerial Corruption and Labor Protests in China", *The China Journal*, vol. 44 (July 2000), pp. 41–43.
17. Wang Shaoguang, "The Social and Political Implications", 2000, op. cit., p. 386; Wang, Chunguang, op. cit., p. 129.
18. Feng Chen, op. cit., pp. 41–63.

Postscript to the Japanese Edition

This book is a revision of my doctoral dissertation submitted to Hitotsubashi University in November 2002, "Factionalization and Collective Violence during the Cultural Revolution." The publication of this book has been made possible through the generous support of the Grant-in-Aid for Scientific Research (2003) from the Japan Society for the Promotion of Science and the invaluable assistance of numerous scholars and friends. It is with deep gratitude for their guidance and support that a young scholar like me was able to publish her first book.

During my time in the master's program at Yokohama City University, where my research journey began, I had the privilege of meeting two mentors who played a crucial role in shaping my academic path. Professor Yuzo Kato instilled in me the importance of imagination, while Professor Kazuko Mori pushed me to study with meticulousness and rigor. The two years I spent completing my master's degree were challenging under Professor Mori's guidance, as I constantly strived to meet her high standards. Looking back on this process, I was amazed at my progress and captivated by the allure of academic research. Even today, I continue to aspire to the standards set by Professor Mori, who remains to be a driving force in my academic pursuits.

Since I entered the doctoral program at Hitotsubashi University, I have received great attention from Professor Takashi Mitani, Professor Hiroko Sakamoto, and my fellow students of Mitani and Sakamoto. Professor

Y. Lijun, *Contest for Citizenship and Collective Violence during China's Cultural Revolution*, IPP Studies in the Frontiers of China's Public Policy, https://doi.org/10.1007/978-981-97-0906-9

Mitani, in particular, patiently tolerated my sometimes unrealistic ideas, granting me the freedom to explore and play with them. He also imparted to me the significance of regional research and empirical analysis, emphasizing the need to investigate our subjects calmly and objectively, free from personal values and subjective assumptions. Professor Hiroko Sakamoto identified the shortcomings in my research through the lens of intellectual history, offering countless insights. Furthermore, my two dissertation reviewers, Professor Osamu Watanabe and Professor Masao Watanabe, contributed valuable guidance from a comparative political science standpoint, transforming the final defense meeting for my Ph.D. into a fresh launchpad for my scholarly pursuits.

I have received guidance from many other scholars throughout my academic journey. Professor Satoshi Amako of Waseda University has been a constant source of encouragement and guidance since our meeting at an academic conference during my master's program. In 1997, I had the unexpected opportunity to meet Professor Ryosei Kokubun of Keio University. We were both hunting for Cultural Revolution material at the Panjiayuan Antique market in Beijing. He was a visiting professor at Peking University then and spoke such eloquent Chinese. The Center for Regional Studies at Keio University (now renamed the Institute of East Asian Studies), which he headed, became an important hub for my research. I also express my gratitude to Professor Shinichi Sato of the University of Tokyo, Professor Yutaka Yoshida of Hitotsubashi University, Professor Tang Liang of Yokohama City University, and Professor Zhao Hongwei of Hosei University for their valuable advice at different stages of writing this book.

Dr. Zheng Yongnian, senior research fellow of the East Asian Institute at the National University of Singapore, has been my life partner and research companion. His extensive knowledge of European and American political theory has greatly influenced my research style, and he has inspired me in countless ways during the completion of this book.

I am also grateful to two incredible couples, Mrs. Keiko Ota (research assistant at Hitotsubashi University) and Mr. Fumon Tsuchida, as well as Mrs. Michio Ueno and Mr. Sejong Oh, both graduate students at Hitotsubashi University. These dear friends supported me throughout my lengthy study abroad journey, sharing the sorrows and joys that came with it. They also undertook the substantial task of revising the Japanese language of this book. Additionally, I extend my gratitude to the graduate students of Hitotsubashi University, including Kouji Hayashi, Masato

Hara, Keji Kato, Akiko Watanabe, Ken Sato, Hirofumi Takase, and Jin Ishizuka (Special Research Fellow of the Japan Society for the Promotion of Science), who dedicated their time and effort to assist in the Japanese language revision. Without their invaluable help, I would not have submitted my doctoral dissertation on schedule.

In addition, I would like to express my sincere thanks to Mr. Morisaku Hashimoto of Ochanomizu Shobo, whose recognition of the value of this book paved the way for its publication. I am also grateful to Mr. Akio Kozakai, who personally dedicated himself from planning to publication, going above and beyond to ensure the success of this book.

The launch of this book project coincided with my father's terminal cancer diagnosis. My father, who lived through the Mao era, was deeply affected by the issues stemming from his non red class origin and my grandfather's history, rendering him unable to shape his own life. Perhaps it was for this reason, he provided me with unwavering encouragement and support. Celebrating my attainment of a doctoral degree was his final wish. For this reason, he fought with great perseverance against his illness, serving as a constant source of inspiration. Sadly, he passed away just as this book was about to be submitted for review as my doctoral thesis. I can only imagine the joy he would have felt had he seen this book during his lifetime, surpassing even my own delight.

September 2003
Yang Lijun

Bibliography

Only the literature directly referenced in this book is listed below.

1 Newspapers and Magazines

《人民日报》1955年, 1966—1969年。

《中国劳工通讯》(电子报) 1999—2002年。

《光明日报》1966—1969年。

《解放军报》1966—1969年。

《解放日报》1966—1969年。

《红旗》1966年。

《河北日报》1978年。

《南华早报》1999年。

2 Historical Materials and Collections

北京电影制片厂毛泽东主义公社编印《战无不胜的毛泽东思想万岁》。
北京函授学院东方红公社编《动态报》第65期, 1967年8月26日。
北京化工学院编《无产阶级文化大革命文件汇编》1967年。
北京首都毛泽东思想兵团《兵团战报》1966年11月。
东方红抗大电校《四三战报》第1期, 1967年6月11日。

Y. Lijun, *Contest for Citizenship and Collective Violence during China's Cultural Revolution*, IPP Studies in the Frontiers of China's Public Policy, https://doi.org/10.1007/978-981-97-0906-9

广东省革命群众批资平反总部编《平反资料汇编》1968年1月。
广西文化大革命大事年表编写组《广西文革大事年表》广西人民出版社, 1992年。
河北北京师范学院《斗争生活》编辑部编《无产阶级文化大革命资料汇编》1966年12月。
红代会井冈山报编辑部《井冈山》1967年。
红代会清华大学井冈山报编辑部《井冈山》1967—1968年。

《江青同志在天桥剧场的讲话》(当时的传单)。
南开大学卫东资料室编《最高指示》1967年。
清华大学井冈山红卫兵宣传队编《清华大学大字报选》1966年。
清华大学井冈山报编辑部《井冈山》1966—1967年。
首都大专院校红卫兵代表大会, 北京政法学院政法公社编《有关无产阶级文化大革命的参考材料》1967年。
首都大专院校红卫兵革命造反总司令部 (第三司令部) 北京政法学院《政法公社》红旗战斗队编《有关无产阶级文化大革命的参考材料》1967年。

《首都红卫兵》第1期, 纪念八．一八专刊, 1967年。
首都中学生革命造反司令部主办《中学文革报》创刊号—第3期。
水电部电研所红色造反总部宣传组《动态资料》1967年。

《无产阶级文化大革命首长讲话汇集》北航红旗印, 1967年4月。
新华公社, 红旗战斗队整理《江青同志重要讲话 (1967年9月5日)》。
中共中国人民大学委员会社会主义思想教育办公室编 《社会主义思想教育参考资料选》第1–4辑, 中国人民大学出版, 1957年。
中国人民解放军国防大学党史党建政工教研室编 《"文化大革命"研究资料》 (上),(中), (下), 人民解放军国防大学出版, 1998年。

《中央负责同志关于无产阶级文化大革命的讲话》(续编), (出处不明), 1967年1月。

《中央领导人接见山东代表团的讲话 (1967年4月27日)》北航《红旗》宣传组翻印, 1967年。

《周恩来, 陈伯达, 康生, 江青, 王力, 关锋, 戚本禹等同志的座谈纪要(1967年1月10日)》北京矿业学院东方红公社整理, 首都职工革命造反总部西城分部印刷, 1967年1月11日。
吴晋增《吴晋增的揭发批判材料》。
中共东方红纺织厂党委会《各单位新班子成员名单》(1966年10月12日)。
中共东方红纺织厂党委会《各单位有关负责人名单》(1966年10月14日)。
东方红纺织厂档案资料《1965年第一季度党员类型》。
东方红纺织厂档案资料《东方红纺织厂档案记录》。

东方红纺织厂档案资料《七个工作队员代表揭发:关于卜虹云在四清运动中所犯的滔天罪恶》(1967年3月15日)。
东方红纺织厂四清运动档案资料《党的基层组织和党员队伍摸底排队情况》(1966年4月28日)。
东方红纺织厂四清运动档案资料《东方红纺织厂文化革命代表名册》(1966年8月)。
东方红纺织厂四清运动档案资料《对敌斗争后各级党组织和党员摸底排队情况》(1966年10月23日)。
东方红纺织厂四清运动档案资料《培养提拔调整干部表》。
东方红纺织厂四清运动档案资料《红卫兵总部临时责任名簿》(1967年1月10日)。
东方红纺织厂革命组织联络站<评我厂红卫兵总部的五大法宝> (1967年1月)。
东方红纺织厂革命组织联络站<假四清、真复辟> (1967年10月26日)。
东方红纺织厂革命组织联络站《沉痛的教训:反戈一击材料汇编之二》(1967年)。
东方红纺织厂革命组织联络站《东方红纺织厂"五八反革命事件":镇压革命干部何文华同志和革命群众的滔天罪行》。
东方红纺织厂革命组织联络站编印<真是一件骇人听闻的政治迫害大冤案> (1967年9月3日)。
东方红纺织厂革命组织、学生联络站<特快新闻> (1967年1月26日)。
东方红纺织厂革命组织联络站宣传部《材料汇辑》(1966–1968)。
东方红纺织厂革命组织联络站宣传部编印《东方红纺织厂无产阶级文化大革命大事记》(1967年12月)。
东方红纺织厂联合指挥部<"东联"的形成及其在晋中"四月黑风"以前> (1967年11月21日)。
东方红纺织厂捍卫毛泽东思想革命造反兵团第十分团<揭开东方红纺织厂"革命组织联络站"一小撮别有用心的人的内幕> (1967年2月27日)。

3 Japanese Literature

天児慧「中国プロレタリア文化大革命」『岩波講座　世界歴史26』岩波書店、1999年。
天児慧『中華人民共和国史』岩波新書、1999年。
天児慧『中国—溶変する社会主義大国』東京大学出版会、1992年。
天児慧《巨竜の胎動——毛沢東vs鄧小平》講談社, 2004年。
安藤正士．古田勝洪．辻康吾著『文化大革命と中国』岩波新書、1986年。
J. R. タウンゼント著、小島朋之訳『現代中国—政治体系の比較分析』慶応通信、1980年(J.R.Townsend, *Politics in China*,1974)。
石塚迅「現代中国法における『四つの基本原則』と思想．言論の自由」『一橋研究』一橋大学大学院、第26巻第1号、2001年4月。
姫田光義等『中国近現代史』(上、下巻)東京大学出版会、1982年。
ウーヴェ・リヒター著、渡部貞昭訳『北京大学の文化大革命』岩波書店、1993年。
加々美光行『歴史のなかの中国文化大革命』岩波、2001年。

加々美光行『逆説として中国革命』田畑書店、1986年。
加々美光行編『現代中国の挫折:文化大革命の省察I』アジア経済研究所、1986年。
加々美光行編『現代中国のゆくえ:文化大革命の省察II』アジア経済研究所、1986年。
国分良成編『中国文化大革命再論』慶応義塾大学出版会、2003年。
新島淳良「中国教育の現状」『中国研究月報』1964年2月号。
谷川真一『中国文化大革命のダイナミクス』御茶ノ水書房、2011年、268頁。
唐亮「改革期の中国における国家=社会関係の変容」『アジア研究』第46号、2000年4月。
竹内実『ドキュメント現代史16 文化大革命』平凡社、1973年。
中兼和津次『中国経済発展論』有斐閣、1999年。
中島嶺雄『中国 歴史．社会．国際関係』中公新書、1982年。
張承志著、小島晋治等訳『紅衛兵の時代』岩波新書、1992年。
福島正夫『人民公社の研究』御茶の水書房、1960年。
古厩忠夫「文化大革命と日本」池田誠等編『世界の中の日中関係(上)』法律文化社、1996年。
前田比呂子「中国における戸籍移転政策」『アジア経済』1996年5月号。
前田比呂子「中華人民共和国における『戸口』管理制度と人口移動」『アジア経済』1993年2月号。
ミハイル．S．ウオスレンスキー著、佐久間穆等訳『ノーメンクラッーラーソヴィエントの赤い貴族』中央公論社、1981年。
毛里和子「中国政治における『幹部』問題」衛藤瀋吉編『現代中国政治の構造』日本国際問題研究所、1982年。
毛里和子『現代中国政治』名古屋大学出版会、1993年。
毛里和子『現代中国政治を読む』山川出版社、1999年。
山本恒人「中国文革初期における労働者階級底辺層—『臨時工．契約工』の造反」『アジア経済』1981年2月。
山本恒人「1960年における労働教育下放の三位一体的政策展開とその破産」加々美光行編『現代中国の挫折:文化大革命の省察I』アジア経済研究所、1986年。
山本恒人「中国文化大革命再考:研究史の側面から」『近きに在りて』2007年6月号。
山本秀夫『中国の農村革命』東洋経済新報社、1975年。
ヨハン・ガルトゥング著、高柳先男ほか訳『構造的暴力と平和』(中央大学出版社、1991年)。
楊麗君「紅衛兵運動の社会的要因:階級と階層の視点から」『現代中国』第74号、2000年。
楊麗君「文革研究の制度的アプローチ:公民権の配分と獲得競争」『アジア研究』第47巻第4号、2001年10月。
楊麗君「文革期における集団的暴力行為:制度論のアプローチから」『アジア研究』第49巻第1号、2003年1月。

楊麗君「文革期の政権再建に関する研究:『三結合政権』における上海モデルの成立」『中国研究月報』2003年11月。
楊麗君「文革期における派閥構造と成因——資源動員論のアプローチ」『アジア研究』第51巻第4号、2005年10月。
楊麗君「文革期における暴力と社会秩序」歴史科学協議会編集『歴史評論』2007年1月号、No.681。
渡辺一衛「湖南文革と"省無聯"」加々美光行『現代中国の挫折:文化大革命の省察I』アジア経済研究所、1986年。
W. ヒントン著、春名徹訳『百日戦争—精華大学の文化大革命』平凡社、1976年。

4 Chinese Literature

安文江〈我不忏悔〉徐友渔编《1966:我们那一代的回忆》中国文联出版公司, 1998年。
白霖(Lynn T.White III)著, 岳经纶译〈解释高潮中的革命〉《社会学学报》香港城市大学出版, 1996年, 第7期。
碧峡〈把回忆的材料留给历史〉徐友渔编《1966:我们那一代的回忆》中国文联出版社, 1998年。
卜大华《"砸烂旧世界"——文化大革命的动乱与浩劫》(中华人民共和国史第六卷:1966–1968), 香港中文大学出版社, 2008年。
陈清泉等著《陆定一传》中共党史出版社, 1999年。
陈晓文〈重庆红卫兵素描〉刘青峰编《文化大革命:史实与研究》香港中文大学出版社, 1996年。
陈祖恩等《上海通史》(第11卷)上海人民出版社, 1999年。
崔之元〈毛泽东"文革"理论的得失与"现代性"的重建〉《社会科学学报》(文革专辑), 香港城市大学出版社, 1996年, 第7期;(修正版)〈毛泽东"文革"理论的得失与"现代性"的重建〉崔之元《第二次思想解放与制度创新》(香港)牛津大学出版社, 1997年。
丁望《中共"文革"运动中的组织与人事问题:1965–1970》当代中国研究所, 1970年。

《当代中国》丛书编集委员会编《当代中国的经济体制改革》中国社会科学出版社, 1984年。

《当代中国》丛书编集委员会编《当代中国经济》中国社会科学出版社, 1987年。
方圆〈自由工人运动的先驱—纪念独立工会全红总成立30周年〉《北京之春》1997年3,4,5月号。
冯骥才《一百个人的十年》江苏文艺出版社, 1997年。
高树华、程铁军《内蒙文革风雷——一位造反派领袖的口述史》香港明镜出版社, 2007年。
龚小夏〈文革中群众暴力行为的起源与发展〉《社会科学学报》(文革专辑),香港城市大学出版, 1996年, 第7期。

国家统计局编《中国统计年鉴1991》中国统计出版社, 1991年。
国家统计局编《中国统计年鉴:1999年》中国统计出版社, 1999年。
何均〈论全业合营—从上海私营工业的七个行业的情况看实行全业合营的作用〉《学习》3月号, 1956年。
胡平〈比赛革命的革命〉《北京之春》1996年, 6月号。
华林山〈文革期间群众性对立派系成因〉刘青峰编《文化大革命:史实与研究》香港中文大学出版社, 1996年。
江沛《红卫兵狂飙》河南人民出版社, 1994年。
金春明《"文化大革命"史稿》四川人民出版社, 1995年。
金观涛、刘青峰《开放中的变迁》香港中文大学出版社, 1993年。
晋夫《文革前十年的中国》中共党史出版社, 1998年。
李培林主编《中国新世纪阶级阶层报告》辽宁人民出版社, 1995年。
李筠〈我与三家村〉《中共党史资料》总第66辑, 中共党史出版社, 1998年6月。
李逊〈文革中发生在上海的"经济主义风"〉罗金义, 郑文龙编《浩劫之外—再论文化大革命》风云论坛出版社, 1997年。
李逊《大崩溃—上海工人造反派兴亡史》时报文化出版企业有限公司, 1996年。
刘国凯编《封杀不了的历史》重新评价文化大革命丛书编辑部出版, 1996年。
刘青峰〈对历史的再发问〉刘青峰编《文化大革命:史实与研究》香港中文大学出版社, 1996年。
刘少奇《在中国共产党第八次全国代表大会上的政治报告》人民出版社, 1956年。
毛泽东《毛泽东著作选读》(上, 下册)人民出版社, 1986年。
毛泽东《毛泽东选集》(1–5卷)人民出版社, 1987年, 1991年。
马克思．韦伯(Max Weber)著, 林荣远译《经济与社会》(上, 下)上午印书馆, 1997年。
麦克法夸尔, 费正清编《剑桥中华人民共和国史》(上, 下)中国社会科学出版社, 1992年。
马继森《外交部文革纪实》香港中文大学出版社, 2003年。
米鹤都《红卫兵这一代》三联书店, 1993年。
米鹤都主编, 卜大华等共著《回忆与反思:红卫兵时代风云人物——口述历史之一》香港中国图书有限公司, 1999年。
张远淦, 林帮光〈试论1955年党内关于农业合作化问题的争论〉《党史研究》1981年1月号。
任知初《红卫兵与嬉皮士》明镜出版社, 1996年。
汝信等编《1998年:中国社会形式分析与预测》社会科学文献出版社, 1998年。
汝信等编《社会蓝皮书:1999年中国社会形式分析与预测》社会科学文献出版社, 2000年。
汝信等编《社会蓝皮书:中国社会形式分析与预测》社会科学文献出版社, 2001年。
上海市总工会《关于集体所有制工厂企业职工要求组织工会的请示报告》1996年8月30日。
孙炳耀等《改革开放与中国的社团组织》中国发展出版社, 1993年。
施拉姆(Stuart R Schram)〈1949年到1976年的毛泽东思想〉麦克法夸尔, 费正清编《剑桥中华人民共和国史》(下)中国社会科学出版社, 1992年。

石云〈红卫兵运动述评〉《中国青运》1989年第1期。
石刚编著,《现代中国的制度与文化》香港社会科学出版社有限公司, 2004年。
史云、李丹慧《难以继续的“继续革命”——从批林到批孔》》(中华人民共和国史第八卷:1972–1976)香港中文大学出版社, 2008年。
史实编著《文革中的检讨书》香港中国历史出版社, 2011年。
苏采青〈“文革”初期三个回合的斗争〉张化, 苏采青编《回首“文革”》(下)中共党史出版社, 2000年。
宋永毅, 孙大进《文化大革命和它的异端思潮》田园书店, 1997年。
宋永毅〈文化大革命中的异端思潮〉刘青峰编《文化大革命:史实与研究》香港中文大学出版社, 1996年。
宋永毅主编《文革大屠杀》香港开放杂志社出版, 2002年。
谭合成《血的神话——公元1967年河南道县文革大屠杀纪实》香港天行健出版社, 2010年。
唐少杰〈红卫兵运动的丧钟:清华大学百日大武斗〉刘青峰编《文化大革命:史实与研究》香港中文大学出版社, 1996年。
唐少杰〈清华井冈山兵团的兴衰>刘青峰编《文化大革命:史实与研究》香港中文大学出版社, 1996年。
唐少杰《一叶知秋——清华大学1968年“百日大武斗”》香港中文大学出版社, 2003年。
王力《现场历史:文化大革命纪事》牛津大学出版社, 1993年。
王年一<评《“文化大革命”十年史》>张化, 苏采青编《回首“文革”》(上)中共党史出版社, 2000年。
王年一<对上海“一月革命”的几点看法>张化, 苏采青编《回首“文革”》(下)中共党史出版社, 2000年。
王年一《大动乱的时代》河南人民出版社, 1988年。
王绍光<拓展文革研究的视野>刘青峰编《文化大革命:史实与研究》香港中文大学出版社, 1996年。
王绍光《理性与疯狂:文化大革命中的群众》牛津大学出版社, 1993年。
王友琴<打老师和大同学之间>刘青峰编《文化大革命:史实与研究》香港中文大学出版社, 1996年。
王友琴<1966:学生打老师的革命>刘青峰编《文化大革命:史实与研究》香港中文大学出版社, 1996年。
王湘<无产阶级文化大革命的经验和教训>刘国凯编《封杀不了的历史》《重新评价文化大革命丛书》编辑部出版, 1996年。

《伟大的十年》人民出版社, 1960年。
文玉山<广州红旗派的兴亡>刘国凯编《封杀不了的历史》《重新评价文化大革命丛书》编辑部出版, 1996年。
文聿《中国左祸》北京朝华出版社, 1992年。
吴文光《革命现场一九六六》时报出版, 1994年。
香港城市大学当代中国研究中心编《社会科学学报》(文革专辑)香港城市大学出版, 1996年春季号, 第7期。
严家其, 高皋《文革十年史》天津人民出版社, 1986年。

杨晓民, 周翼虎《中国单位制度》中国经济出版社, 1989年。
叶永烈《王洪文兴衰录》时代文艺出版社, 1989年。
印红标<红卫兵运动的主要流派>《青年研究》1997年, 第4期。
印红标〈文革的"第一张马列主义大字报"〉刘青峰编《文化大革命:史实与研究》香港中文大学出版社, 1996年。
印红标,《失踪者的足迹——文化大革命期间的青年思潮》, 香港中文大学出版社, 2009年。
萧喜东〈「两个文革」, 或「一个文革」?〉罗金义、郑文龙编《浩劫之外——再论文化大革命》(台湾)风云论坛出版社有限公司、1997年。
徐贲〈论文革政治文化中的恐惧和暴力〉罗金义, 郑文龙编《浩劫之外—再论文化大革命》风云论坛出版社, 1997年。
徐友渔〈文化大革命中红卫兵行为动因的调查与分析〉《中国社会科学季刊》香港城市大学出版, 1996年夏季卷, 第15期。
徐友渔《自由的言说》长春出版社, 1999年。
徐友渔,《形形色色的造反——红卫兵精神素质的形成和演变》, 香港中文大学出版社, 1999年。
遇罗克<出身论(1966年11月)>《首都中学文革报》创刊号, 1967年1月18号。
曾慧燕<采访高皋>《九十年代月刊》1987年4月。
张庆五编《户口登记常识》法律出版社, 1983年。
张乐天《告别理想—人民公社制度研究》东方出版中心, 1998年。
郑义<清华附中, 红卫兵与我>《北京之春》1996年11月号。
郑永年《中国民族主义的复兴:民族国家向何处去?》(香港)三联书店, 1997年。
中共中央组织部编写组《2000—2001中国调查报告:新形势下的人民内部矛盾研究》中央编译出版社, 2001年。
中共中央党史研究室《中共党史大事年表》人民出版社, 1981年。
中共中央文献研究室编《关于建国以来党的若干历史问题的决议 注释本》人民出版社, 1985年。

《中国统计年鉴:1999年》中国统计出版社。
仲维光<"清华附中红卫兵小组"诞生史实>《北京之春》1996年10月号。
周恩来《关于发展国民经济的第二个五年计划的建议的报告》人民出版社, 1956年。
周恩来《历史在这里沉思》(第五卷)华夏出版社, 1986年。
总政治部干部部, 军事科学院军制研究部编《人民解放军干部制度概要》军事科学出版社, 1988年。

5 English Literature

Barnes, Samuel J., Max Kaase, et al., *Political Action: Mass Participation in Five Western Democracies* (Beverly Hills, CA: Sage Publications, 1979).
Beijing Review, May 22, 2000.
Bendix, Reinhard, *Kings, or People: Power and the Mandate to Rule* (Berkeley, CA: University of California Press, 1978).

Bendix, Reinhard, *Max Weber, An Intellectual Portrait* (New York: Garden City, 1960). (折原浩訳『マックス . ウェーバー:その学問の全体像』中央公論社、1966年)。

Bergesen, Albert J., "A Durkheimian Theory of 'Witch-Hunts' with the Chinese Cultural Revolution of 1966–1969 as an Example," *Journal of the Scientific Study of Religion*, vol. 17, no. 1 (1978), pp. 19–29.

Button, James, *Black Violence: Political Impact of the 1960s Riots* (Princeton, NJ: Princeton University Press, 1978).

Chan, Anita, *Children of Mao: Personality Development and Political Activism in the Red Guard Generation* (Seattle: University of Washington Press, 1985).

Chan, Anita, Stanley Rosen and Jonathan Unger, "Students and Class Warfare: The Social Roots of the Red Guard Conflict in Canton," *The China Quarterly*, no. 83 (September 1980), pp. 397–446.

Chen-Feng, "Subsistence Crisis, Managerial Corruption and Labor Protests in China," *The China Journal*, 44 (July 2000), pp. 41–43.

China Daily, April 22, 2000.

Colin, Crouch and Alessandro Pizzorno, eds., *The Resurgence of Class Conflict in Western Europe Since 1968*, vol. II (London: Macmillan, 1978).

Dahrendorf, Ralf, *Class and Class Conflict in Industrial Society* (London: Routledge and Kegan Paul, 1959). (富永健一訳『産業社会における階級および階級闘争』ダイヤモンド社、1964年)。

Dittmer, Lowell, *Liu Shao-ch'i and Chinese Cultural Revolution: The Politics of Mass Criticism* (Berkeley: University of California Press, 1987).

Djilas, Milovan, *The New Class: An Analysis of the Communist System* (New York: Praeger, 1957). (原子林二郎訳『新しい階級:共産主義制度の分析』時事通信社、1957年)。

Douglas A. Hibbs, "Industrial Conflict in Advanced Industrial Societies," *American Political Science Review*, vol. 70 (1976), pp. 1033–1058.

Eisinger, Peter, "The Conditions of Protest Behavior in American Cities, "*American Political Science Review*, vol. 67 (1973), pp. 11–28.

Evans, Peter B., Dietrich Rueschemeyer, and Theda Skocpol, eds., *Bringing the State Back In* (New York: Cambridge University Press, 1985).

Gamson, William A., *The Strategy of Social Protest* (Homewood, III: Dorsey, 1975).

Gerth, H. H and C.Wright. Mills, eds., *From Max Weber: Essays in Sociology* (New York: Oxford University Press, 1958). (山口和男等訳『マックス . ウェーバー:その人と業績』ミネルヴァ書房、1962年)。

Gurr, Ted Robert, "A Causal Model of Civil Strife: A Comparative Analysis Using New Indices," *American Political Science Review*, vol. 62 (1968), pp. 1104–1124.

Graham, Hugh Davis and Gurr, Ted Robert, eds., *Violence in America: Historical and Comparative Perspectives* (Washington, DC: US Government Printing Office, 1969).

Gurr, Ted Robert, *Why Men Rebel* (Princeton, NJ: Princeton University Press, 1970).

Guoqiang, Dong and Walder, Andrew G., "Factions in a Bureaucratic Setting: The Origins of Cultural Revolution Conflict in Nanjing." *China Journal*, no. 65 (January 2011).

Guoqiang, Dong and Walter, Andrew G., "Foreshocks: Local Origins of Nanjing's Qingming Demonstrations of 1976." *China Quarterly*, vol. 220 (2014).

Halsey, A.H., "T.H. Marshall: Past and Present, 1893–1981," *Sociology*, vol. 18 (1984).

Harding, Harry, *Organizing China: The Problem of Bureaucracy, 1949–1976* (Stanford: Stanford University Press, 1981).

Henderson, A.M. and Talcott, Parsons. eds., *Max Weber: The Theory of Social and Economic Organization* (New York: The Free Press, 1964).

Hibbs, Douglas, *Mass Political Violence: A Causal Analysis* (New York: Wiley, 1973).

Hirschman, Albert O., *Exit, Voice, and Loyalty: Responses to Decline in Firms, Organizations, and States* (Cambridge, MA: Harvard University Press, 1970). (三浦隆之訳『組織社会の論理構造:退出．告発．ロイヤルティ』ミネルヴァ書房, 1975年。

Hirschman, Albert O., *Shifting Involvements: Private Interest and Public Action* (Princeton, NJ: Princeton University Press, 1982). (佐々木毅等訳『失望と参画の現象学:私的利益と公的行為』法政大学出版局、1988年)。

Hobsbawm, Eric J., *Primitive Rebels: Studies in Archaic Forms of Social Movement in the 19th and 20th Centuries* (Manchester, England: Manchester University Press, 1959).(青木保編訳『反抗の原初形態—千年王国主義と社会運動』中公新書239、1971年;水田洋等訳『素朴な反逆者たち—思想の社会史』社会思想社、1989年)。

Yang, Lijun and Shan, Wei, eds., *Governing Society in Contemporary China* (London and Singapore: World Scientific Publishing, 2016).

Yang, Lijun, "From the 'Four Cleans' Movement to the Cultural Revolution: The Origins of Factional Conflicts." *China: An International Journal,* 13(3) (2015), pp. 1–24.

Katzenstein, Mary F. and Carole Mueller, *The Women's Movements of the United States and Western Europe: Consciousness, Political Opportunity, and Public Policy* (Philadelphia: Temple University Press, 1987).

Kitschelt, Herbert, "Political Opportunity Structures and Political Protest: Anti-Nuclear Movements in Four Democracies," *British Journal of Political Science*, vol. 16 (1986), pp. 57–85.

Klandermans, Bert, Hanspeter Kriesi, and Sidney Tarrow, eds., *From Structure to Action: Comparing Social Movement Research Across Cultures* (Greenwich, Conn.: JAI, 1988).

Korpi, Walterand Micheal Shalev, "Strikes, Power, and Politics in the Western Nations," in I.M. Zeitlin, ed., *Political Power, and Social Theory* (Greenwich, Conn.: JAI, 1980).

Leese, Daniel, *Mao Cult: Rhetoric and Ritual in China's Cultural Revolution* (Cambridge: Cambridge University Press, 2011). (中文版:丹尼尔・里斯、秦禾声等译,《崇拜毛——文化大革命中的言辞崇拜与仪式崇拜》,香港:香港中文大学出版社, 2017年)。

Lee, Hong Yung, "The Radical Students in Kwangtung during the Cultural Revolution," *The China Quarterly*, no. 64 (December 1975), pp. 645–683.

Lee, Hong Yung, *The Politics of the Chinese Cultural Revolution: A Case Study* (Berkeley, CA: University of California Press, 1978).

Leeds, Anthony and Elizabeth Leeds, "Accounting for Behavioral Difference: The Political Systems and the Responses of Squatters in Brazil, Peru and Chile," in John Walton and Louis Masotti, eds., *The City in Comparative Perspective* (New York: Wiley, 1976), pp. 193–248.

Li, Lianjiang and Kevin J.O' Brien, "Villagers and Popular Resistance in Contemporary China," *Modern China*, vol. 22, no. 1 (January 1996), pp. 28–61.

Lifton, Robert J., *Revolution Immortality: Mao Tse-tung and the Chinese Culture Revolution* (New York: Vintage Books, 1968). (井上勇訳『革命の不死性:毛沢東と中国文化大革命』時事通信社、1970年)。

Lipset, Seymour Martin, "Tom Marshall-Man of Wisdom," *British Journal of Sociology*, vol. 24 (1973).

Lipsky, Micheal, "Protest as a Political Resource," *American Political Science Review*, vol. 62 (1968).

Lockwood, David, "For T.H. Marshall," *Sociology*, vol. 8 (1974).

MacFarquhar, Roderick and John K. Fairbank, *The Cambridge History of China, volume 14: The People's Republic, Part 1: The emergence of revolutionary China, 1949–1965, part 2: Revolution within the Chinese Revolution, 1966–1982* (Cambridge University Press, 1987, 1991). (马可法夸尔, 费正清编《剑桥中华人民共和国史》(上, 下), 中国社会科学出版社, 1990, 1992年)。

MacFarquhar, Roderick, *The Origins of the Cultural Revolution*, three volumes (New York: Columbia University Press, 1974, 1983, 1996). (麦克法夸尔《文化大革命的起源》第1–2卷, 河北人民出版社, 1989, 1991年)。

Mann, Micheal, "Ruling Class Strategies and Citizenship," in Mann, *States, War, and Capitalism: Studies in Political Sociology* (Oxford: Blackwell Publishers, 1992), pp. 188–209.

Mann, Micheal, "The Autonomous Power of the State: Its Origins, Mechanism, and Results," in John A. Hall, ed., *State in History* (Oxford: 【xix】 Blackwell Publishers, 1986).

Marshall, T.H. and Tom Bottomore, *Citizenship and Social Class* (Doubleday, Garden City NY, 1965). (岩崎信彦等訳『シティズンシップと社会的階級』法律文化社、1993年)。

McAdam, Doug, *Political Process and the Development of Black Insurgency, 1930–1970* (Chicago: University of Chicago Press, 1982).

McCarthy, John D. and Mayer N. Zald, "Resource Mobilization and Social Movements: A Partial Theory," *American Journal of Sociology*, vol. 82 (1977), pp. 1212–1241.

Micheals, Robert, *Political Parties: A Sociological Study of the Oligarchical Tendencies of Modern Democracy* (New York: The Free Press, 1962).

Milwertz, Cecilia, "Control as Care: Interaction between Urban Women and Birth Planning Workers," in Kjeld Erik Brodsgaard and David Strand, eds., *Reconstructing Twentieth Century China: State Control, Civil Society, and National Identity* (Oxford: Clarendon Press, 1998), pp. 92–112.

Moore, Barrington, Jr., *Social Origins of Dictatorship and Democracy* (Boston: Beacon, 1966). (宮崎隆次等訳『独裁と民主政治の社会的起源:近代世界形成過程における領主と農民』岩波書店、1986年)。

Morris, Aldon D. and Carol McClurg Mueller, eds., *Frontiers in Social Movement Theory* (New Heaven: Yale University Press, 1992).

Oberschall, Anthony, *Social Conflict and Social Movements* (Englewood Cliffs, NJ: Prentice Hall, 1973).

Olson, Mancur, Jr., *The Logic of Collective Action* (Cambridge, CA: Harvard University Press, 1965). (依田博等訳『集合行為論』ミネルヴァ書房, 1983年)。

Pareto, Vilfredo, *The Rise and Fall of Elites* (New York: Arno Press, 1979).

Parkin, Frank, *Middle Class Radicalism* (Cambridge: Cambridge University Press, 1968).

Pang, Lai Kwan, *The Art of Cloning: Creative Production During China's Cultural Revolution* (Verso, 2017) 。(中文版:彭丽君著、李祖乔译《复制的艺术——文革期间的文化生产及实践》, 香港:香港中文大学出版社, 2017年) 。

Perry, Elizabeth J. and Li Xun, *Proletarian Power: Shanghai in the Cultural Revolution* (Boulder: Westview Press, 1997).

Pfeiffer, Richard, "The Pursuit of Purity: Mao's Cultural Revolution," *Problems of Communism*, vol. 18, no. 6 (1969).

Piven, France Fox and Richard A. Cloward, *Poor People's Movements: Why They Succeed, How They Fail* (New York: Vintage, 1979).

Popkin, Samuel L., *The Rational Peasant: The Political Economy of Rural Society in Vietnam* (Berkeley and Los Angeles: University of California Press, 1979).

Pye, Lucian W., *The Dynamics of Chinese Politics* (Cambridge, MA: 【xx】 Oelgeschlager, Gunn & Hain, 1981).

Pye, Lucian W., *The Spirit of Chinese Politics: A Psychocultural Study of The Authority Crisis in Political Development* (Cambridge, MA: MIT Press, 1992).

Roddock, David, *Political Behavior of Adolescents in China: The Cultural Revolution in Kwangchow* (Tuscon, Ariz: The University of Arizona Press).

Rosen, Stanley, *Red Guard Factionalism, and the Cultural Revolution in Guangzhou* (Boulder, Colorado: Westview Press, 1982).

Rozman, Gilbert, *The Modernization of China* (London: New York: Free Press, 1981). (罗兹曼《中国的现代化》江苏人民出版社, 1988年)。

Su, Yang, *Collective Killings in Rural China during the Cultural Revolution* (New York: Cambridge University Press, 2011). (中文版:宋熙译《文革时期中国农村的集体杀戮》香港:香港中文大学出版社, 2017年)。

Shan, Wei and Yang, Lijun, eds., *Changing State-Society Relations in Contemporary China* (Singapore: World Scientific Publishing, 2016).

Snyder, David and Charles Tilly, "Hardship and Collective Violence in France: 1830–1960," *American Sociological Review*, vol. 37 (1972), pp. 520–532.

Tarrow, Sidney, *Peasant Communism in Southern Italy* (New Haven: Yale University Press, 1967).

Tarrow, Sidney, *Power in Movement: Social Movements, Collective Action and Politics* (New York: Cambridge University Press, 1994).

Tarrow, Sidney, *Struggle, Politics, and Reform: Collective Action, Social Movements, and Cycles of Protest*, Western Societies Program Occasional Paper No. 21, 2nd edition (Center for International Studies, Cornell University, 1991).

Teiwes, Frederick, *Leadership, Legitimacy and Conflict in China: From a Charismatic Mao to the Politics of Succession* (Armond, New York: Sharpe, 1984).

Terrill, Ross, *Mao: A Biography* (New York: Harper and Row, 1980).

Thompson, E.P., "The Moral Economy of the English Crowd in the Eighteenth Century," *Past and Present*, vol. 50 (1971), pp. 76–136.

Thurston, Anne F., *Enemies of the People: The Order of the Intellectuals in China's Great Cultural Revolution* (New York: Knopf, 1987).

Tilly, Charles, "Collective Violence in European Perspective," in Hugh D. Graham and Ted R. Gurr, eds., *Violence in America: Historical and Comparative Perspectives* (Washington, DC: US Government Printing Office, 1969).

Tilly, Charles, ed., *The Formation of National States in Western Europe* (Princeton, NJ: Princeton University Press, 1975).

Tilly, Charles, *From Mobilization to Revolution* (Englewood Cliffs, NJ: Prentice-Hall, 1978). (堀江湛訳『政治変動論』芦書房、1984年)。

Tilly, Charles, *The Contentious French* (Cambridge, MA: Harvard University Press, 1986).

Tilly, Charles, *The Vendee* (Cambridge: Harvard University Press, 1964).

Tong, Yanqi and Shaohua Lei, *Social Protest in Contemporary China, 2003–2010: Transitional Pains and Regime Legitimacy* (London: Routledge, 2014).

Turner, B.S., *Citizenship and Capitalism: The Debate Over Reformism* (London: Allen and Unwin, 1986).

Unger, Jonathan, *Education Under Mao: Class and Competition in Canton Schools, 1960–1980* (New York, NY: Columbia University Press, 1982).

Walder, Andrew G., "The Chinese Cultural Revolution in the Factories: Party-State Structures and Patterns of Conflict," in Elizabeth J. Perry, ed., *Putting Class In Its Place: Worker Identities In East Asia* (Berkeley: University of California Institute of East Asian Studies, 1996).

Walder, Andrew G., *Communist Neo-Traditionalism: Work and Authority in Chinese Industry* (Berkeley, CA: University of California Press, 1986).

Walder, Andrew G., *Fractured Rebellion: The Beijing Red Guard Movement* (Cambridge: Harvard University Press, 2009).

Walton, John, "Urban Political Movements and Revolutionary Change in the Third World," *Urban Affairs Quarterly*, vol. 15 (1979).

Wang, Shaoguang and Hu Angang, *The Political Economy of Uneven Development: The Case of China* (Armonk, NY: M.E. Sharpe, 1999).

Wang, Shaoguang, "The Social and Political Implications of China's WTO Membership," *Journal of Contemporary China*, vol. 9, no. 25 (2000), pp. 379–380.

Wang, Shaoguang, *Failure of Charisma: The Cultural Revolution in Wuhan* (Hong Kong: Oxford University Press, 1995). (中文版:王绍光《理性与疯狂:文化大革命中的群众》(香港)牛津大学出版社, 1993年)。

Welch, Susan, "The Impact of Urban Riots on Urban Expenditures," *American Journal of Political Science*, vol. 19 (1975), pp. 741–760.

White, Lynn T. III, *Policies of Chaos: The Organizational Causes of Violence in China's Cultural Revolution* (Princeton, NJ: Princeton University Press, 1989).

World Bank, *China 2020: Sharing Rising Incomes* (Washington, DC: The World Bank, 1997).

Zald, Mayer N. and John D. McMarthy, *Social Movements in an Organizational Society: Collected Essays* (New Brunswick, NJ: Transaction, 1987).

Index of Names

Y. Lijun, *Contest for Citizenship and Collective Violence during China's Cultural Revolution*, IPP Studies in the Frontiers of China's Public Policy, https://doi.org/10.1007/978-981-97-0906-9

Index of Subject

Y. Lijun, *Contest for Citizenship and Collective Violence during China's Cultural Revolution*, IPP Studies in the Frontiers of China's Public Policy, https://doi.org/10.1007/978-981-97-0906-9

Q

R

GPSR Compliance
The European Union's (EU) General Product Safety Regulation (GPSR) is a set of rules that requires consumer products to be safe and our obligations to ensure this.

If you have any concerns about our products, you can contact us on

ProductSafety@springernature.com

In case Publisher is established outside the EU, the EU authorized representative is:

Springer Nature Customer Service Center GmbH
Europaplatz 3
69115 Heidelberg, Germany

www.ingramcontent.com/pod-product-compliance
Ingram Content Group UK Ltd.
Pitfield, Milton Keynes, MK11 3LW, UK
UKHW020833290926
15470UKWH00004B/1326
9789819709052